BULLETIN OF THE OHIO BIOLOGICAL SURVEY
New Series

Volume 8	Number 1

THE VASCULAR FLORA OF THE GLACIATED ALLEGHENY PLATEAU REGION OF OHIO

BARBARA K. ANDREAS

Cuyahoga Community College
Department of Natural Sciences

Kent State University
Department of Biological Sciences

D1200056

PUBLISHED BY
COLLEGE OF BIOLOGICAL SCIENCES
THE OHIO STATE UNIVERSITY

COLUMBUS, OHIO 43210   1989

OHIO BIOLOGICAL SURVEY

BULLETIN NEW SERIES: ISSN 0078-3994
BULLETIN NEW SERIES VOLUME 8 NUMBER 1:
ISBN 0-86727-104-3
LIBRARY OF CONGRESS NUMBER: 87-072905

EDITOR

Veda M. Cafazzo

CITATION

Andreas, Barbara K. 1989. The Vascular Flora of the Glaciated Allegheny Plateau Region of Ohio. Ohio Biol. Surv. Bull. New Series Vol. 8 No. 1 viii + 191 p.

COVER PHOTOGRAPH—Autumn in a tamarack bog. Photograph by Gary Meszaros. The cover design was created by David M. Dennis, Biological Illustrator, College of Biological Sciences, The Ohio State University.

COLLEGE OF BIOLOGICAL SCIENCES
THE OHIO STATE UNIVERSITY
COLUMBUS, OHIO 43210
1989

12-89—1.5M

ABSTRACT

The flora of the Glaciated Allegheny Plateau region of Ohio encompasses all of seven and parts of 16 counties. The vascular flora of the region includes 2002 species and 27 interspecific hybrids, belonging to 706 genera and 148 families. Approximately 72% of the taxa are native and 28% are alien. Information concerning frequency, habitat, and distribution of each taxon is presented as known in 1988. One hundred and fifty-one taxa are, in Ohio, more or less confined to the Glaciated Allegheny Plateau region. Of these definitive taxa of this phytogeographical region, 50% are confined to wetlands, especially peatlands. Another 18% are confined to hemlock-white pine-hardwood forests, and 7% occur on eroding slopes found where post-Pleistocene streams cut through lacustrine deposits and shales. The remaining 25% of these definitive taxa grow in a variety of forested communities. Eighteen types of plant communities are described. Three of these—semi-ombrotrophic bogs, boggy forests, and hemlock-white pine-hardwood forests—are restricted, in Ohio, to this phytogeographical region. The introductory section of this text briefly describes the climate, topography, bedrock and glacial geology, and vegetation of the Glaciated Allegheny Plateau region.

iii

DEDICATION

This book is dedicated to Dr. Tom S. Cooperrider in honor of his many contributions to Ohio floristics. Under his influence, throughout the past 30 years, the floristic knowledge of Ohio has expanded greatly. In addition to serving as Chair of the Ohio Flora Committee (1969 to present) and publishing numerous scientific articles on the topic, he has, to date, trained 22 graduate students, each of whom has made a personal contribution to Ohio floristics—Joyce E. Amann, Sandra M. Anderson, Barbara K. Andreas, Gregory D. Bentz, Judy S. Bradt-Barnhart, Bruce L. Brockett, James F. Burns, Robert J. Cline, Tammy E. Cook, Barbara H. Costelloe, Allison W. Cusick, David P. Emmitt, Edward J. P. Hauser, William D. Hawver, George A. McCready, Lynne D. Miller, Paul L. Pusey, Robert F. Sabo, Gene M. Silberhorn, Lewis W. Tandy, Paula L. Van Natta, and Hugh D. Wilson.

Tom has demonstrated integrity and professionalism to all who have been fortunate to have worked with him. He has encouraged me to strive for excellence as a botanist and a teacher.

CONTENTS

THE CATALOGUE OF THE VASCULAR PLANT

LIST OF FIGURES

ACKNOWLEDGEMENTS

The compilation of a study such as the Vascular Flora of The Glaciated Allegheny Plateau region of Ohio requires the assistance of many individuals. I thank the curators of the following herbaria for their assistance in examining herbarium specimens:

James K. Bissell, Cleveland Museum of Natural History
Philip D. Cantino, Ohio University
Carl F. Chuey, Youngstown State University
Tom S. and Miwako K. Cooperrider, Kent State University
John J. Furlow, The Ohio State University
George T. Jones, Oberlin College
Warren P. Stoutamire, The University of Akron.

I am grateful to the past botanical collectors, especially Almon N. Rood, Roscoe J. Webb, Erwin M. Herrick, Floyd Bartley, Leslie L. Pontius, and Charles Goslin, whose legacy made it possible for me to compare the modern flora with that of the past. To supplement my own collections, those of numerous additional modern collectors have provided the herbarium records on which the bulk of this document is based. Of special importance was the work of graduate students from Kent State University who prepared floras for various counties within the Glaciated Allegheny Plateau: Joyce E. Amann (1961), Stark County; Sandra M. Anderson (1969), Medina County; James F. Burns (1980), Trumbull County; Robert J. Cline (1977), Perry County; David P. Emmitt (1981), Ashland County; William D. Hawver (1961), Geauga County; Paul L. Pusey (1976), Knox County; and Hugh D. Wilson (1972), Holmes County. The floristic efforts of Allison W. Cusick have added a wealth of information to this document.

Special thanks are extended to the landowners who gave me permission to survey their property. Their insights into the significance of their parcels have saved many remaining pieces of natural vegetation in Ohio.

John F. Gwinn, Alexander J. Karlo, and Warren P. Stoutamire accompanied me on various field trips throughout the Glaciated Allegheny Plateau region of Ohio. I am especially indebted to Jeffrey D. Knoop whose companionship on field trips provided an awareness of the forests and large trees remaining in Ohio.

The portions of the Ohio and Michigan floras prepared to date by Tom S. Cooperrider and Edward G. Voss, respectively, simplified decisions regarding nomenclature and circumscription. Dr. Voss' habitat information was always helpful.

Charles C. King, Executive Director, Ohio Biological Survey; Dennis M. Anderson, former Plant Ecologist, Division of Natural Areas and Preserves, Ohio Department of Natural Resources; Jane L. Forsyth, Professor of Geology, Bowling Green State University, and Tom S. Cooperrider, Professor of Biological Sciences, Kent State University, provided guidance and technical advice throughout the research and writing of this study. Comments and suggestions made by anonymous peer reviewers enhanced greatly the final version of the manuscript. Harriet S. Boker provided valuable copy-editing services. I wish to thank Veda M. Cafazzo, Editor, Ohio Biological Survey, for her editorial work, and Tom S. and Miwako K. Cooperrider, Jeffrey D. Knoop, and David P. Emmitt, for reading early versions of the manuscript. Tom S. and Miwako K. Cooperrider also reviewed early galley proofs.

Cuyahoga Community College granted to me two professional leaves to work on this project. The National Science Foundation, through Grant Number SPI-816510, provided financial support for the field work conducted during 1981 and 1982.

And finally, a sincere appreciation is expressed to my sons, Jason and Aaron. They were but five and four years old, respectively, when this endeavor began. Forever I will have fond memories of them playing in roadside streams, running up and down slopes, and inspecting rock outcrops to occupy their time while I botanized. It is for them and their children that documenting the flora of Ohio in the 1980's will have value.

Barbara K. Andreas

Cuyahoga Community College, Eastern Campus
Division of Natural Sciences
4250 Richmond Road
Warrensville Township, OH 44122

Kent State University
Department of
Biological Sciences
Kent, OH 44242

PUBLISHER'S NOTE

Publication of this Bulletin was assisted by financial contributions from the following:

Barbara K. Andreas; Anonymous; Audubon Society of Greater Cleveland; John E. Berger; Burroughs Nature Club; Davey Tree Company Foundation; The Davey Tree Expert Company in honor of Dr. Roger C. Funk; Mr. and Mrs. Robert Dix; William D. Ginn; J. Arthur Herrick; The Holden Arboretum — Horvath Forbes Education and Research Fund; Jack and Elsie Joy; Jeffrey D. Knoop; Lake Parks Foundation; Mr. and Mrs. Richard E. Leppo; Barbara A. Lipscomb; The Lubrizol Foundation; The Lillian and Thomas Mastin Foundation; Lucia S. Nash; Native Plant Society of Northeast Ohio; Evan W. Nord; The Nord Family Foundation; Gretta S. and Hugh D. Pallister, Jr. in memory of Emma M. Pallister; The Sears Family Foundation; The Bill and Edith Walter Foundation, James E. Lane, President; and The Wilderness Center Botanizers.

INTRODUCTION

The Glaciated Allegheny Plateau region of Ohio, as defined in this study, is a physiographic area comprised of all or parts of 23 counties. This area encompasses (1) all of the "northeastern or glaciated Allegheny Plateau" region, (2) the western section of the "Miami region", and (3) the northwestern portion of the "southern unglaciated area of the Allegheny Plateau" proposed by Schaffner (1932). As detailed here, the vascular plant flora of the region consists of 2002 species and 27 interspecific hybrids belonging to 706 genera and 148 families. Approximately 72% of the species and hybrids are native (indigenous) and 28% are alien (non-indigenous). One hundred fifty-one of the species are, in Ohio, more or less confined to this region, and are considered indicator, or definitive, taxa. Thirty-two of these species are among the 110 listed by the Division of Natural Areas and Preserves (1988) as presumably extirpated from Ohio.

With a renaissance taking place in Ohio floristics, primarily promoted by the work of the Ohio Biological Survey (e.g. Cusick and Silberhorn, 1977; Cooperrider, 1982); the Ohio Flora Committee of the Ohio Academy of Science (Cooperrider, 1984a); and the Ohio Natural Heritage Program of the Division of Natural Areas and Preserves, Ohio Department of Natural Resources (McCance and Burns, 1984), a significant amount of knowledge regarding Ohio's flora is being added each year. In addition, separate floras have been prepared for at least 24 individual Ohio counties (Roberts and Stuckey, 1974; Pusey, 1976; Cline, 1977; Burns, 1980; Andreas, 1980; Emmitt, 1981). To cite one example of the increased knowledge gained from such studies, 28 years ago Cooperrider (1961) estimated that 217 vascular plant species had been collected from Holmes County. After intensive field investigations, Wilson (1972) reported a total of 1,056 species collected from the county.

In order to assemble information on the flora of the Glaciated Allegheny Plateau region of Ohio, field surveys were made by the author from 1976 to 1988. Major Ohio herbaria with substantial collections from the region were examined for herbarium specimens collected from the 23 counties included in this study. Fifteen of the 23 counties included in the region were surveyed floristically at the county level by other investigators prior to this study. The purpose of this text is three-fold: 1) to compile the floras of the remaining eight counties; 2) to integrate the floras of all 23 counties into a comprehensive, modern flora of this phytogeographic region; and 3) to prepare a document on the glaciated portion of the Allegheny Plateau to complement the work of Silberhorn (1970a; 1970b) and Cusick and Silberhorn (1977), thereby completing the flora of the entire Plateau in Ohio.

The catalogue of vascular plant species for the Glaciated Allegheny Plateau region of Ohio presents information for each taxon as it is known in 1988. In the future, additional county records, including those overlooked in the field to date, those added by plant migrations, and those added by accidental and intentional introductions, will enlarge the flora. Disappearing habitats caused by manipulation of the land through human activity probably will result in deletions from the flora through species extirpation.

DEFINITION OF THE STUDY AREA

The study area includes that portion of Ohio which lies within the Glaciated Allegheny Plateau region of the Appalachian Plateau Province. It is bordered on the east by the political boundary separating Ohio and Pennsylvania. The southern boundary of the Glaciated Allegheny Plateau region is delimited by the southern extension of the Pleistocene glacial advances (Goldthwait et al, 1967). The northern boundary is marked by the Portage Escarpment (Stout et al, 1943) which separates the Glaciated Allegheny Plateau region from the Great Lakes region (Lake Plains) of the Central Lowland Province (Fenneman, 1938). The western boundary of the Glaciated Allegheny Plateau region is somewhat arbitrarily defined. Fenneman stated that the western boundary is fixed "empirically by the contrast of its topography with the smoother surface of the Till Plains in northern Ohio." Anderson (1983), following the map of

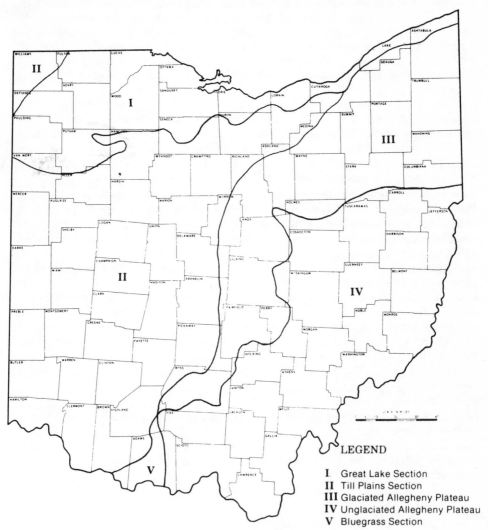

LEGEND

I Great Lake Section
II Till Plains Section
III Glaciated Allegheny Plateau
IV Unglaciated Allegheny Plateau
V Bluegrass Section

Figure 1. Physiographic sections of Ohio (Anderson, 1983).

2

the "Physiographic Sections of Ohio" (Division of Geological Survey, no date), excluded the northwestern corner of Medina County and all of Lorain County from the Glaciated Allegheny Plateau region (Figure 1). White (1982), Lafferty (1979), Stout et al (1943), and Fenneman (1938) primarily used the west-facing escarpment of the Mississippian rocks (the Allegheny Front Escarpment) as the western boundary. This boundary includes all of Medina County and most of Lorain and Huron counties. For the purpose of this study, the western boundary follows the "Physiographic Sections of Ohio" map with modifications to include Medina and Lorain Counties (Figure 2).

The Glaciated Allegheny Plateau region of Ohio includes all of seven counties (Geauga, Mahoning, Medina, Portage, Summit, Trumbull, and Wayne) and parts of 24 counties (Ashland, Ashtabula, Columbiana, Coshocton, Cuyahoga, Delaware, Fairfield, Franklin, Highland, Hocking, Holmes, Huron, Knox, Lake, Licking, Lorain, Morrow, Muskingum, Perry, Pickaway, Richland, Ross, Stark,

Figure 2. Outline of the study area.

3

Figure 3. Northern third of the study area including counties and townships.

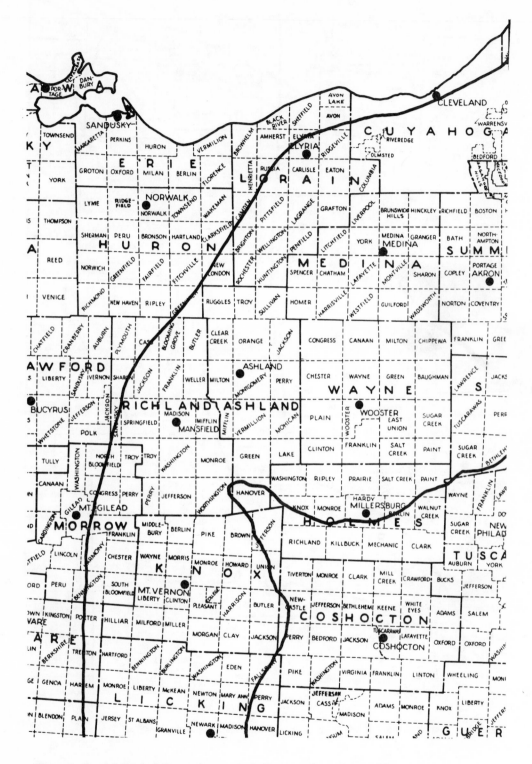

Figure 4. Central third of the study area including counties and townships.

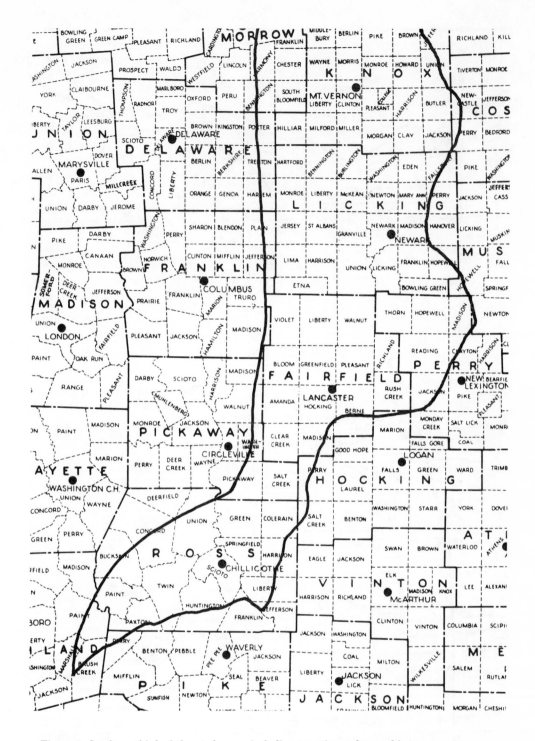

Figure 5. Southern third of the study area including counties and townships.

Table 1. Counties and townships constituting the study area, the physiographic and phytogeographical area of the Glaciated Allegheny Plateau region of Ohio.

County	Townships Included in Study Area
Ashland	Clear Creek, Green, Jackson, Lake, Mifflin, Milton, Mohican, Montgomery, Orange, Perry, Ruggles, Troy, Sullivan, Vermillion
Ashtabula	Andover, Austinburg, Cherry Valley, Colebrook, Denmark, Dorset, Harpersfield, Hartsgrove, Jefferson, Lenox, Monroe, Morgan, New Lyme, Orwell, Pierpont, Plymouth, Richmond, Rome, Sheffield, Trumbull, Wayne, Williamsfield, Windsor
Columbiana	Butler, Center, Elk Run, Fairfield, Hanover, Knox, Middleton, Perry, Salem, Unity, West
Cuyahoga	Lower 2/3 of county (not divided into townships)
Fairfield	Amanda, Berne (north 1/2), Bloom, Clear Creek, Greenfield, Hocking, Liberty, Madison, Pleasant, Richland, Rush Creek, Violet, Walnut
Geauga	All townships
Holmes	Berlin, Hardy, Knox, Monroe, Paint, Prairie, Ripley, Salt Creek, Washington
Knox	Berlin, Brown, Butler, Clay, Clinton, College, Harrison, Hilliar, Howard, Jackson, Liberty, Middlebury, Milford, Miller, Monroe, Morgan, Morris, Pike, Pleasant, Union, Wayne
Lake	Concord, Leroy, Kirtland, Madison (south 1/2)
Licking	Bennington, Bowling Green, Burlington, Eden, Etna, Fallsbury, Franklin, Granville, Harrison, Hartford, Hopewell, Jersey, Liberty, Licking, Lima, Madison, Mary Ann, McKean, Monroe, Newark, Newton, St. Albans, Union, Washington
Lorain	Brighton, Camden, Carlisle, Columbia, Eaton, Elyria, Grafton, Henrietta, Huntington, LaGrange, Penfield, Pittsfield, Ridgeville, Rochester, Russia, Wellington
Mahoning	All townships
Medina	All townships
Morrow	Bennington, Chester, Congress, Franklin, Harmony, North Bloomfield, Perry, South Bloomfield, Troy

Perry	Hopewell, Jackson, Madison, Reading, Thorn
Pickaway	Pickaway, Salt Creek, Walnut, Washington
Portage	All townships
Richland	Butler, Franklin, Jackson, Jefferson, Madison, Mifflin, Monroe, Perry, Sandusky, Sharon, Springfield, Troy, Washington, Weller, Worthington
Ross	Colerain, Green, Harrison, Huntington, Liberty, Paxton, Scioto, Springfield, Twin, Union
Stark	Bethlehem, Canton, Jackson, Lake, Lawrence, Lexington, Marlboro, Nimishillen, Osnaburg, Paris, Perry, Plain, Sugar Creek, Washington, Tuscarawas
Summit	All townships
Trumbull	All townships
Wayne	All townships

and Tuscarawas). Eight of the latter counties have been excluded from the study. Delaware and Franklin Counties have been omitted because the western boundary of the Allegheny Plateau transects only a part of the easternmost tier of townships within these counties. Coshocton, Highland, Hocking, Huron, Muskingum, and Tuscarawas Counties have been eliminated because each has an area of less than a township within the Glaciated Allegheny Plateau region. Counties and townships which constitute the study area are illustrated in Figures 3, 4, and 5 and listed in Table 1.

A flora inclusive of the Glaciated Allegheny Plateau region of Ohio is difficult to define due to the area's irregular and nebulous natural boundaries. In addition, incomplete location data provided on some herbarium specimens make it difficult to determine if the specimen was collected within the phytogeographical region. During the preparation of this study, data from herbarium records for all 23 counties were collected. Records from portions of the counties outside the boundaries of the region are presented in Appendix A. Where county records represent taxa that are definitive to a neighboring phytogeograhic region, such as the Great Lakes region (Anderson, 1983) or the Unglaciated section (Cusick and Silberhorn, 1977), a notation is made in Appendix A.

Schaffner (1932) divided Ohio into nine phytogeographic regions (Figure 6). Two of these regions, the "central non-glaciated region" and the "southern non-glaciated region" were studied by Silberhorn (1970a), and Cusick and Silberhorn (1977). However, Schaffner's phytogeographic regions do not co-incide with the natural boundaries of the Unglaciated Allegheny Plateau. For example, most of glaciated Perry and Ross Counties is included in Schaffner's unglaciated region. Because of the way Schaffner drew his regional boundaries, several counties included by Cusick and Silberhorn (1977) are also part of the present study.

8

Figure 6. Phytogeographic regions of Ohio (Schaffner, 1932).

All or parts of the northernmost sixteen counties (Ashland, Ashtabula, Columbiana, Cuyahoga, Geauga, Holmes, Lake, Lorain, Mahoning, Medina, Portage, Richland, Stark, Summit, and Wayne) in the study area are contained in Schaffner's "northeastern or glaciated Allegheny Plateau region." These counties form a coherent phytogeographic unit characterized by species with wetland and boreal affinities. The glacial geology of these counties is similar and is discussed by White (1982).

Schaffner (1932) did not separate the south-central glaciated counties (Fairfield, Knox, Licking, Morrow, and Pickaway) as a distinct region and incorporated them within the "Miami glaciated" phytogeographic region. These south-central counties are primarily a transitional region separating the calcareous substrates of western Ohio from the acidic substrates of eastern Ohio, and include representative taxa from both.

9

In Ohio, 151 taxa are more or less confined to the Glaciated Allegheny Plateau region of Ohio (Appendix B). This list of definitive taxa includes 22 of the 38 taxa listed by Schaffner (1932) as characteristic of the Glaciated Allegheny Plateau region. These vascular plants typically have a northern affinity (Thompson, 1939; Stuckey and Denny, 1981), with approximately 50% of the taxa occurring in wetlands, especially peatlands (bogs and fens). Another 18% occur in the hemlock-white pine-hardwood communities of extreme northeastern Ohio. Approximately 7% of the taxa occur on eroding slopes found where post-Pleistocene streams have cut through thick lacustrine deposits (Stout et al, 1943) and shales. The remaining 25% occur in a variety of woodland communities, e.g., beech-maple forests, hemlock slopes, oak-hickory forests, and mixed mesic forests. Forty-four of the 151 definitive taxa are known within the region from only one or two collections. Apparently, in Ohio, these taxa always have been rare. Examples include *Lonicera oblongifolia* (swamp fly honeysuckle), *Oxalis montana* (white wood-sorrel), and *Streptopus roseus* (twisted-stalk).

DESCRIPTION OF THE STUDY AREA
CLIMATE

The climate of the Glaciated Allegheny Plateau region of Ohio is relatively uniform, with cooler average temperatures occurring in the northern tier of counties, and warmer average temperatures occurring in the southern portion due to differences in latitude and topography of the area. Changes in the weather occur every few days from passing cold or warm fronts, which generally approach from the northwest or southwest (Miller, 1969).

The growing season for most of the region ranges from 176 days in Ross County to 133 days in Portage and southern Ashtabula Counties. Much longer growing seasons occur in the Lake Plains region of the northern counties where Lake Erie modifies the climate. In areas along Lake Erie in Lake and Lorain Counties, for example, the average growing season is 193 days and 184 days, respectively. The moderating effect of Lake Erie is felt beyond the line separating the Lake Plains from the Allegheny Plateau, but this effect is diminished quickly.

Average annual precipitation in the northern counties varies from a mean high of 111.4 cm in Geauga County to a low of 86.6 cm in Mahoning and Trumbull Counties. In the southernmost counties, Ross and Pickaway, the mean high is approximately 96.5 cm. Annual mean snowfall is highest in Geauga County with 269.2 cm and lowest in Ross County with 54.9 cm. While the northeastern counties bordering Lake Erie are in the snowbelt region of Ohio, the effect of that snowfall is confined mostly to the area near the Portage Escarpment. Short periods of drought may occur yearly, but severe droughts are infrequent.

Daily mean temperatures range from a high of 18.2° C and 23.8° C in Ross and Pickaway Counties, respectively, to a low of 3.0° C and 2.4° C in Portage and Ashtabula Counties, respectively.

The overall effect of the higher daily mean temperatures, lower snowfall, and longer growing season are factors that affect the differences in overall vegetation seen in the extreme southern counties of the Glaciated Allegheny Plateau region of Ohio. The above climatic data are summarized from Miller (1969) and Ruffner (1980).

TOPOGRAPHY

The topography of the Glaciated Allegheny Plateau region of Ohio has been influenced by three main factors: 1) glaciation, 2) bedrock elevational highs, and 3) stream erosion. The overall topography may be described as broad, gently rolling uplands with moderate relief. This is typical of glaciated topography where stream valleys have been filled with glacial deposits and the uplands have been abraded by ice.

The highest elevation in the study area is 448 m above sea level near Mansfield, Richland County. This upland extends into northeastern Morrow County where elevation is as high as 440 m above sea level. Other regions with relatively high elevation include northern Geauga County with an elevation of 408 m above sea level; eastern Stark County, 415 m above sea level; and an area near the Summit County-Medina County line, 403 m above sea level. These elevational highs are formed by underlying bedrock.

The lowest elevation in the region, 180 m above sea level, is in Ross County where the Scioto River crosses the glacial border. Another low area occurs in Cuyahoga County where the Cuyahoga River enters the Lake Plains. Here the elevation is 183 m above sea level.

Local relief is moderate except where major modern streams have carved out valleys. At these areas local relief can be as much as 60 m. Due to continuing stream erosion, these slopes are usually subject to slumping.

Fenneman (1938) describes a "sag" occurring in the upland area of northeastern Ohio. This sag corresponds to the pre-Illinoian Grand River valley which extends south through Ashtabula and northern Portage and Trumbull Counties. Fuller (1965) and Winslow (1957) consider this area to be a continuation of the Lake Plains, but its flora is included in this study.

The earliest evidences of drainage patterns for Ohio are from the late Tertiary period. The major Ohio stream of that period was the Teays River (Ver Steeg, 1946). The term "Teays River System", as used by Stout et al (1943), refers to the Teays River, its tributaries, and all contemporary streams. The Teays River System was formed in the general period of erosion that was terminated by the Pleistocene.

During the late Tertiary period the study area was separated by a major divide that traversed northern Morrow County, central Richland County, northern Ashland County, central Wayne County, then turned southeast to southeastern Belmont County, then directly northward to the southeastern corner of Ashtabula County (this eastern edge of the divide corresponds to the present-day Flushing Escarpment, Stout et al, 1943). That portion of the study area to the north of this divide primarily drained northward into the Erigan River (Spencer, 1894) by the Dover and Ravenna Rivers (Stout et al, 1943). The remainder of the study area drained south and southwest through the Groveport and Cambridge Rivers and their tributaries, and eventually into the Teays River.

The Teays River System within the study area was obliterated with the advent of the first Pleistocene ice advance. No river drainage within the Glaciated Allegheny Plateau region of Ohio was left untouched by these advances. Drainage evidence of these glacial advances include stream reversal, ponded finger lakes and new valleys formed by the spilling over of ponded water.

Drainage patterns changed with each Pleistocene glacial advance. Stout et al (1943) and Cummins (1959) present maps showing hypothetical drainage patterns during the various ice advances. Andreas (1985) discussed how pre-Pleistocene drainage patterns influenced the distribution of Ohio peatlands.

The Wisconsinan ice advances and their moraines have had the most recent significant impact on drainage patterns and have formed the present drainage systems in glaciated Ohio. When the Wisconsinan ice melted, present-day streams became established. A continental divide now separates the study area into the St. Lawrence River and Mississippi River drainage systems. The area to the north of the divide primarily drains into Lake Erie through six major rivers: Ashtabula River, Grand River, Chagrin River, Cuyahoga River, Rocky River, and Black River.

Drainage from the region south of the continental divide flows into the Ohio River via eight major streams: Mahoning River, Tuscarawas River, Killbuck Creek, Mohican River, Kokosing River, Licking River, Hocking River, and Scioto River.

'The Mahoning River joins Beaver River upstream from the confluence with the Ohio River. The Mohican and Kokosing Rivers become the Walhonding River. Killbuck Creek joins the Walhonding which, along with the Licking River, flow into the Muskingum River which empties into the Ohio River. The Hocking and Scioto Rivers drain directly into the Ohio River.

Present-day drainage has had two major effects on the vegetation of the Glaciated Allegheny Plateau region of Ohio. First, many marshes and peat deposits are found on poorly drained areas along the continental divide. Secondly, where contemporary streams have created steep ravines and eroding slopes, distinctive plant communities have developed. In narrow ravines, especially in those facing north and northeast, hemlocks (*Tsuga canadensis*) frequently dominate the vegetation. Where the dissected valleys are broad with south- or west-facing slopes, eroding slope communities often form. These slopes are frequently dominated by dry oak woodlands on the uplands.

GEOLOGY

The seas that covered Ohio during the Paleozoic era laid down sedimentary rock systems that are distinguished as Ordovician, Silurian, Devonian, Mississippian, Pennsylvanian, and Permian (Figure 7). In the Glaciated Allegheny

Plateau region of Ohio, only Devonian, Mississippian, and Pennsylvanian deposits are found above stream levels. Tills deposited during Pleistocene glaciation cover the surface bedrock. Bedrock is observed within the region only in areas of rock exposures such as cliffs, ledges, ravines, and bluffs where tills have been eroded away, and in areas where tills have been removed by roadcuts and quarries.

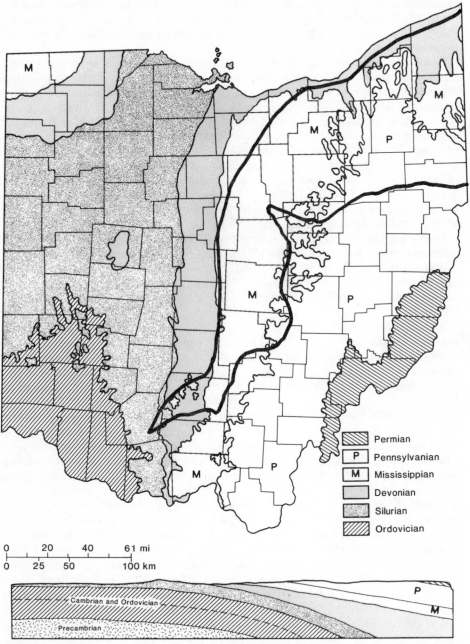

Figure 7. Geological map and cross-section of Ohio. Modified from Ohio Division of Geological Survey.

13

Devonian rocks are the major bedrock strata in Ashtabula, Lake, Pickaway, and Ross Counties. These rocks, primarily marine-deposited limestones and shales (Szmuc, 1970a), are exposed along the banks of major rivers, such as the Grand River, Chagrin River, and Cuyahoga River, and their tributaries.

Mississippian-age rocks were formed from marine, transitional, and non-marine sediments (Szmuc, 1970b). These strata, comprised of shales and sandstones, are the dominant bedrocks in 11 counties: Ashland, Geauga, Fairfield, Knox, Licking, Lorain, Medina, Richland, Summit, Trumbull, and Wayne. Mississippian-age rocks occur in stream and river valleys throughout these counties.

Pennsylvanian-age bedrock is the dominant type in five counties: Columbiana, Holmes, Mahoning, Portage, and Stark. In addition, Pennsylvanian-age rocks make up the elevational highs in Geauga, Medina, and Summit Counties. According to Rau (1970), these rocks were formed from 1) river deposited sand, 2) floodplain deposited silts and clays, and 3) coastal plain or floodplain deposited coals. The oldest, and botanically most important, Pennsylvanian-age rock in Ohio is Sharon Conglomerate. It occurs as massive rock outcrops in Geauga, Portage, and Summit Counties at well-known natural areas such as Thompson Ledges, Geauga County; Nelson Ledges, Portage County; and Virginia Kendall Park and Whipps Ledges, Summit County. After sedimentation and rock formation during the Pennsylvanian period, depositional seas never covered the study area again.

Following the ice advances during the Pleistocene (Figure 8), bedrock and bedrock-derived soils no longer were the major factor influencing the vegetation of the study area. While no depositional signs of the pre-Illinoian ice advances are evident within the Glaciated Allegheny Plateau region (White, 1934; Goldthwait et al, 1967), no surface within the study area was left untouched by them.

The Illinoian ice advance is well documented within the Glaciated Allegheny Plateau region, and it is the Illinoian margin that forms the eastern and southern boundaries of three-fourths of the study area. The Illinoian till is exposed in narrow bands ranging in width from 3-6 km in Columbiana and Stark Counties, to 16-22 km in Knox and Perry Counties (Goldthwait et al, 1967).

Illinoian tills have had little effect on the vegetation of the region. Where Illinoian till is exposed, much of it has washed away (Goldthwait, 1959), or is so thin that the parent sandstones and shales have a strong influence on soil formation. Illinoian till supports the same vegetation type (mixed mesophytic and mixed oak forests) as that occurring in the unglaciated portions of these counties (Gordon, 1966).

The Wisconsinan, the last ice advance to enter Ohio, reached its maximum extension between 24,600 and 16,600 years BP (before present). Its tills have had a significant impact on the present vegetation of Ohio. According to Goldthwait (1959), the Wisconsinan added as much as 27 m of tills to those already laid down by previous advances. Tills in the Teays-age buried rivers are as much as 180-200 m in depth (Stout et al, 1943). Bedrock areas above 395 m, primarily the highs in Richland, Morrow, Ashland, and Geauga Counties, were covered with only thin layers of tills. Many of these have been washed away, or the soils formed are heavily influenced by parent bedrock. Mixed mesophytic woodlands have developed on these areas.

The areas of thicker tills throughout much of northern Fairfield, western Licking, Knox, Morrow, Richland, Lorain, Medina, Geauga, Ashtabula, and

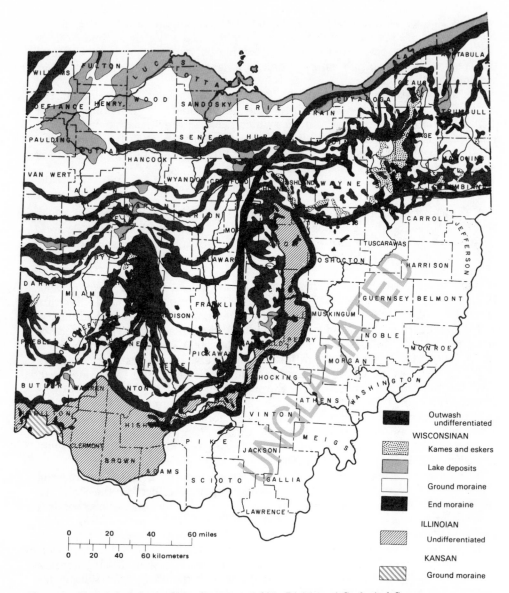

Figure 8. Glacial deposits in Ohio. Courtesy of Ohio Division of Geological Survey.

northeast Portage and northwest Trumbull Counties, are dominated by beech-maple woodlands. A somewhat nebulous line that separates areas occupied by beech-maple forests from those occupied by mixed oak and mesophytic forests more or less corresponds to the border of the maximum Wisconsinan advance (Gordon, 1966; Forsyth, 1970).

The Wisconsinan advance also influenced the distribution of Ohio peatlands. With the exception of several fens found on outwash deposits in the valleys of Rush and Clear Creeks in Fairfield County, peatlands in Ohio are confined to the area glaciated by the Wisconsinan advance (Andreas, 1985).

The flora of the southern portion of the Glaciated Allegheny Plateau region of Ohio, both in community type and species present, includes plants from

15

the Till Plains, as well as from unglaciated Ohio. This area differs from the remainder of the region in that tills are thinner and are higher in lime content (its origins having been from limestone-dominated western Ohio). In addition, this region has a warmer climate and a longer growing season than other portions of the study area.

Overall, Pleistocene glaciation has affected the present vegetation of the section by 1) modifying topography and local relief, 2) providing tills of different origins, thicknesses, and ages which result in the formation of different soil types, and 3) changing drainage patterns. Forsyth (1970) discusses the relationship of glacial geology to vegetation. White (1982) presents a detailed study of glacial geology in northeastern Ohio.

VEGETATION

A flora, presumably comprised of Arcto-Tertiary elements (Braun, 1955) occurred throughout the region prior to the advance of Pleistocene glaciers. As glaciers melted, colonization of the exposed bedrock and glacial deposits began and eventually led to the vegetation present at the time of the arrival of the European settlers. This flora resulted from 1) the northward migration of plant species which had survived south of the glacial moraine (Braun, 1950; 1955; Delcourt and Delcourt, 1981); 2) the establishment in suitable habitats of northern plants which had migrated into Ohio in front of the glacial advances; 3) the eastward migration of prairie plants and plants more typical of drier areas during the Xerothermic Period 8,000 to 5,000 years BP (Benninghoff, 1964); and 4) the westward migration of coastal species via eastward drainage channels that formed in the St. Lawrence lowlands as the ice front retreated.

At the time of the arrival of the European settlers about 98% of Ohio was forested (Gordon, 1966; Cooperrider, 1982). Forest types included beech, mixed oak, oak-sugar maple, elm-ash-maple swamp forests, and mixed mesophytic. The remaining 2% of the surface of Ohio was in open areas such as freshwater marshes, bogs, fens, prairies, oak barrens, and oak openings (Sears, 1926; Transeau, 1935; Gordon, 1966; 1969).

While Native Americans modified the vegetation to a small extent, the arrival of the European settlers had a monumental effect on the flora. In addition to changing the landscape by building roads and canals, these settlers cleared large areas of forests and converted the land to agricultural fields. Peatlands, once widespread (Dachnowski, 1912), were mined for peat and quarried for the underlying calcareous marl which was used for fertilizer. Wetlands were further destroyed through drainage (Gordon, 1969). Streams and rivers were channelized to control flooding, and dammed to provide feeder systems for canals and later for power and water sources.

Another major effect that Europeans had on the flora was the introduction of non-native taxa. Cooperrider (1982) estimates that 900 of the 2700 species of vascular plants in Ohio are alien.

While none of the major vegetation types found in the Glaciated Allegheny Plateau region of Ohio at the time of the earliest European settlers has disappeared, the amount of land occupied by each type has been reduced greatly.

16

PLANT COMMUNITIES

Throughout the Glaciated Allegheny Plateau region of Ohio are groups of plants that usually grow together. These plant associations, or plant communities, share similar edaphic factors such as the availability of moisture, parent geologic material, topography, and direction of slope. These communities do not occur at random; their distributions are predictable on the basis of geology and topography. Forsyth (1970) stated that the "correlation between the vegetation and the associated geology is expectable [because] the original vegetation [has had] thousands of years to adjust to the underlying substrate." To date there is no complete study of the plant communities of Ohio. Anderson (1982), however, prepared a preliminary list of community types and presented descriptions and species lists for many of them.

This section highlights the dominant plant communities of the Glaciated Allegheny Plateau region of Ohio. In general, the names of naturally occurring plant communities follow Anderson (1982). The brief descriptions and lists of taxa are intended to acquaint the reader with common community types and some of their prevalent vascular plant species. The lists, arranged alphabetically by tree, shrub, and herbaceous vegetation layer, are not all-inclusive. Dominance is not implied unless it is so stated. It should also be remembered that plant communities do not have abrupt boundaries but instead often merge gradually into one another. Predominant plant communities of the section are: wet forests, mesic forests, dry forests, prairie remnants, eroding slope communities, wetlands, and communities of disturbed sites.

Wet Forests

BOGGY FORESTS. Boggy forests develop at the edge of kettle lakes or acidic bogs in wet depressions where soils have high organic content. In Ohio, this plant community is confined to the Glaciated Allegheny Plateau region. Gordon (1969) referred to this community type as developing in the absence of *Picea mariana* (black spruce), *Larix larcina* (tamarack), *Pinus strobus* (white pine), and *Tsuga canadensis* (hemlock). Trees characteristic of boggy forests include *Acer rubrum* (red maple), *Acer saccharinum* (silver maple), *Betula alleghaniensis* (yellow birch), *Nyssa sylvatica* (sour gum), *Quercus palustris* (pin oak), and *Ulmus americana* (American elm). Shrubs include *Ilex verticillata* (winterberry), *Lindera benzoin* (spicebush), *Rubus hispidus* var. *obovalis* (small dewberry), *Viburnum lentago* (nannyberry), and *Viburnum recognitum* (arrowwood). The herbaceous layer is dominated by the following ferns: *Dryopteris* x *boottii* (Boott's fern), *Dryopteris cristata* (crested wood fern), *Dryopteris marginalis* (marginal shield fern), *Dryopteris spinulosa* (spinulose shield fern), *Osmunda cinnamomea* (cinnamon fern), and *Osmunda regalis* (royal fern). Other herbaceous taxa include *Brachyelytrum erectum*, *Carex canescens* (glaucous sedge), *Carex howei* (Howe's sedge), *Carex seorsa* (weak bog sedge), *Carex trisperma* (three-seeded sedge), *Circaea lutetiana* (enchanter's nightshade), *Hydrocotyle americana* (American pennywort), *Maianthemum canadense* (Canada mayflower), *Pilea pumila* (clearweed), *Trillium flexipes*, and *Viola cucullata* (marsh violet).

HEMLOCK-WHITE PINE-HARDWOOD FORESTS. This community type is restricted to, and is rare in the Glaciated Allegheny Plateau region of Ohio, entering northeastern Ohio in Ashtabula County, with small outliers in Geauga,

Portage, and Trumbull Counties (Braun, 1950). This forest type occurs on more or less flat topography, such as river terraces and broad pre-Wisconsinan buried river valleys. Tree species include *Acer pensylvanicum* (striped maple), *Acer rubrum* (red maple), *Betula alleghaniensis* (yellow birch), *Fagus grandifolia* (beech), *Magnolia acuminata* (cucumber-tree), *Prunus serotina* (wild cherry), and *Tsuga canadensis* (hemlock). *Pinus strobus* (white pine) was, prior to lumbering, a frequent component of this community type (Hicks, 1933). Shrubs include *Lonicera canadensis* (fly honeysuckle), *Sambucus pubens* (red-berried elder), *Viburnum alnifolium* (hobble-bush), and *Viburnum lentago* (nannyberry). *Aralia nudicaulis* (wild sarsaparilla), *Claytonia caroliniana* (Carolina spring beauty), *Dryopteris spinulosa* (spinulose shield fern), *Maianthemum canadense* (Canada mayflower), *Tiarella cordifolia* (foamflower), and *Viola rotundifolia* (round-leaved yellow violet) are common understory taxa. Pools of standing water frequently occur in these forests. Typical species of the pools include *Nyssa sylvatica* (sour gum), *Cephalanthus occidentalis* (buttonbush), *Vaccinium corymbosum* (blueberry), *Viburnum lentago* (nannyberry), *Viburnum recognitum* (arrow-wood), *Carex canescens* (glaucous sedge), *Carex seorsa* (weak bog sedge), *Coptis trifolia* (goldthread), *Cyperus engelmannii* (Engelmann's umbrella sedge), *Lycopus* spp. (water horehound), *Osmunda cinnamomea* (cinnamon fern), *Trientalis borealis* (star-flower), *Viola pallens* (northern white violet), *Woodwardia virginica* (chain-fern), and *Sphagnum* spp. (sphagnum moss).

SWAMP FORESTS. According to Gordon (1966), this community type occurred throughout the region on flat areas of poor drainage such as floodplains, old lake beds, and post-glacial flats. Forsyth (1970) listed lacustrine clays and silts as parent materials on which these communities develop. Dominant trees are *Acer rubrum* (red maple), *Acer saccharinum* (silver maple), *Fraxinus americana* (white ash), *Ulmus americana* (American elm), and occasionally *Carya laciniosa* (big shellbark hickory), *Carya ovata* (shagbark hickory), *Quercus bicolor* (swamp white oak), and *Quercus palustris* (pin oak). Understory species include *Sambucus canadensis* (elderberry), *Bidens* spp., *Carex* spp., *Cinna arundinacea* (wood reedgrass), *Laportea canadensis* (wood nettle), *Leersia oryzoides* (rice cutgrass), *Onoclea sensibilis* (sensitive fern), and *Phalaris arundinacea* (reed canary grass). Sampson (1930a) described this community type as it occurs in northwestern Ohio.

FLOODPLAIN FORESTS. Floodplain forests have developed within the region on thick silt deposits along floodplains of major rivers such as the Scioto, Killbuck, and Cuyahoga. Trees in this community type include *Acer negundo* (box-elder), *Acer saccharinum* (silver maple), *Aesculus glabra* (Ohio buckeye), *Fraxinus pennsylvanica* (red ash), *Juglans nigra* (black walnut), *Platanus occidentalis* (sycamore), *Populus deltoides* (cottonwood), and *Ulmus americana* (American elm). Understory species are *Sambucus canadensis* (elderberry), *Athyrium thelypteroides* (silvery spleenwort), *Hystrix patula* (bottlebrush grass), *Impatiens capensis* (jewelweed), *Leersia oryzoides* (rice cutgrass), *Leersia virginica* (white grass), *Matteuccia struthiopteris* (ostrich fern), *Parthenocissus quinquefolia* (Virginia creeper), *Polygonum hydropiper* (common smartweed), *Polygonum virginianum* (jumpseed), *Rhus radicans* (poison ivy), *Rudbeckia laciniata* (cut-leaf coneflower), *Verbesina alternifolia* (wingstem), and *Viola striata* (striped white violet). *Alliaria petiolata* (garlic mustard) and *Hesperis matronalis* (dame's-rocket), both alien taxa, are becoming increasingly

18

well-established in floodplain forests throughout the study area. Hardin and Wistendahl (1983) described floodplain forests in the unglaciated Allegheny Plateau region.

Mesic Forests

Mesic forests, as used in this study, include mixed mesophytic, beech-maple, and hemlock slope forests. These community types are treated together because 1) the topography on which they develop is similar, 2) members of the tree layer are similar, although they vary in percentage of species composition, and 3) the shrub and herbaceous understory layers are similar. Mesic forests are common on the slopes of ravines, stream terraces, and gentle slopes. Large areas of mesic deciduous forests occur throughout the region.

BEECH-MAPLE FORESTS. Beech-maple forests occur in areas covered with ground and end moraines (Forsyth, 1970). The southern extent of beech-maple *(Fagus grandifolia-Acer saccharum)* forests is generally that of Wisconsinan glaciation (Braun, 1950). This community type occurs on areas of gently rolling topography and on gentle north- and northeast-facing slopes. In relatively undisturbed beech-maple communities studied throughout the region, Pell and Mack (1977) found, with the exception of *Podophyllum peltatum*, few common taxa. Andreas (1980) and Gordon (1969), listed several taxa typically found in beech-maple forests. Trees include *Acer rubrum* (red maple), *Acer saccharum* (sugar maple), *Carpinus caroliniana* (American hornbeam), *Liriodendron tulipifera* (tulip-tree), *Magnolia acuminata* (cucumber-tree), *Prunus serotina* (wild cherry), *Quercus alba* (white oak), and *Quercus rubra* (red oak). Other taxa are *Hamamelis virginiana* (witch-hazel), *Viburnum acerifolium* (maple-leaved viburnum), *Carex* spp., *Disporum lanuginosum* (yellow mandarin), *Dryopteris intermedia* (intermediate shield fern), *Epifagus virginiana* (beech-drops), *Erythronium americanum* (yellow fawn lily), *Festuca obtusa* (nodding fescue), *Floerkea proserpinacoides* (false mermaid-weed), *Hepatica acutiloba* (sharp-lobed hepatica), *Milium effusum* (millet grass), *Osmorhiza claytonii* (sweet jervil), *Phlox divaricata* (woodland phlox), *Poa* spp., *Polygonatum biflorum* (smooth Solomon's seal), *Smilacina racemosa* (false Solomon's seal), *Smilax glauca* (greenbrier), *Sanicula* spp., and *Viola* spp. In northeastern Ohio, occasionally well-drained areas support nearly homogeneous stands of *Fagus grandifolia*. Temporary pools and small streams are frequent in beech-maple forests. Members of the herbaceous flora associated with these moister areas include *Erythronium albidum* (white fawn lily), *Lilium canadense* (Canada lily), *Mertensia virginica* (Virginia bluebells), *Onoclea sensibilis* (sensitive fern), *Panax trifolius* (dwarf ginseng), *Ranunculus septentrionalis* (swamp buttercup), and *Viola cucullata* (marsh violet).

MIXED MESOPHYTIC FORESTS. These forests contain a wide assortment of trees, including *Acer rubrum* (red maple), *Acer saccharum* (sugar maple), *Carpinus caroliniana* (American hornbeam), *Carya cordiformis* (bitternut hickory), *Carya ovata* (shagbark hickory), *Fraxinus americana* (white ash), *Fagus grandifolia* (beech), *Liriodendron tulipifera* (tulip-tree), *Magnolia acuminata* (cucumber-tree), *Ostrya virginiana* (hop-hornbeam), *Prunus serotina* (wild cherry), *Quercus alba* (white oak), and *Quercus rubra* (red oak). *Amelanchier arborea* (serviceberry), *Cornus alternifolia* (alternate-leaved dogwood), *Cornus florida* (flowering dogwood), *Hamamelis virginiana* (witch-hazel), *Lindera ben-*

zoin (spicebush), *Ribes cynosbati* (gooseberry), and *Viburnum acerifolium* (maple-leaved viburnum) are frequent shrubs. The mixture of trees and shrubs varies from locality to locality and is related to drainage and slope. According to Forsyth (1970), this community type is common to areas of resistant sandstone, conglomerate ridges, and gravel kames. Understory species include *Actaea pachypoda* (white baneberry), *Anemonella thalictroides* (rue-anemome), *Bromus ciliatus* (fringed brome), *Carex cephalophora, Carex rosea, Viola pubescens* (common yellow violet), *Viola rostrata* (long-spurred violet), and many species that are associated with beech-maple forests. Sampson (1930b) discussed mixed mesophytic forests in Ohio.

HEMLOCK SLOPE FORESTS. With few exceptions, the distribution of hemlock in Ohio is limited to the study area (Gordon, 1969). Black and Mack (1977) conducted an ecological study of hemlock ravines throughout the Glaciated Allegheny Plateau region. This community type occurs in cool microhabitats on steep north- or northeast-facing slopes and in narrow gorges. Rock ledges, rock outcrops, and occasionally waterfalls are associated with these communities. Taxa associated with hemlock *(Tsuga canadensis),* are *Betula alleghaniensis* (yellow birch), *Betula lenta* (cherry birch), *Acer spicatum* (mountain maple), *Cornus florida* (flowering dogwood), *Sambucus pubens* (red-berried elder), *Taxus canadensis* (Canada yew), *Viburnum acerifolium* (maple-leaved viburnum), *Aralia racemosa* (spikenard), *Carex blanda, Carex debilis* var. *rudgei* (weak sedge), *Carex pedunculata* (long-stalked sedge), *Cinna arundinacea* (wood reedgrass), *Circaea alpina* (small enchanter's nightshade), *Cystopteris fragilis* (fragile fern), *Dryopteris marginalis* (marginal shield fern), *Dennstaedtia punctilobula* (hay-scented fern), *Hieracium paniculatum, Maianthemum canadense* (Canada mayflower), *Lycopodium lucidulum* (shining clubmoss), *Solidago flexicaulis* (zigzag goldenrod), *Viola blanda* (sweet white violet), *Viola rotundifolia* (round-leaved yellow violet), and *Viola hastata* (halberd-leaf violet). Rock outcrops and ledges within the hemlock ravines often support *Camptosorus rhizophyllus* (walking-fern), *Cystopteris bulbifera* (bulblet fern), *Cystopteris fragilis* (fragile fern), *Lycopodium lucidulum* (shining clubmoss), *Pilea pumila* (coolwort), *Polypodium virginianum* (polypody fern), and *Thelypteris phegopteris* (long beech-fern).

Dry Forests

OAK-HICKORY FORESTS. Throughout the Glaciated Allegheny Plateau region, dry forests form in thin soils over sandstone bedrock, on well-drained ridges formed by stream erosion (Andreas, 1980), and on large areas of gravel deposits such as kames and gravel outwash (Forsyth, 1970). Trees in this plant community are *Carpinus caroliniana* (American hornbeam), *Carya ovalis* (sweet pignut hickory), *Carya ovata* (shagbark hickory), *Ostrya virginiana* (hop-hornbeam), *Quercus alba* (white oak), *Quercus muhlenbergii* (chinquapin oak), *Quercus rubra* (red oak), and *Quercus velutina* (black oak). *Amelanchier arborea* (serviceberry), *Cornus florida* (flowering dogwood), *Gaylussacia baccata* (huckleberry), and *Vaccinium pallidum* (sweet early blueberry) are common shrubs. Frequent herbaceous taxa include *Aureolaria virginica* (downy false foxglove), *Baptisia tinctoria* (wild indigo), *Carex artitecta, Carex pensylvanica, Danthonia spicata* (poverty grass), *Hieracium paniculatum, Hieracium venosum* (rattlesnake weed), *Helianthus hirsutus, Houstonia caerulea* (bluets), *Krigia biflora* (dwarf dandelion), *Lespedeza* spp., *Panicum lanuginosum,* and *Panicum linearifolium.* Where the canopy is open, areas frequently are dominated by members of the Ericaceae. This plant association includes *Chimaphila maculata* (spotted wintergreen), *Gaultheria procumbens* (wintergreen), *Mitchella repens* (partridge-berry), *Vaccinium angustifolium* (low sugarberry), and *Vaccinium pallidum* (sweet early blueberry).

Prairie Remnants

PRAIRIE REMNANTS. Prairie remnants are a type of treeless plant community within the Glaciated Allegheny Plateau region of Ohio. According to Sears (1926), prairies in Ohio are more or less confined to glaciated areas, and are best developed in the Till Plains in Madison, Union, and Wyandot Counties. Sears listed prairies and "oak barrens" in Ashland, Holmes, Lorain, Pickaway, Portage, Stark, Trumbull, and Wayne Counties. Transeau (1935) mapped prairies as occurring in Fairfield, Ross, and Summit Counties. To the best of the author's knowledge, no true prairies remain in the study area. Remnants, including dry areas rich in prairie indicators, still occur in Richland, Stark, Wayne (Cusick and Troutman, 1978), and Summit Counties (Andreas, 1980). Plants in these remnants include *Corylus americana* (hazel-nut), *Salix humilis* (upland willow), *Allium cernuum* (nodding onion), *Andropogon gerardii* (big bluestem), *Andropogon scoparius* (little bluestem), *Anemone virginiana* (thimbleweed), *Coreopsis tripteris* (coreposis), *Desmodium paniculatum, Euphorbia corollata* (flowering spurge), *Lespedeza hirta* (pubescent bush-clover), *Linum virginianum* (Virginia flax), *Rudbeckia hirta* (black-eyed Susan), *Silphium trifoliatum* (whorled rosinweed), *Sorghastrum nutans* (Indian grass), and *Tradescantia ohiensis* (Ohio spiderwort).

Eroding Slopes

ERODING SLOPES. Eroding slope communities occur on steep, eroding sides of major streams. In the Glaciated Allegheny Plateau region, these communities occur along the Grand River, Chagrin River, Conneaut River, Furnace Run, Rocky River, Tinkers Creek, and Cuyahoga River. According to Gardner (1972), rivers continuously undercut the "toe" of the banks causing the upper slope to slump. Parent materials of these slopes are usually shales, or lacustrine deposits that have been laid down by glacial lakes. These deposits may be as much as 33 m thick (Stout et al, 1943). Plants typically found on eroding slopes are *Corylus americana* (hazel-nut), *Salix humilis* (upland willow), *Shepherdia canadensis* (buffalo-berry), *Andropogon gerardii* (big bluestem), *Andropogon scoparius* (little bluestem), *Andropogon virginicus* (brome grass), *Apocynum androsaemifolium* (spreading dogbane), *Asclepias quadrifolia* (whorled milkweed), *Aster laevis* (smooth aster), *Carex aurea* (golden-fruited sedge), *Carex eburnea* (bristle-leaved sedge), *Carex granularis, Carex pensylvanica, Coreopsis tripteris* (coreopsis), *Erigeron pulchellus* (Robin's plantain), *Equisetum arvense* (field horsetail), *Euphorbia corollata* (flowering spurge), *Gentiana quinquefolia* (stiff gentain), *Hypoxis hirsuta* (yellow-eyed grass), *Krigia biflora* (dwarf dandelion), *Lobelia spicata* (pale spike lobelia), *Silphium trifoliatum* (whorled rosinweed), *Sorghastrum nutans* (Indian grass), and *Taenidia integerrima.* Many of these eroding slopes have the aspect of a prairie. The uplands above eroding slopes are usually dry oak-hickory forests. Andreas (1980) presented an extensive list of species found on an eroding slope in Summit County.

Wetlands

Wetlands are frequent within the Glaciated Allegheny Plateau region of Ohio, especially in the northern counties. Wetlands include the following plant communities: rivers, streams, ponds, lakes, marshes, shrub swamps, bogs, and fens. The type of plant community and the degree of vascular plant diversity occurring in wetlands are determined by 1) the rate of water movement, 2) the depth of water, 3) the chemistry of the substrate, 4) the source of water entering the wetland, 5) the amount of disturbance, and 6) the impact of run-off from surrounding uplands. Wetlands generally are open areas where tree species may be present but are usually stunted and not part of the dominant vegetation.

RIVERS AND STREAMS. Rivers and streams are abundant throughout the Glaciated Allegheny Plateau region of Ohio. With the exception of slow-moving rivers, streams in full sunlight, or oxbows, these communities do not exhibit high vascular plant diversity. The following taxa are associated with rivers and streams and their margins: *Alnus rugosa* (speckled alder), *Alnus serrulata* (common alder), *Salix* spp., *Athyrium thelypteroides* (silvery spleenwort), *Carex torta* (twisted sedge), *Cyperus* spp., *Elodea canadensis* (Canadian waterweed), *Elymus riparius* (river wild-rye), *Equisetum hyemale* (scouring rush), *Glyceria striata* (fowl manna grass), *Heteranthera dubia* (mud-plantain), *Justicia americana* (water-willow), *Peltandra virginica* (arrow-arum), *Phalaris arundinacea* (reed canary grass), *Poa palustris* (marsh bluegrass), *Poa trivialis* (rough bluegrass), *Polygonum* spp., *Pontederia cordata* (pickerelweed), and *Potamogeton* spp.

PONDS AND LAKES. Glacial lakes and well-established artificial lakes are frequent in the region. With the exception of a few kettle lakes surrounded by extensive peatlands and a few remaining alkaline kettle lakes, most of these lakes have been subject to some degree of disturbance. They vary in depth of water and surface area size. Typical vascular plants associated with these lakes are *Ceratophyllum demersum* (coontail), *Elodea canadensis* (Canada waterweed), *Elodea nuttallii* (Nuttall's waterweed), *Lemna minor* (duckweed), *Lemna trisulca* (star duckweed), *Najas flexilis* (slender naiad), *Nuphar advena* (spatter-dock), *Nymphaea odorata* (water-lily), *Polygonum amphibium* (water smartweed), *Potamogeton foliosus, Potomogeton nodosus,* and *Utricularia vulgaris* (bladderwort). *Myriophyllum spicatum* var. *spicatum* (water-milfoil), *Najas minor,* and *Potamogeton crispus* (curly-leaf pondweed) are characteristic alien taxa occurring in ponds and lakes.

MARSHES. Marshes are wet, treeless communities dominated by graminoid-like plants. Within the region, marshes occur in the floodplains of major rivers, in areas near the continental divide, and in pre-Wisconsinan buried river valleys. The vegetation of marshes is fairly uniform throughout the region. The degree of moisture and amount of open water determines dominant species. Areas with standing water are dominated by *Acorus calamus* (sweetflag), *Carex* spp., *Sparganium americanum* (bur-reed), and *Typha latifolia* (cat-tail). Moist areas with infrequent standing water support sedge meadows dominated by *Carex stricta, Carex* spp., *Scirpus cyperinus* (wool-grass), and *Thelypteris palustris* (marsh fern). Marshes, where the water table fluctuates, are dominated by *Typha angustifolia* (narrow-leaved cat-tail), *Typha latifolia* (common cat-tail), and several species of grasses including *Glyceria septentrionalis* (floating manna grass), *Glyceria striata* (fowl manna grass), *Leersia oryzoides* (cut rice-grass), and *Phalaris arundinacea* (reed canary grass). Andreas (1980) provided a list of addditional taxa associated with marsh communities.

SHRUB SWAMPS. Shrub swamps are wetlands that occur in areas inundated for part of the year. Frequent taxa include *Aronia prunifolia* (black chokeberry), *Cephalanthus occidentalis* (buttonbush), *Ilex verticillata* (winterberry), *Salix discolor* (pussy willow), *Salix* spp., *Vaccinium corymbosum* (northern highbush blueberry), and *Viburnum recognitum* (arrow-wood). When *Sphagnum* hummocks develop at the bases of shrubs, taxa such as *Carex canescens* (glaucous sedge), *Carex trisperma* (three-seeded sedge), *Coptis trifolia* (goldthread), and *Osmunda cinnamomea* (cinnamon fern) may occur. Additional herbaceous species associated with buttonbush swamps include *Bidens frondosa* (beggar ticks), *Bidens tripartita* (Spanish-needles), *Carex grayi*, *Leersia oryzoides* (cut rice-grass), *Lycopus* spp. (water-horehound), *Polygonum amphibium* (water smartweed), and *Polygonum arifolium* (halberd-leaved tearthumb). In her study of buttonbush swamps in central Ohio, Tyrrell (1987) found, in addition to *Cephalanthus occidentalis*, *Bidens tripartita* [*B. comosus*] (bur-marigold), *Boehmeria cylindrica* (false nettle), *Scutellaria lateriflora* (mad-dog skullcap), and *Sium suave* (water-parsnip) present in each area surveyed. Although Tyrrell found no indicator taxa for buttonbush swamps, members of the Cyperaceae and Polygonaceae were well represented within the swamps. Taxa frequent in non-*Cephalanthus* dominated shrub swamps are *Ribes americanum* (wild black currant), *Rosa palustris* (swamp rose), *Spiraea alba* (meadow-sweet), *Alisma plantago-aquatica* (mud plantain), *Asclepias incarnata* (swamp milkweed), *Carex stipata*, *Decodon verticillatus* (swamp loosestrife), *Galium asprellum* (rough bedstraw), *Polygonum punctatum*, *Scirpus cyperinus* (woolgrass), and *Sium suave* (water-parsnip).

BOGS. According to Andreas (1985), general characteristics for Ohio bogs are these: 1) bogs develop in areas where drainage is blocked or impeded and there is little or no circulation of water; 2) bogs contain a *Sphagnum*-dominated ground layer that accumulates to form a more or less continuous mat; 3) bogs have shrubby vegetation dominated by members of the Ericaceae and a herbaceous layer primarily dominated by members of the Cyperaceae; and 4) bogs have a water pH of between 3.5 and 5.5. Semi-ombrotrophic bogs (Andreas and Bryan, 1990) are confined to the Glaciated Allegheny Plateau region. Dominant shrub taxa include *Chamaedaphne calyculata* (leatherleaf), *Gaylussacia baccata* (huckleberry), *Nemopanthus mucronatus* (mountain holly), and *Vaccinium corymbosum* (northern highbush blueberry). Trees, including *Betula alleghaniensis* (yellow birch), *Larix laricina* (tamarack), and *Nyssa sylvatica* (sour gum), occur along the outer edge of the floating mat where the mat becomes drier and firmer. Herbaceous taxa include *Carex canescens* (glaucous sedge), *Carex trisperma* (three-seeded sedge), *Decodon verticillatus* (swamp loosestrife), *Eriophorum virginicum* (tawny cotton-grass), *Hypericum virginicum* (pink St. John's-wort), *Lycopus virginicus* (water-horehound), and *Sarracenia purpurea* (pitcher-plant). Aldrich (1943), Gordon (1969), and Andreas (1985) list additional representative associated species. Dachnowski (1912) and Andreas (1985) present distributional information.

In Geauga, Portage, Stark, and Summit Counties a special type of bog occurs in shallow kettle-hole depressions. These "leatherleaf" bogs are dominated by an almost continuous mat of *Sphagnum* spp. and *Chamaedaphne calyculata* (leatherleaf). Other associated taxa include *Cephalanthus occidentalis* (buttonbush), *Ilex verticillata* (winterberry), *Carex howei* (Howe's sedge), *Carex oligosperma* (few-seeded sedge), *Carex* spp., *Drosera rotundifolia* (round-leaved sundew), and *Eriophorum virginicum* (tawny cotton-grass). A moat, or lagg, surrounding the more or less floating mat of *Chamaedaphne* is vegetated with *Calla palustris* (wild calla), *Glyceria canadensis* (rattlesnake grass), *Glyceria grandis* (tall manna grass), *Leersia oryzoides* (cut rice-grass), and *Polygonum* spp. A similar moat is associated with most Ohio bogs.

FENS. Fens have: 1) relatively clear water coming from artesian sources that surface as springs or seeps; 2) a wet, springy calcareous substrate which supports bryophytes which do not accumulate to form a continuous mat; 3) vegetation dominated by members of the Cyperaceae, Compositae, Rosaceae, and Gramineae and with *Potentilla fruticosa* almost always present, and; 4) water pH between 5.5 and 8 (Andreas, 1985). The fens of the Glaciated Allegheny Plateau region are primarily weakly minerotrophic peatlands (Andreas and Bryan, 1990). Fens exhibit much greater species diversity than bogs. Knoop and Andreas (1987) found as many as 30 taxa in a square meter. Taxa associated with fens include *Potentilla fruticosa* (shrubby cinquefoil), *Physocarpus opulifolius* (ninebark), *Rhamnus alnifolia* (alder-leaved buckthorn), *Salix candida* (hoary willow), *Salix serissima* (autumn willow), *Carex atlantica* (prickly bog sedge), *Carex cephalantha* (little prickly sedge), *Carex cryptolepis* (little yellow sedge), *Carex prairea* (prairie sedge), *Carex stricta* (hummock sedge), *Carex tetanica* (slender millet sedge), *Lobelia kalmii* (Kalm's lobelia), *Bromus kalmii* (Kalm's brome grass), *Parnassia glauca* (grass-of-Parnassus), *Sanguisorba canadensis* (Canada burnet), *Solidago ohioensis* (Ohio goldenrod), and *Solidago uliginosa* (marsh goldenrod). Occasionally, *Androgopgon gerardii* (turkey-foot grass) and *Sorghastrum nutans* (Indian grass) occur in fens on mossy mats or sedge peat. Where peaty hummocks accumulate at the base of shrubs, species more typical of bogs, such as *Drosera rotundifolia* (round-leaf sundew) and *Vaccinium macrocarpon* (cranberry), occur. *Rhamnus frangula* (European buckthorn), an alien taxon, is becoming more frequent in wetlands, especially fens. Gordon (1969), Stuckey and Denny (1981), and Andreas (1985) list common fen taxa. Dachnowski (1912), Stuckey and Denny (1981), and Andreas (1985) present distributions of these communities in Ohio.

Common Communities of Disturbed Sites

Most of the land surfaces in Ohio, with the possible exception of a few wetlands, have been subject to some degree of disturbance since the arrival of the European settlers. The degree of disturbance ranges from mild (careful selective cutting) to severe (removing native vegetation and plowing). Some areas remain in perpetual disturbance and bear no resemblance to native plant communities. Curtis (1959) refers to such areas as "weed communities." According to Curtis, they have in common: 1) the presence of bare soil; 2) exposure to full sun; and, 3) a wide variation in soil temperature and moisture regimes. In addition, there is a preponderance of alien species. Alien taxa are indicated by an asterisk (*) in the species lists in the following descriptions.

WASTE PLACES. Waste places are areas where the top soil has been disturbed severely and frequently. Curtis (1959) refers to these areas as "ruderal" habitats. Examples include borrow pits, quarries, dumps, parking lots, railroad yards, clearings, barnyards, cultivated fields, farm roads, abandoned gardens, and strip mines. Taxa found in dry waste areas include *Ambrosia artemisiifolia* (common ragweed), **Capsella bursa-pastoris* (shepherd's purse), **Cichorium intybus* (chicory), *Conyza canadensis* (horseweed), **Digitaria sanguinalis* (crab-grass), **Dipsacus fullonum* (teasel), *Euphorbia maculata,* **Festuca arundinacea* (alta fescue), **Melilotus officinalis* (yellow sweet clover), *Oenothera biennis* (common evening primrose), **Polygonum aviculare* (knotweed), **Polygonum persicaria* (lady's thumb), **Ranunculus acris* (tall buttercup), **Saponaria officinalis* (bouncing-bet), **Tussilago farfara* (coltsfoot), and **Verbascum thapsus* (common mullein). Moist waste areas include **Echinochloa crusgalli* (barnyard grass), *Equisetum arvense* (false horsetail), *Juncus effusus* (common rush), *Phragmites australis* (reed grass), *Polygonum lapathifolium* (dock-leaved smartweed), *Scirpus cyperinus* (wool-grass), *Typha angustifolia* (narrow-leaved cat-tail), *Typha latifolia* (common cat-tail), and *Urtica dioica* (stinging nettles). In recent years, *Elaeagnus umbellata* (oleaster) has become increasingly frequent in waste places.

OLD FIELDS. Old fields include abandoned agricultural fields, pastures and broad areas of clearings such as abandoned quarry floors. Shrubs common to old fields are *Cornus racemosa* (gray dogwood), *Crataegus* spp., *Prunus serotina* (wild cherry), **Rhamnus frangula* (European buckthorn), *Rhus glabra* (smooth sumac), *Rhus typhina* (staghorn sumac), and *Rubus occidentalis* (black raspberry). Herbaceous taxa frequently found are **Achillea millefolium* (common yarrow), **Agropyron repens* (quack grass), **Amaranthus hybridus* (pigweed), *Arctium minus* (great burdock), *Asclepias syriaca* (common milkweed), *Aster novae-angliae* (New England aster), *Aster umbellatus,* **Barbarea vulgaris* (wild mustard), **Cirsium arvense* (Canada thistle), **Dactylis glomerata* (orchard grass), **Daucus carota* (wild carrot), *Desmodium paniculatum, Euthamia graminifolia* (flat-topped goldenrod), **Festuca pratensis* (meadow fescue), *Juncus tenuis* (path rush), *Lobelia inflata* (Indian tobacco), **Poa compressa* (Canada bluegrass), **Potentilla recta* (upright cinquefoil), **Rumex crispus* (yellow dock), *Sisyrinchium angustifolium* (blue-eyed grass), *Solanum carolinense* (horse-nettle), *Solidago canadensis* (field goldenrod), and *Vernonia gigantea* (ironweed).

A distinct plant association develops in old fields in well-drained acidic soils in the northern portion of the Glaciated Allegheny Plateau region. Characteristic species are *Quercus imbricaria* (shingle oak), *Aronia prunifolia* (black chokeberry), *Rubus flagellaris* (dewberry), *Rubus hispidus* var. *obovalis* (small dewberry), *Spiraea tomentosa* (steeple-bush), *Vaccinium corymbosum* (tall blueberry), *Andropogon virginicus* (broom-sedge), *Danthonia spicata* (poverty grass), **Hieracium aurantiacum* (devil's paintbrush), **Hieracium florentinum* (king devil), *Lespedeza hirta* (pubescent bush-clover), *Lycopodium clavatum* (trailing clubmoss), *Lycopodium flabelliforme* (ground pine), *Polygala sanguinea* (red milkwort), and *Pteridium aquilinum* (bracken fern).

Taxa common to cultivated fields and abandoned gardens are **Abutilon theophrasti* (velvet-leaf), *Acalypha virginica* (three-seeded mercury), **Brassica* spp., **Convolvulus arvensis* (field bindweed), **Glecoma hederacea* (gill-over-the-ground), **Hordeum jubatum* (squirrel-tail barley), **Lotus corniculata* (birdsfoot trefoil), **Malva neglecta* (cheeses), **Matricaria matricarioides* (pineapple-weed), **Mollugo verticillata* (carpet-weed), *Oxalis dillenii* (wood-sorrel), **Portulaca oleracea* (common pigweed), *Rumex acetosella* (sheep sorrel), **Se-

taria glauca (yellow foxtail grass), *Setaria viridis* (green foxtail grass), *Solanum dulcamara* (deadly nightshade), *Solanum nigrum,* and *Stellaria media* (common chickweed). *Veronica filiformis* (blue-eyed speedwell) is an increasingly frequent lawn weed in northeastern counties.

ARTIFICIAL IMPOUNDMENTS. Artificial bodies of water often occur in association with dwellings, pastures, roads, or agricultural fields. Plants associated with these impoundments include *Bidens* spp., *Carex* spp., *Ceratophyllum demersum* (coontail), *Eleocharis* spp., *Elodea canadensis* (Canada waterweed), *Galium tinctorium, Impatiens capensis* (jewelweed), *Lemna minor* (duckweed), *Ludwigia alternifolia* (seedbox), *Myriophyllum spicatum* var. *spicatum* (water-milfoil), *Najas minor, Nuphar advena* (spatter-dock), *Polygonum* spp., *Potamogeton crispus* (curly-leaved pondweed), *Potamogeton nodosus, Sagittaria latifolia* (duck potato), and *Typha latifolia* (common cattail). In recent years, *Iris pseudacorus* (yellow water flag), *Lythrum salicaria* (purple loosestrife), and *Phragmites australis* (reed grass) have become more frequent in wetlands in general, and in artificial wetlands, in particular.

THE CATALOGUE OF THE VASCULAR PLANT FLORA

Explanation of the Catalogue

This catalogue contains a list of 2,002 species and 27 interspecific hybrids of vascular plants that occur within the Glaciated Allegheny Plateau region of Ohio. When used in combination with Appendix A, the catalogue provides a list of the flora of 23 Ohio counties: Ashland, Ashtabula, Columbiana, Cuyahoga, Fairfield, Geauga, Holmes, Knox, Lake, Licking, Lorain, Mahoning, Medina, Morrow, Perry, Pickaway, Portage, Richland, Ross, Stark, Summit, Trumbull, and Wayne.

The arrangement of the catalogue is in the taxonomic hierarchy proposed by Cronquist, Takhtajan, and Zimmerman (1966) as modified by Cronquist (1973). The taxa listed are a part of the Subkingdom Embryobionta (embryophytes), in which the vascular plants are assigned to the following Divisions: Lycopodiophyta (clubmosses), Equisetophyta (horsetails and scouring rushes), Polypodiophyta (ferns), Pinophyta (gymnosperms), and Magnoliophyta (angiosperms). The Magnoliophyta are divided into the classes Magnoliopsida (dicotyledons) and Liliopsida (monocotyledons). Families within these major groups are arranged alphabetically. Appendix C lists the vascular plant families found in the Glaciated Allegheny Plateau region in the Englerian sequence used by Fernald (1950) and Cusick and Silberhorn (1977). Genera and species are listed alphabetically within families.

Nomenclature and circumscription at the family level generally follow Fernald (1950). For seed plants, departures from Fernald at the family level include the use of the name Betulaceae instead of Corylaceae (International Code of Botanical Nomenclature, 1972); the inclusion of the Amaryllidaceae within the Liliaceae and the Pyrolaceae within the Ericaceae, and the separation of Menyanthaceae from the Gentianaceae (Thorne, 1968); the separation of the Fumariaceae from the Papaveraceae, and of the Cupressaceae and Taxodiaceae from the Pinaceae (Gleason and Cronquist, 1963); and the separation of the Ruppiaceae, Zannichelliaceae, and Potamogetonaceae from the Zosteraceae (Voss, 1972). For pteridophytes, following the treatment proposed by Crabbe, Jermy, and Mickel (1975) and Mickel (1979), the Adiantaceae, Aspleniaceae, Blechnaceae, Dennstaedtiaceae, and Thelypteridaceae are separated from the Polypodiaceae *sensu lato*, and the Azollaceae from the Salviniaceae *sensu lato*.

Nomenclature and circumscription for taxa at the generic and species levels in the Division Pinophyta and Class Liliopsida essentially follow Voss (1972). Exceptions include the genera *Platanthera* L. C. Richard, *Coeloglossum* Hartman, and *Spiranthes* L. C. Richard, where the treatment proposed by Luer (1975) is followed. Where a name differs from one used by Braun (1961; 1967), the latter is given in synonymy.

Nomenclature and circumscription for the Class Magnoliopsida follow Voss (1985) for all families in the Apetalae, with the exception of the genus *Crataegus* L. where Gleason and Cronquist (1963) is followed, and those families in the Polypetalae not covered by Cooperrider (1985a). Cooperrider (1985a) is followed for nomenclature and circumscription for portions of the Polypetalae and all of the Sympetalae, with the exception of the family Compositae, where Cronquist (1980) is followed. For taxa in the Division Magnoliophyta, in instances where a name differs from one used by Fernald (1950), the latter is given in synonymy. Statistical summaries of the numbers of families, and native and alien genera, species, and hybrids in each family are presented in Table 2 and Table 3.

Table 2. Statistical summary indicating numbers of native and alien species and interspecific hybrids for families of vascular plants within the Glaciated Allegheny Plateau region of Ohio. If a single species of a family or genus is native, that family or genus is designated as native. An asterisk (*) indicates an alien family.

Family	Genera		Species and Hybrids	
	Native	Alien	Native	Alien
Division Lycopodiophyta (Clubmosses)				
Isoetaceae	1		2	
Lycopodiaceae	1		10	
Selaginellaceae	1		2	
Division Equisetophyta (Horsetails and Scouring Rushes)				
Equisetaceae	1		8	
Division Polypodiophyta (Ferns)				
Adiantaceae	2		2	
Aspleniaceae	10		26	
*Azollaceae		1		1
Blechnaceae	1		2	
Dennstaedtiaceae	2		2	
Ophioglossaceae		2	7	
Osmundaceae	1		3	
Polypodiaceae	1		1	
Thelypteridaceae	1		4	
Division Pinophyta (Gymnosperms)				
Cupressaceae	1	1	2	1
Pinaceae	3		6	2
Taxaceae	1	1		
*Taxodiaceae		1		1
Division Magnoliophyta (Angiosperms)				
Class Magnoliopsida (Dicotyledons)				
Acanthaceae	2		3	
Aceraceae	1		6	1
*Aizoaceae		1		1
Amaranthaceae	1	1	1	9
Anacardiaceae	1	1	6	1
Annonaceae	1		1	

28

Family	Genera		Species and Hybrids	
	Native	Alien	Native	Alien
	Division Magnoliophyta (Angiosperms) Class Magnoliopsida (Dicotyledons) (cont.'d)			
Apocynaceae	1	2	4	2
Aquifoliaceae	2		2	1
Araliaceae	2		5	1
Aristolochiaceae	2		2	
Asclepiadaceae	3		15	1
Balsaminaceae	1		2	1
Berberidaceae	3	1	3	2
Betulaceae	5		11	2
*Bignoniaceae		2		4
Boraginaceae	6	5	9	15
Cactaceae	1		1	
Callitrichaceae	1		3	
Campanulaceae	3		8	1
Cannabaceae	1	1	1	2
Capparaceae	1	2	1	2
Caprifoliaceae	7		23	10
Caryophyllaceae	6	10	12	30
Celastraceae	2		3	3
Ceratophyllaceae	1		2	
Chenopodiaceae	2	5	5	12
Cistaceae	2		8	
Compositae	37	29	152	84
Convolvulaceae	3	1	9	7
Cornaceae	1		8	
Crassulaceae	1		1	4
Cruciferae	10	24	23	46
Cucurbitaceae	2	3	2	3
*Dipsacaceae		1		3
Droseraceae	1		2	
Ebenaceae	1		1	
Elaeagnaceae	1	1	1	2
Ericaceae	13	1	24	1
Euphorbiaceae	3	1	10	11
Fagaceae	3		16	
Fumariaceae	3	1	5	1
Gentianaceae	5	1	9	1
Geraniaceae	1	1	4	5
Guttiferae	1		14	1
Haloragaceae	2		2	1
Hamamelidaceae	1		1	
Hippocastanaceae	1		2	1
Hydrophyllaceae	2		6	
Juglandaceae	2		8	
Labiatae	18	13	36	25

Family	Genera		Species and Hybrids	
	Native	Alien	Native	Alien
	Division Magnoliophyta (Angiosperms) Class Magnoliopsida (Dicotyledons) (cont.'d)			
Lauraceae	2		2	
Leguminosae	23	6	52	29
Lentibulariaceae	1		5	
Limnanthaceae	1		1	
Linaceae	1		3	1
Lythraceae	5		5	2
Magnoliaceae	2		2	
Malvaceae	1	7	2	10
*Martyniaceae		1		1
Melastomataceae	1		1	
Menispermaceae	1		1	
Menyanthaceae	1		1	
Moraceae	1	1	1	2
Myricaceae	2	2		
*Nyctaginaceae		1		2
Nymphaeaceae	4	1	4	1
Nyssaceae	1		1	
Oleaceae	1	2	4	4
Onagraceae	5		18	3
Orobanchaceae	3		3	
Oxalidaceae	1		5	1
Papaveraceae	2	2	2	5
Passifloraceae	1		1	
Phrymaceae	1		1	
Phytolaccaceae	1		1	
Plantaginaceae	1		3	4
Platanaceae	1		1	
Polemoniaceae	2		6	1
Polygalaceae	1		5	
Polygonaceae	2	1	18	12
Portulacaceae	1	1	2	1
Primulaceae	4	2	10	4
Ranunculaceae	16		34	6
*Resedaceae		1		1
Rhamnaceae	2		3	2
Rosaceae	21	5	64	28
Rubiaceae	5	1	19	5
Rutaceae	2		2	
Salicaceae	2		21	8
Santalaceae	1		1	
Sarraceniaceae	1		1	
Saururaceae	1		1	
Saxifragaceae	9	1	14	4
Scrophulariaceae	16	7	28	18

Family	Genera		Species and Hybrids	
	Native	Alien	Native	Alien

	Native	Alien	Native	Alien
Division Magnoliophyta (Angiosperms) Class Magnoliopsida (Dicotyledons) (cont.'d)				
*Simaroubaceae		1		1
Solanaceae	2	5	5	12
Staphyleaceae	1		1	
Thymelaeaceae	1		1	
Tilaceae	1		1	
Ulmaceae	2		4	
Umbelliferae	17	6	25	7
Urticaceae	5		6	
Valerianaceae	2		4	3
Verbenaceae	2		8	
Violaceae	2		25	3
Viscaceae	1		1	
Vitaceae	2		6	
*Zygophyllaceae		1		1
Class Liliopsida (Monocotyledons)				
Alismataceae	3		7	
Araceae	5		7	
Commelinaceae	1	1	3	3
Cyperaceae	12		173	1
Dioscoreaceae	1		1	1
Eriocaulaceae	1		1	
Gramineae	42	19	117	59
Hydrocharitaceae	2		3	
Iridaceae	2		9	2
Juncaceae	2	24		
Juncaginaceae	2		3	
Lemnaceae	4		7	
Liliaceae	22	12	43	16
Najadaceae	1		3	1
Orchidaceae	17	1	43	1
Pontederiaceae	2		2	
Potamogetonaceae	1		21	1
*Ruppiaceae		1		1
Sparganiaceae	1		3	
Typhaceae	1		2	
Xyridaceae	1		1	
Zannichelliaceae	1		1	

Table 3. Statistical summary indicating numbers of native and alien families, genera, species, and interspecific hybrids for the major taxonomic groups of vascular plants as recorded in the catalogue for the Glaciated Allegheny Plateau region of Ohio.

Major Taxonomic Group	Families		Genera		Species and Hybrids	
	Native	Alien	Native	Alien	Native	Alien
Lycopodiophyta	3		3		14	
Equisetophyta	1		1		8	
Polypodiophyta	8	1	20	1	47	1
Pinophyta	3	1	5	2	9	4
Magnoliophyta						
Magnoliopsida	101	8	356	160	908	479
Liliopsida	21	1	124	34	473	86
Totals	137	11	509	197	1459	570
Grand Totals	148		706		2029	

A number symbol (#) preceeding a scientific name indicates that the taxon is definitive of the Glaciated Allegheny Plateau region of Ohio. These taxa are also listed separately in Appendix B.

An asterisk (*) preceeding the scientific name indicates that the taxon is alien to this region of Ohio. The status of these taxa was determined from Fernald (1950), Braun (1961; 1967), Gleason and Cronquist (1963), Voss (1972; 1985), Cronquist (1980), and Cooperrider (1985a). Based on information provided by the above authors and my own field experience, an attempt is made to determine the approximate degree of establishment of alien taxa. The term "adventive" is used for those species that either do not reproduce spontaneously or do so for a short time and then disappear, and the term "naturalized" is used for those species that are more permanently established in the flora (Lawrence, 1951; Radford et al, 1974). Common names, where provided after the scientific names, are consistent with the above authors and with the "Rare Species of Native Ohio Wild Plants" list (Division of Natural Areas and Preserves, 1988).

For each taxon, a frequency rating is given. These ratings are based on the following criteria: common, 23 to 16 counties; frequent, 15 to 11 counties; infrequent, 10 to 5 counties; and rare, 4 to 1 counties. Some native taxa are known only from historical records, defined as those based on collections made prior to 1968 (Division of Natural Areas and Preserves, 1988). In these cases, the phrase "presumably extirpated" is inserted after the frequency rating.

Following the frequency rating, habitat information is provided. These habitat data have been gathered from herbarium records and my field notes.

For information on county distribution, the following two-letter code is used for the counties in the study area. This is consistent with that used by

Silberhorn (1970) and Cusick and Silberhorn (1977):

AS	Ashland	LK	Lake	PO	Portage
AB	Ashtabula	LI	Licking	RI	Richland
CL	Columbiana	LR	Lorain	RO	Ross
CU	Cuyahoga	MH	Mahoning	ST	Stark
FA	Fairfield	MD	Medina	SU	Summit
GE	Geauga	MO	Morrow	TR	Trumbull
HL	Holmes	PE	Perry	WA	Wayne
KN	Knox	PC	Pickaway		

For rare taxa, abbreviated habitat information, general location, name of collector, collection number, date, and herbarium accession number are given. For the remaining taxa, county distribution, arranged alphabetically, is provided.

The abbreviation "s.n.", used throughout the catalogue, represents *sine numero* and is used when there is no collection number or no herbarium accession number.

The following herbaria were surveyed for information: The University of Akron, Cleveland Museum of Natural History (CLM), Kent State University (KE), Oberlin College (OC), The Ohio State University (OS), and Ohio University (BHO). In addition, several records were obtained from specimens at other institutions through loans. In those situations, the full name of the institution is given. Occasional records are taken from the literature, and in such instances, the author is cited after the record. Occasionally another author, for example Braun (1961; 1967), cites a record that could not be located. These records are listed as taken from the literature.

The following references were consulted for aid in specimen identification and for data on county distribution: Bigelow (1841), Shupe (1930), Hicks (1933), Crowl (1937), Heffner (1939), Fernald (1950), Amann (1961), Braun (1961; 1967), Hawver (1961), Cruden (1962), Gleason and Cronquist (1963), Hauser (1963; 1964; 1965), Easterly (1964), Sabo (1965), Fisher (1966), Valley and Cooperrider (1966), Pringle (1967), McCready (1968), Anderson (1969), Bentz (1969), Brockett (1969), Cooperrider and Sabo (1969), Andreas (1970; 1980), Blackwell (1970), Weishaupt (1971), Chuey (1972), Noelle and Blackwell (1972), Voss (1972; 1985), Wilson (1972; 1974), Kerrigan and Blackwell (1973), Haynes (1974), O'Connor and Blackwell (1974), Luer (1975), Miller (1976), Pusey (1976), Tandy (1976), Cline (1977), Bentz and Cooperrider (1978), McCready and Cooperrider (1978; 1984), Mickel (1979), Andreas and Cooperrider (1979; 1980; 1981), Burns (1980; 1986), Cronquist (1980), Riehl and Ungar (1981), Aiken (1981), Emmitt (1981), Arbak and Blackwell (1982), Bayer and Stebbins (1982), Cooperrider (1982; 1984b; 1985a; 1985b; 1986a; 1986b), Stuckey and Roberts (1982), Adams (1982), Brockett and Cooperrider (1983), Cusick (1983; 1984), Spooner (1983), Spooner and Shelly (1983), Ungar and Loveland (1982), Loconte and Blackwell (1984), McCance and Burns (1984), Bissell (1984), Bissell et al (1985), Hardin (1985), and Lellinger (1985).

The Catalogue

SUBKINGDOM EMBRYOBIONTA

For explanation of the use of the number symbol (#) and of the asterisk (*), see page 32.

Division Lycopodiophyta

ISOETACEAE (Quillwort Family)

1 # *Isoetes echinospora* Dur.   Spiny-spored Quillwort
Once rare, now presumably extirpated from Ohio; clear shallow water. PO: Lake Brady, Shilling s.n., 16 Now 1935 (OS 11400); Sandy Lake, Hopkins 1052, 2 Aug 1919 (KE 11192 & 1399); East Twin Lake, Hopkins 3140, 20 Jul 1915 (OS s.n.); Lake Brady, Hopkins 3139, 10 Jul 1915 (OS s.n.); East Twin Lake, Webb 23 Jul 1915 (KE 11193).

2 # *Isoetes engelmannii* A. Braun   Appalachian Quillwort, Engelmann's Quillwort
Rare; shallow waters, pond shores, marshes. MH: Berlin Twp., Vickers 2390, 1938 (KE 11191). PO: submerged in water, Lake Brady, Hopkins s.n., 3 Jul 1913 (OS 1396). TR: mud. Hartford Twp., Vujevic s.n., 14 Jun 1976 (Youngstown State University Herbarium 10722).

LYCOPODIACEAE (Clubmoss Family)

3 *Lycopodium clavatum* L.   Trailing Clubmoss, Ground Pine
Frequent; dry sandy fields, open secondary woods, sandstone quarries. AB, CL, CU, GE, FA, KN, LK, LI, LR, MH, PO, ST, SU, TR.

4 *Lycopodium dendroideum* Michx.   Flat-branch Tree Clubmoss
Frequent; open mesic woods, thickets, old fields. AB, FA, GE, HL, LK, LR, MD, PO, ST, SU, TR.

5 *Lycopodium flabelliforme* (Fern.) Blanchard   Ground Pine
Lycopodium complanatum L. var. *flabelliforme* Fern.
Common; moist, mesic and dry woods, fields, thickets. All counties except PC.

6 *Lycopodium* x *habereri* House *(L. flabelliforme* x *L. tristachyum)*
Rare; weedy fields, sandy openings. CL: weedy field, Unity Twp., Cusick 3169, 15 Sep 1966 (KE 28709). SU: border of sandy woods, Green Twp., Cusick 10100, 4 Jul 1969 (KE 28899).

7 # *Lycopodium inundatum* L.   Bog Clubmoss
Infrequent; wet sandy soils, sphagnous mats, abandoned sandstone quarries. CU, GE, KN, LK, PE, PO, SU, TR.

8 *Lycopodium lucidulum* Michx.   Shining Clubmoss
Common; moist and mesic woods, hemlock slopes. AS, AB, CL, CU, FA, GE, HL, KN, LK, LR, MD, PE, PO, ST, SU, TR, WA.

9 *Lycopodium lucidulum* Michx. x *L. porophilum* Lloyd & Underw.
Rare; sandstone ledges. FA: Berne Twp., Goslin s.n., 3 Nov 1940 (BHO 20926). MD: hemlock gorge, Sharon Twp., Cooperrider & Herrick 7486, 27 Jul 1960 (KE 7935). PO: rocky side of ravine, Woodsworth Glen, Webb 684, 19 Aug 1909 (KE 11020).

10 *Lycopodium obscurum* L.   Tree Clubmoss
Frequent; mesic and moist woods, fields, damp thickets. AB, CL, CU, FA, GE, LI, MD, PE, PO, ST, SU, TR, WA.

11 *Lycopodium porophilum* Lloyd & Underw.   Rock Clubmoss
Infrequent; sandstone ledges. AS, AB, FA, LI, PO.

12 # *Lycopodium tristachyum* Pursh   Ground Cedar
Infrequent; dry sandy fields, woods, sandstone quarries. AB, CU, FA, GE, LK, LI, LR, PO, SU, WA.

SELAGINELLACEAE (Spikemoss Family)

13 *Selaginella apoda* (L.) Spring   Meadow Spikemoss
Infrequent; damp soils along stream margins, wooded floodplains, wet woods, meadows, fens. AB, LK, LR, PC, PO, RO, TR.

14 *Selaginella rupestris* (L.) Spring   Rock Spikemoss
Rare; exposed sandstone rocks, packed sands. FA: Reese's Knob, Hocking Twp., Goslin s.n., 6 Apr 1941 (OS 1937); Kettle Hill, Pontius s.n., 8 Aug 1938 (OS 49436). PO: sandstone rocks, Nelson Twp., Andreas 1739, 10 May 1978 (KE 41931); rocks, Nelson Twp., Rood s.n., 16 Jul 1939 (KE 14422); Newell's Ledges, Nelson Twp., Rood 678, 18 Sep 1921 (KE 16249). Adams (1982) indicates a record for LI.

Division Equisetophyta

EQUISETACEAE (Horsetail Family)

15 *Equisetum arvense* L.   Field Horsetail
Common; roadsides, railroads, stream and pond margins, fields, waste places. ALL COUNTIES.

16 *Equisetum* x *ferrissii* Clute *(E. hyemale* var. *affine* x *E. laevigatum)*
     Ferriss's Scouring Rush
Infrequent; stream banks, fields, roadsides, fens. AS, AB, GE, HL, LK, LI, MD, PO, ST, WA.

17 *Equisetum fluviatile* L.   Water Horsetail
Frequent; marshes, fens, thickets, wet woods. AS, AB, CL, CU, GE, HL, LK, LR, MH, PC, PO, RO, ST, SU, WA.

18 *Equisetum hyemale* L. var. *affine* (Engelm.) A. A. Eaton   Common Scouring Rush
Common; floodplains, stream banks, abandoned quarries, marshes, fens, waste places. ALL COUNTIES.

19 *Equisetum laevigatum* A. Braun   Smooth Scouring Rush
Frequent; fens, marshes, wet meadows. CU, GE, HL, KN, LK, LR, PE, PO, RO, ST, TR.

20 *Equisetum* x *nelsonii* (A. A. Eaton) Schaffner *(E. laevigatum* x *E. variegatum)*
Rare; marshy grounds, open floodplains, sandy soils. SU: low marshy ground, Boston Twp., Cooperrider & Herrick 7499, 27 Jul 1960 (KE 6960).

21 # *Equisetum sylvaticum* L.   Woodland Horsetail
Infrequent; weedy thickets, open wet woods, shaded roadsides. AB, CU, GE, LK, LR, PO, ST, SU, TR.

22 *Equisetum variegatum* Schleich.   Variegated Horsetail
Rare; calcareous sandy soils, marl fens. GE: sand quarry, Newbury Twp., Bissell 1981:195, 5 Nov 1981 (KE 45087). ST: seepage area, Jackson Twp., Amann 264, 4 Jun 1960 (KE 8628). SU: sandy soil, Deep Lock Quarry, Peninsula, Andreas 6991, 29 May 1986 (KE 50267); sandy flats, Akron, Stoutamire 7523, 5 May 1968 (University of Akron Herbarium s.n.).

Division Polypodiophyta

ADIANTACEAE (Maidenhair Family)

23 *Adiantum pedatum* L.   Northern Maidenhair Fern
Common; mesic and moist rich woods, moist ravines, shaded rocky slopes. ALL COUNTIES.

24 *Vittaria lineata* (L.) J. E. Smith   Appalachian Gametophyte, Shoestring Fern
Rare; dark crevices, cavern openings. GE: Sharon Conglomerate, Chardon Twp., Cusick 22000, 24 Aug 1982 (OS s.n.).

ASPLENIACEAE (Spleenwort Family)

25 *Asplenium montanum* Willd.   Mountain Spleenwort
Rare; moist shaded sandstone outcrops, crevices and ledges. FA: camp, Hocking Twp., Goslin s.n., 8 Jul 1940 (BHO 21005). MH: Standing Rock, Mahoning River, Vickers s.n., 22 Sep 1909 (OS 711). SU: Cuyahoga Falls, Bogue s.n., 24 Jun 1894 (OS 709).

26 *Asplenium pinnatifidum* Nutt.   Pinnatifid Spleenwort
Infrequent; sandstone outcrops, cliffs, ledges. AS, AB, FA, KN, LI, MH, PE, RO, SU.

27 **Asplenium platyneuron** (L.) Oakes   Ebony Spleenwort
Common; dry and mesic woods, roadbanks, steep eroding slopes, dry hillsides, sandstone
outcrops. ALL COUNTIES.

28 **Asplenium trichomanes** L.   Maidenhair Spleenwort
Frequent; calcareous rock crevices, cliffs and ledges. AS, AB, CU, FA, GE, HL, KN, LK,
LR, PO, RI, RO, SU.

29 **Asplenium** x **trudellii** Wherry **(A. montanum** x **A. pinnatifidum)**
Rare; sandstone exposures. FA: Greenfield Twp., Goslin s.n., 25 Nov 1956 (BHO 21023);
cliff near Hamburg, Bartley s.n., 25 Feb 1945 (OS 44423).

30 **Athyrium felix-femina** (L.) Roth   Northern Lady Fern
Athyrium angustum (Willd.) Presl
Common; moist and mesic woods, wooded slopes, thickets, wooded floodplains, stream
terraces. ALL COUNTIES.

31 **Athyrium pycnocarpon** (Sprengel) Tidestrom   Glade-fern, Narrow-leaved Spleenwort
Common; rich mesic slopes, ravines, wooded floodplains, stream terraces. AS, AB, CU, FA,
GE, HL, KN, LK, LI, LR, MH, MD, PE, PO, RI, RO, SU, WA.

32 **Athyrium thelypteroides** (Michx.) Desv.   Silvery Spleenwort, Silvery Glade-fern
Common; moist slopes, floodplains, rich woods. All counties except PC.

33 **Camptosorus rhizophyllus** (L.) Link   Walking Fern
Asplenium rhizophyllum L.
Common; shaded rock exposures and boulders. AS, AB, CL, CU, FA, GE, HL, KN, LK,
LI, LR, MH, PE, PO, RI, RO.

34 **Cystopteris bulbifera** (L.) Bernh.   Bulblet Bladder Fern
Frequent; moist sandstone outcrops, cliffs, talus slopes. AS, AB, CL, CU, FA, GE, KN, LK,
LI, LR, PO, RO, SU, TR, WA.

35 **Cystopteris fragilis** (L.) Bernh.   Fragile Fern, Brittle Fern
Common; moist sandstone outcrops, shady banks, talus slopes, constructed sandstone walls.
All counties except MD, MO, PC.

36 **Cystopteris protrusa** (Weath.) Blasdell   Lowland Fragile Fern
Cystopteris fragilis (L). Bernh. var. *protrusa* Blasdell
Common; moist woods, stream terraces, wooded floodplains, rocky slopes. All counties
except SU.

37 **# Dryopteris** x **boottii** (Tuckerman) Undrew. **(D. cristata** x **D. intermedia)**
                                                                        Boott's Wood Fern
Infrequent; boggy woods, marsh margins, wooded seeps, thickets. AB, GE, LR, PO, RO,
ST, SU, TR, WA.

38 **# Dryopteris clintoniana** (D.C. Eaton) Dowell   Clinton's Wood Fern
Dryopteris cristata (L.) Gray var. *clintoniana* (D.C. Eaton) Underw.
Infrequent; rich moist woods, boggy thickets, damp woods. AB, GE, LK, LR, WA.

39 **Dryopteris cristata** (L.) Gray   Crested Wood Fern
Common; calcareous fen margins, boggy thickets, marshes, wooded seeps, stream terraces.
AS, AB, CL, CU, FA, GE, HL, LK, LI, LR, MH, PO, RO, ST, SU, TR, WA.

40 **Dryopteris goldiana** (Hooker) Gray   Goldie's Wood Fern
Frequent; slightly calcareous moist woods, stream terraces, floodplains. AB, CL, CU, FA,
GE, HL, LK, LI, LR, MD, PE, PO, RO, SU, WA.

41 **Dryopteris intermedia** Gray   Intermediate Shield Fern
Dryopteris spinulosa (O. F. Muell.) Watt. var. *intermedia* (Muhl.) Underw.
Common; rich mesic woods, thickets, stream terraces, rocky slopes. All counties except MO,
PC, RO.

42 **Dryopteris marginalis** (L.) Gray   Marginal Shield Fern
Common; rich mesic woods, secondary woods, hemlock slopes, rocky slopes. All counties
except PC.

43 **Dryopteris** x **neo-wherryi** Wagner **(D. goldiana** x **D. marginalis)**
                                                                        Wherry's Wood Fern
Rare; moist woods. SU: moist alluvial woods, Boston Twp., Cooperrider & Herrick 7529, 27
Jul 1960 (KE 6938 & 6939).

44 *Dryopteris spinulosa* (O. F. Muell.) Watt. Spinulose Shield Fern
Common; rich mesic woods, thickets, wooded floodplains. All counties except RO.

45 *Dryopteris* x *triploidea* Wherry *(D. intermedia* x *D. spinulosa)*
Frequent; shaded roadside banks, moist woods, thickets. AS, AB, CL, CU, GE, HL KN, LK, LI, LR, PO, ST, SU.

46 # *Gymnocarpium dryopteris* (L.) Newman Oak Fern
Infrequent; cool woods, rocky slopes, hemlock slopes. AB, CL, LK, Li, LR, WA.

47 *Matteuccia struthiopteris* (L.) Todaro Ostrich Fern
Pteretis pensylvanica (Willd.) Fern.
Frequent; moist ravine terraces, wooded floodplains, wet mature woods. AS, AB, CU, GE, HL, LK, LR, MD, PE, PO, RI, SU, TR, WA.

48 *Onoclea sensibilis* L. Sensitive Fern, Bead Fern
Common; floodplains, lake and pond margins, stream terraces, roadside ditches, fens. ALL COUNTIES.

49 *Polystichum acrostichoides* (Michx.) Schott Christmas Fern
Common; moist and mesic woods, rocky slopes, shaded roadside banks. ALL COUNTIES.

50 *Woodsia obtusa* (Sprengel) Torrey Blunt-lobed Woodsia
Infrequent; rocky woods, sandstone outcrops, constructed sandstone walls. AS, CL, FA, LN, LI, PE, RO.

AZOLLACEAE (Mosquito Fern Family)

51 **Azolla caroliniana* Willd. Water Fern
Rare, naturalized; quiet waters, canals. CU: pond, Hunting Valley, Bissell 1981:65, 22 Jul 1981 (CLM 18377). LK: near Painesville, Werner s.n., 1879 (CLM 7267). ST: canal, Jackson Twp., Roberts 4086, 10 Aug 1973 (KE 37176). SU: Portage Lakes, Akron, Blaydes s.n., Oct 1929 (OS 1386).

BLECHNACEAE (Chain Fern Family)

52 *Woodwardia areolata* (L.) Moore Netted Chain Fern
Lorinseria areolata (L.) Presl
Rare; peat bogs, swamps, moist woods. AB: hemlock-hardwood swamp, Austinburg Twp., Bissell 1982-313, 9 Sep 1982 (CLM 17337). LK: Lake County, Beardslee s.n., 27 May 1927 (OS 576). Adams (1982) indicates a record for PO.

53 # *Woodwardia virginica* (L.) Small Virginia Chain Fern
Anchistea virginica (L.) Presl
Frequent; sphagnous bogs, swamps, wet hemlock woods. AB, CU, FA, GE, LK, LI, LR, PO, ST, SU, WA.

DENNSTAEDTIACEAE (Bracken Family)

54 *Dennstaedtia punctilobula* (Michx.) Moore Hay-scented Fern
Common; mesic woods, wooded ravines, rocky slopes, shaded road banks. All counties except MO, PC.

55 *Pteridium aquilinum* (L.) Kuhn Bracken Fern
Common; dry woods, sandstone outcrops, burned fields, railroads, disturbed areas. All counties except MO, PC.

OPHIOGLOSSACEAE (Adder's-tongue Family)

56 *Botrychium dissectum* Sprengel (including *B. dissectum* var. *obliquum* (Muhl.) Clute)
Common Grape-fern
Common; mesic woods, pastures, disturbed woodlots. ALL COUNTIES.

57 # *Botrychium lanceolatum* (Gmel.) Angstr. Lance-leaved Grape-fern, Lance-leaved Moonwort
Once rare, now presumably extirpated from Ohio; mature beech-maple woods, peaty slopes, meadows. AB: East Conneaut, Hicks & Chapman s.n., 17 Jun 1933 (OS 59 & 60); beech-maple woods, Monroe Twp., Hicks s.n., 18 Jul 1929 (OS 61). GE: Thompson, Howells & Simms s.n., 16 Aug 1907 (OS 63). PO: rich woods, Freedom Twp., Rood 143, 8 Jun 1902 (KE 11044). Adams (1982) indicates records for LK, LR, TR.

58 *Botrychium matricariifolium* A. Braun Daisy-leaf Grape-fern, Daisy-leaved Moonwort
Frequent; moist and mesic open woods, thickets. AS, AB, CL, CU, FA, GE, LK, LR, MH, ME, PO, ST, SU, TR.

59 # *Botrychium multifidum* (Gmel.) Rupr. var. *intermedium* (D. C. Eaton) Farw.
Leathery Grape-fern
Infrequent; gravelly slopes, thickets, abandoned fields. AB, CU, LR, MH, MD, PO, SU.

60 *Botrychium oneidense* (Gilbert) House Blunt-lobed Grape-fern
Frequent; disturbed woodlots, floodplains, meadows, abandoned fields. AS, CL, FA, GE, KN, LI, LR, MD, PO, TR, WA.

61 *Botrychium virginianum* (L.) Swartz Rattlesnake Fern
Common; floodplains, rich woods, wooded terraces, rocky slopes, disturbed woodlots. ALL COUNTIES.

62 *Ophioglossum vulgatum* L. Adder's Tongue
Frequent; rich mesic woods, pastures, thickets, secondary woodlots. AB, CU, FA, LK, MH, MO, PC, PO, RI, RO, SU, WA.

OSMUNDACEAE (Royal Fern Family)

63 *Osmunda cinnamomea* L. Cinnamon Fern
Common; floodplains, fens, bogs, wet woods, marshes. All counties except MO, PC, RO.

64 *Osmunda claytoniana* L. Interrupted Fern
Common; mesic and dry woods, thickets, stream banks, wooded roadside banks. All counties except PC.

65 *Osmunda regalis* L. Royal Fern, Flowering Fern
Common; marshes, thickets, floodplains, wet woods, pastures, lake margins, sphagnous thickets. AS, AB, CL, CU, FA, GE, HL, LK, LI, LR, MH, MO, PO, RI, RO, ST, SU, TR, WA.

POLYPODIACEAE (Polypody Family)

66 *Polypodium virginianum* L. Rock-cap Polypody
Common; sandstone outcrops and ledges, shaded tree trunks, rocky slopes. All counties except MO, PC, ST.

THELYPTERIDACEAE (Marsh Fern Family)

67 *Thelypteris hexagonoptera* (Michx.) Weath. Southern Beech Fern, Broad Beech Fern
Dryopteris hexagonoptera (Michx.) Christens.
Common; rich mesic woods, roadside banks, wooded floodplains. All counties except PC.

68 *Thelypteris noveboracensis* (L.) Nieuwl. New York Fern
Dryopteris noveboracensis (L.) Gray
Common; dry, mesic and moist mature woods, disturbed woodlots, wooded stream terraces. All counties except MO, PC.

69 *Thelypteris palustris* Schott Marsh Fern
Dryopteris thelypteris (L.) Gray
Common; marshes, fens, moist woods, wet meadows. AS, AB, CL, CU, GE, HL, KN, LK, LI, LR, MH, MD, PO, RI, RO, ST, SU, TR, WA.

70 # *Thelypteris phegopteris* (L.) Slosson Long Beech Fern, Northern Beech Fern
Dryopteris phegopteris (L.) Christens.
Phegopteris connectilis (Michx.) Watt.
Frequent; moist sandstone outcrops and ledges, wooded hemlock slopes. AB, CL, CU, GE, KN, LK, LI, LR, MD, PO, SU, TR.

Division Pinophyta (Gymnosperms)

CUPRESSACEAE (Cypress Family)

71 *Juniperus communis* L. Ground Juniper
Infrequent; dry eroding calcareous slopes. CU, GE, LK, LR, PO, SU, WA.

72 *Juniperus virginiana* L. Red Cedar
Common; dry slopes, fields, disturbed woodlots. ALL COUNTIES.

38

73 *Thuja occidentalis* L. White Cedar
Rare, adventive; abandoned homesites, waste places. KN: shaley roadbank below cemetery, Union Twp., Pusey 214, 27 Aug 1974 (KE 37890).

PINACEAE (Pine Family)

74 # *Larix laricina* (DuRoi) K. Koch Tamarack, Larch
Infrequent; bogs, sphagnous fens, sandy kames. AB, CU, GE, LK, LR, PO, ST, SU, TR, WA.

75 *Pinus echinata* Miller Yellow Pine
Rare; dry slopes, oak ridges. PE: hillside, Clayton Twp., Cline 643, 5 Jun 1975 (KE 43247). RO: Colerain Twp., Bartley & Pontius s.n., 29 Jun 1929 (OS 2062 & 2063); sandstone hills, Harrison and Colerain Twps., Dean s.n., 8 Jul 1922 (OS 2060).

76 *Pinus nigra* Arnold Austrian Pine
Rare, adventive; disturbed areas. AB: Phelps Creek, Hicks s.n., 16 Jun 1932 (OS 106655). PE: disturbed woods, Clayton Twp., Cline 784, 27 Jun 1975 (KE 43248).

77 *Pinus rigida* Miller Pitch Pine
Infrequent; dry ridges, oak woods. CL, FA, HL, LI, PE, RO.

78 *Pinus strobus* L. White Pine
Common, often escaping from cultivation; dry oak ridges, woodland borders, sphagnous bogs, wet woods. AS, AB, CL, CU, GE, HL, KN, LK, LI, LR, MD, PE, PO, RI, ST, SU, TR, WA.

79 *Pinus sylvestris* L. Scotch Pine
Infrequent, adventive; fields, woodland borders, roadside banks. AB, FA, LK, KN, MD, PE, TR.

80 *Pinus virginiana* Miller Virginia Pine, Scrub Pine
Infrequent, adventive in northern counties; dry ridges, oak woods. AS, CL, FA, LI, PE, RO, TR.

81 *Tsuga canadensis* (L.) Carr. Hemlock
Common; shaded ravines, rocky north-facing slopes, moist woods, swamps. All counties except MO, PE, PC.

TAXACEAE (Yew Family)

82 *Taxus canadensis* Marsh. American Yew
Frequent; shaded ravines, rocky slopes, moist hemlock woods. AB, CL, CU, GE, HL, KN, LK, LI, LR, MH, PE, SU.

TAXODIACEAE (Bald-cypress Family)

83 *Taxodium distichum* (L.) Richard Bald-cypress
Rare, adventive; wet fields, buttonbush swamps, sphagnous seeps. PE: margin of railroad ballast, Jackson Twp., Cline 526, 12 Oct 1974 (KE 43274). PO: buttonbush swamp along railroad, Windham Twp., Andreas & Cooperrider 2200, 30 Jun 1978 (KE 47497). SU: wet sphagnous mat, Twinsburg Twp., Andreas 4293, 25 Jul 1979 (KE 47498).

Division Magnoliophyta (Angiosperms)
CLASS MAGNOLIOPSIDA (DICOTYLEDONS)

ACANTHACEAE (Acanthus Family)

84 *Justicia americana* (L.) Vahl Water-willow
Infrequent; shallow sunny streams, river margins. AS, CL, CU, FA, KN, LI, MH, MO, RO.

85 *Ruellia humilis* Nutt. var. *calvescens* Fern.
Rare; rocky slopes, open woods. FA: Hocking Twp., Goslin s.n., 10 Sep 1935 (OS 62343). RO: Colerain Twp., Bartley & Pontius s.n., 12 Aug 1937 (OS 32519).

86 *Ruellia strepens* L.
Infrequent; wooded floodplains, woodland borders. FA, LI, PC, RI, RO.

ACERACEAE (Maple Family)

87 *Acer negundo* L. Box-elder, Ash-leaved Maple
Common; marshes, floodplains, river banks. ALL COUNTIES.

88 *Acer pensylvanicum* L. Striped Maple, Moosewood
Rare, but locally abundant; rich moist woods, boggy hemlock woods. AB: wooded slope, Conneaut Twp., Bissell 1974:25, 19 May 1974 (CLM 8227).

89 **Acer platanoides* L. Norway Maple
Infrequent, adventive; disturbed woods, abandoned fields, river slopes. CU, LK, MD, PO, ST, SU.

90 *Acer rubrum* L. Red Maple
Common; dry, mesic and moist woods, sphagnous thickets, floodplains, marshes, old fields. ALL COUNTIES.

91 *Acer saccharinum* L. Silver Maple
Common; wooded floodplains, stream terraces, swamp forests, boggy woods. All counties except PC.

92 *Acer saccharum* Marsh. (including *A. nigrum* Michx. f.) Sugar Maple
Common; rich dry and mesic woods, woodland borders, fence rows. ALL COUNTIES.

93 *Acer spicatum* Lam. Mountain Maple
Common; shaded cliffs, ravine bottoms, stream margins. AS, AB, CL, CU, GE, HL, KN, LK, LI, MR, MD, MH, PE, PO, RI, SU, TR, WA.

AIZOACEAE (Carpet-weed Family)

94 **Mollugo verticillata* L. Carpet-weed, Indian Chickweed
Common, adventive; gardens, waste places, sandy roadsides, railroads. AB, CL, CU, FA, GE, HL, KN, LK, LI, MH, MD, PE, PC, PO, RO, ST, SU, TR, WA.

AMARANTHACEAE (Amaranth Family)

95 *Amaranthus albus* L. Tumbleweed
Common; waste places, cultivated fields, roadsides. AB, CU, FA, GE, HL, KN, LK, LR, MH, MD, PC, PO, RO, ST, SU, TR, WA.

96 **Amaranthus blitoides* S. Watson Tumbleweed
Amaranthus graecizans L.
Common, adventive; railroads, roadsides, fields, waste places. AB, CL, CU, GE, KN, LK, LR, MH, MO PE, PO, RO, ST, SU, TR, WA.

97 **Amaranthus hybridus* L. Pigweed, Wild Beet
Common, adventive; waste places, roadsides, cultivated fields. All counties except LI, PE, RI.

98 **Amaranthus lividus* L.
Rare, adventive; waste places. PO: weed, Kent, Cusick 13549, 27 Jul 1974 (KE 40225). SU: near Manchester, Wilson s.n., 12 Aug 1929 (OS 93727). TR: mudflat, Mecca Twp., Rood & Davis 2906, 2 Oct 1955 (KE 12251).

99 **Amaranthus palmeri* S. Watson Palmer's Amaranth
Rare, adventive; waste places, railroads. RO: railroad, Chillicothe, Bartley s.n., 12 Sept 1965 (OS 104690).

100 **Amaranthus retroflexus* L. Green Amaranth
Common, adventive; cultivated fields, waste places. AB, CL, CU, GE, HL, KN, LK, LI, LR, MH, MD, PC, PO, RI, RO, SU, TR, WA.

101 **Amaranthus spinosus* L. Thorny Amaranth
Rare, adventive; waste places. RO: Chillicothe, Kellerman s.n., 12 Aug 1898 (OS 18098).

102 **Amaranthus tamariscinus* Nutt. Water-hemp
Acnida tamariscinus (Nutt.) Wood
Infrequent, adventive; alluvial fields, open floodplains, waste places. CU, HL, LK, PC, RO, TR, WA.

103 **Amaranthus tuberculatus* (Mog.) Sauer Water-hemp
Acnida altissima Riddell
Frequent, adventive; river banks, pond margins, wet waste places. FA, GE, HL, KN, LK, LI, MH, PC, PO, RO, ST, SU, WA.

104 *Froelichia gracilis* (Hooker) Moq.   Slender Cottonweed
Rare, adventive; sandy soil. RO: Chillicothe, Carr 2057, 20 Aug 1979 (KE 45628); Chillicothe, Bartley s.n., 12 Sep 1965 (OS 80088).

ANACARDIACEAE (Cashew Family)

105 *Cotinus coggygria* Scop.   Smoke-tree
Rare, adventive; waste places, roadsides. AB: Wayne Twp., Hicks 1394, 26 Aug 1928 (OS 24888). TR: escaped, Hicks s.n., Aug 1928 (OS 24384).

106 **Rhus aromatica** Aiton var. **aromatica**   Fragrant Sumac
Infrequent; rocky hillsides, open woods, woodland borders. AS, CU, LI, LR, MO, PC, RO, ST, SU.

107 **Rhus copallina** L.   Shining Sumac, Winged Sumac
Common; dry roadside banks, fields, thickets. All counties except MO, PC, RI.

108 **Rhus glabra** L.   Smooth Sumac
Common; dry roadside banks, fields, thickets. ALL COUNTIES.

109 **Rhus radicans** L.   Poison Ivy
Common; dry, mesic and moist woods, thickets, waste places. ALL COUNTIES.

110 **Rhus typhina** L.   Staghorn Sumac
Common; thickets, railroads, dry fields. All counties except MO, PE, PC.

111 **Rhus vernix** L.   Poison Sumac
Common; sphagnous thickets, fens, bogs, buttonbush swamps, marshes. All counties except MO, PE, PC.

ANNONACEAE (Custard-apple Family)

112 **Asimina triloba** (L.) Dunal   Pawpaw
Common; fields, wooded floodplains, woodland borders. All counties except MO, RI.

APOCYNACEAE (Dogbane Family)

113 *Amsonia tabernaemontana* Walter var. *salicifolia* (Pursh) Woodson
Rare, adventive; fields, roadsides. PC: bog, Pickaway Twp., Adams s.n., 16 May 1964 (Muskingum College Herbarium 6340).

114 **Apocynum androsaemifolium** L.   Spreading Dogbane
Common; shaded roadsides, rocky slopes, open woods, woodland borders. AS, AB, CL, CU, FA, GE, HL, KN, LK, LI, LR, MH, MD, PO, RI, ST, SU, TR, WA.

115 **Apocynum cannabinum** L.   Indian Hemp
Common; woodland borders, dry fields, roadsides, waste places. All counties except MO.

116 **Apocynum** x **medium** Greene **(A. androsaemifolium** x **A. cannabinum)**
Frequent; waste places, railroads, dry fields. AB, CU, GE, KN, LR, PO, RI, ST, SU, TR, WA.

117 **Apocynum sibiricum** Jacq.   Clasping-leaf Dogbane
Infrequent; sandy lake shores, rocky woodland banks. AB, GE, LK, LR, MD.

118 *Vinca minor* L.   Periwinkle, Myrtle
Common, naturalized; rich mesic and moist woods, roadside banks, floodplains, abandoned homesites, cemeteries. All counties except MO, PC, RO.

AQUIFOLIACEAE (Holly Family)

119 *Ilex opaca* Aiton   American Holly
Rare, adventive; disturbed woodlots, waste places. FA: Berne Twp., Goslin s.n., 8 Jul 1945 (OS 44260). LI: Cranberry Island, Snodgrass 63, no date (OS s.n.).

120 **Ilex verticillata** (L.) Gray   Winterberry
Common; sphagnous thickets, wet thickets, bogs, wooded floodplains, moist woods. All counties except MO.

121 # **Nemopanthus mucronatus** (L.) Trel.   Mountain-holly
Frequent; sphagnous thickets, bogs. AS, AB, CU, GE, LK, LR, PO, ST, SU, TR, WA.

ARALIACEAE (Ginseng Family)

122 **# Aralia hispida** Vent.   Bristly Sarsaparilla
Infrequent; dry open woods, sandstone ledges, sandy fields. AB, CU, GE, LK, LI, WA.

123 **Aralia nudicaulis** L.   Wild Sarsaparilla
Common; rich moist and mesic woods, wooded floodplains, ravines. All counties except AS, MO, PC.

124 **Aralia racemosa** L.   Spikenard
Common; shaded cliffs, ravines, hemlock slopes, moist woods. All counties except PC.

125 ***Aralia spinosa** L.   Devil's Walking Stick
Infrequent, native in south-central Ohio, adventive elsewhere; wooded floodplains, ravines, mesic and dry woods. AB, CU, MD, PE, PO, ST, SU, WA.

126 **Panax quinquefolius** L.   Ginseng
Common; rich mesic and moist woods, hemlock slopes. All counties except KN, PC, TR.

127 **Panax trifolius** L.   Dwarf Ginseng
Common; rich moist woods, hemlock ravines, wooded floodplains, stream terraces. AS, AB, CL, CU, GE, HL, KN, LK, LR, MH, MD, MO, PO, RI, ST, SU, TR, WA.

ARISTOLOCHIACEAE (Birthwort Family)

128 **Aristolochia serpentaria** L.   Virginia Snakeroot
Frequent; dry wooded slopes, dry oak woods. AS, AB, CL, CU, FA, HL, LI, PO, ST, SU, TR, WA.

129 **Asarum canadense** L.   Wild Ginger
Common; rich mesic and moist woods, wooded floodplains, hemlock slopes. All counties except PC.

ASCLEPIADACEAE (Milkweed Family)

130 **Asclepias amplexicaulis** Smith   Wavy-leaf Milkweed
Rare; sandy soil in open woods, roadsides. FA: dry soil, Hocking Twp., Cusick 16744, 20 Jun 1977 (KE 39342); sandstone, Hocking Twp., Terrell 636, 2 Jul 1941 (OS 109532); Amanda, Kellerman s.n., 20 Aug 1895 (OS 30625). LR: railroad, Kipton, Dick s.n., 18 Jun 1894 (OC 75159); Oberlin, Ricksecker s.n., 11 Jun 1894 (OC 75149). RO: Green Twp., Bartley & Pontius s.n., 4 Jun 1933 (OS 30618).

131 **Asclepias exaltata** L.   Poke Milkweed
Common; rich mesic and moist woods, shaded roadside banks. AB, CL, CU, FA, GE, HL, KN, LK, LI, LR, MH, MD, PE, PO, RI, RO, ST, SU, WA.

132 **Asclepias hirtella** (Pennell) Woodson
Rare; moist meadows and sandy fields. PC: Pickaway Twp., Bartley & Pontius s.n., 13 Sep 1931 (OS 30441).

133 **Asclepias incarnata** L.   Swamp Milkweed
Common; wet ditches, meadows, marshes, fens, open floodplains. ALL COUNTIES.

134 **Asclepias purpurascens** L.   Purple Milkweed
Infrequent; roadsides, fields, pastures, woodland borders. FA, LK, PC, PO, RO, ST, SU, TR.

135 **Asclepias quadrifolia** Jacq.   Four-leaf Milkweed
Common; dry oak woods, wooded hillsides, dry eroding slopes. All counties except MD, TR.

136 **Asclepias sullivantii** Engelm.   Prairie Milkweed
Rare; roadsides, prairie remnants, fence rows, railroads. FA: Fairfield County, Kellerman s.n., 2 Jul 1892 (OS 30605). LR: roadbank, Camden Twp., Cusick 11838, 6 Jul 1971 (KE 26476).

137 **Asclepias syriaca** L.   Common Milkweed
Common; fence rows, fields, waste places, roadsides, railroads. All counties except MO.

138 **Asclepias tuberosa** L.   Butterfly-weed, Pleurisy-root
Common; dry roadside banks, sandy fields, prairie remnants, railroads. AB, CL, CU, FA, GE, HL, KN, LK, LI, LR, MD, PE, PC, PO, RI, RO, ST, SU, WA.

42

139 *Asclepias variegata* L. White Milkweed
Rare; dry woodland margins, shaded roadside banks. FA: roadside, Wolfe et al, s.n., 31 May 1941 (OS 43641); Hocking Twp., Goslin s.n., 18 Jun 1936 (OS 30639 & OS 30640). SU: Akron, Werner s.n., 3 Jun 1892 (OS 30630).

140 *Asclepias verticillata* L. Whorled Milkweed
Infrequent; dry roadside banks, fields. AB, CU, FA, LK, RO, SU.

141 *Asclepias viridiflora* Raf. Green-flowered Milkweed
Rare; dry woodland borders, fields, roadsides. AB: below Harpersfield, Hicks s.n., 20 Jun 1932 (OS 30400). RI: Marshfield, Selby 2163, 17 Jun 1899 (OS 7052).

142 *Asclepias viridis* Walter Spider Milkweed
Asclepiodora viridis (Walter) Gray
Rare; dry woodland borders, fields, prairies. RO: Scioto Twp., Thomas s.n., 4 Jun 1935 (OS 81297).

143 *Cynanchum laeve* (Michx.) Persoon Sandvine
Ampelamus albidus (Nutt.) Britton
Infrequent; fence rows, waste places, railroads, river banks. FA, KN, LI, PC, RO, ST.

144 *Cynanchum nigrum* (L.) Persoon Swallow-wort
Rare, naturalized; waste places, roadsides. PC: Circleville Twp., Bartley s.n., 20 Aug 1958 (OS 68713).

145 *Matelea obliqua* (Jacq.) Woodson Angle-pod
Gonolobus obliquus (Jacq.) Schultes
Rare; thickets, fence rows, woodland borders. RO: Colerain Twp., Bartley & Pontius s.n., 22 Jun 1932 (OS 30827).

BALSAMINACEAE (Touch-me-not Family)

146 *Impatiens balsamina* L. Garden Balsam
Rare, adventive; waste places. HL: refuse dump, Monroe Twp., Wilson 2468, 20 Aug 1971 (KE 29256).

147 *Impatiens capensis* Meerb. Spotted Touch-me-not
Common; wet woods, swamps, fens, shaded ditches, thickets. All counties except PC.

148 *Impatiens pallida* Nutt. Pale Touch-me-not, Yellow Jewelweed
Common; woodland borders, shaded ditches, thickets, moist woods. All counties except MO & PC.

BERBERIDACEAE (Barberry Family)

149 *Berberis thunbergii* DC. Japanese Barberry
Common, adventive; mesic woods, fields. ALL COUNTIES.

150 *Berberis vulgaris* L. Common Barberry
Once rare, now presumably extirpated from Ohio, adventive; fields, roadsides. LK, LI, LR, PE, PO.

151a *Caulophyllum thalictroides* (L.) Michx. var. *thalictroides* Blue-cohosh
Common; rich mesic and moist woods, ravines. All counties except PC.

151b *Caulophyllum thalictroides* (L.) Michx. var. *giganteum* Farw.
Common; rich mesic and wet woods, wooded floodplains. All counties except PC, RI.

152 *Jeffersonia diphylla* (L.) Persoon Twinleaf
Frequent; rich mesic woods, wooded floodplains, eroding slopes. AB, CU, GE, LK, LR, MD, MO, PE, PO, RO, SU, WA.

153 *Podophyllum peltatum* L. May-apple
Common; rich dry, mesic and moist woods, floodplains, pastures, shaded roadsides. All counties except PC.

BETULACEAE (Birch Family)

154 *Alnus glutinosa* (L.) Gaertner Black Alder
Infrequent, naturalized; waste places, moist fields. AB, CL, CU, HL, LK, MO, PO.

155 **#*Alnus rugosa*** (DuRoi) Sprengel   Speckled Alder
Frequent; thickets, fen margins, marshes, open floodplains. AS, AB, CL, CU, GE, HL, LK, LR, MD, PO, RI, ST, SU, TR, WA.

156 **Alnus serrulata** (Aiton) Willd.   Common Alder
Common; marshes, wet thickets, fen margins, open floodplains. All counties except MO, PC.

157 **Betula alleghaniensis** Britton   Yellow Birch
Betula lutea Michx. f.
Common; shaded ravines, sphagnous bogs, wet woods, buttonbush swamps. AS, AB, CL, CU, FA, GE, HL, LK, LR, MH, MD, PE, PO, ST, SU, TR, WA.

158 **Betula lenta** L.   Cherry Birch, Sweet Birch
Infrequent; rocky ledges, ravines, rich moist slopes. AS, CL, CU, FA, GE, LI, LR, PO, SU.

159 **Betula nigra** L.   River Birch
Rare; river and stream margins. FA: Rush Creek Twp., Goslin s.n., 22 Jul 1956 (BHO 21425); Bremen, Kellerman s.n., 24 Oct 1895 (OS 26340). PE: Junction City, Kellerman s.n., 26 Apr 1901 (OS 26333).

160 ***Betula pendula*** Roth   European Birch
Rare, adventive; fields, disturbed thickets, rocky areas. MD: roadside, Chippewa Lake, Lafayette Twp., Mutz 1360, 8 Aug 1965 (KE 15751). PO: disturbed peat bog, Hiram Twp., Andreas 3502, 24 May 1979 (KE 49079); rocky field, Nelson Twp., Rood 1786, 13 May 1941 (KE 10821). SU: wet quarry floor, Boston Twp., Andreas & Stoutamire 4377, 1 Aug 1979 (KE 49081).

161 **Betula populifolia** Marsh.   Gray Birch
Infrequent; abandoned stone quarries, sphagnous thickets, disturbed sandy areas. AB, CU, GE, LK, PO, ST, SU, TR.

162 **Betula pumila** L.   Swamp Birch
Rare; calcareous marshes, tamarack fens, sphagnous thickets. PO: shrubby fen, Suffield Twp., Andreas 5570, 2 Jun 1981 (KE 44034); Suffield Bog, Rood et al, 16 Sep 1923 (OS 26359). ST: Jackson Twp., Brown s.n., 17 Dec 1939 (OS 26357); Congress Lake, Kellerman s.n., 28 Jun 1898 (OS 26358). SU: bog, Norton Twp., Andreas 2500, 3 Aug 1978 (KE 42003); tamarack bog, Coventry Twp., Andreas 4081, 11 Jul 1979 (KE 42297); Long Pond, Hicks s.n., 15 Jun 1933 (OS 25355).

163 **Carpinus caroliniana** Walter   American Hornbean, Ironwood
Common; rich mesic woods, wet thickets, pastures, bog and fen margins. ALL COUNTIES.

164 **Corylus americana** Walter   American Hazelnut
Common; woodland borders, wet and mesic thickets, stream margins, pastures. ALL COUNTIES.

165 **Corylus cornuta** Marsh.   Beaked Hazel
Once rare, now presumably extirpated from Ohio; thickets. AB: gorge, Windsor Twp., Hicks and Bartley 2115, 30 Jul 1954 (US).

166 **Ostrya virginiana** (Miller) K. Koch   Hop Hornbean
Common; mesic and dry woods, roadside banks. ALL COUNTIES.

BIGNONIACEAE (Trumpet Creeper Family)

167 ***Campsis radicans*** (L.) Seem.   Trumpet-creeper
Common, native to southern counties, naturalized elsewhere (Cooperrider, 1985a); roadside banks, railroads, thickets, fence rows. AS, CU, FA, GE, HL, KN, LI, LR, MD, PE, PO, RO, ST, SU, TR, WA.

168 ***Catalpa bignonioides*** Walter   Southern Catalpa
Infrequent, naturalized; roadsides, pastures. AS, CL, MD, PO, ST.

169 ***Catalpa ovata*** G. Don   Chinese Catalpa
Rare, naturalized; waste places, disturbed areas, roadsides. AB: Pine Run, Hicks s.n., summer 1929 (OS 32374). GE: waste area, Bainbridge Twp., Hauser 265, 3 Aug 1960 (KE 6587). MD: roadside, Chatham Twp., Mutz 1216, 10 Jul 1965 (KE 15399). SU: parking lot, Akron, Dumke 654, 16 Jun 1966 (KE 15714).

170 ***Catalpa speciosa*** Warder   Cigar-tree
Frequent, naturalized; fields, pastures, roadsides, waste places. AS, GE, HL, KN, LI, MH, MD, PE, PC, RO, ST, TR, WA.

BORAGINACEAE (Borage Family)

171 *Anchusa azurea* Miller   Alkanet
Rare, adventive; waste places. AB: escape near Austinburg, Hicks 1782, 17 Aug 1930 (OS 70788).

172 *Cynoglossum officinale* L.   Common Hound's Tongue
Common, naturalized; roadsides, pastures, waste places. AB, CL, CU, GE, HL, KN, LK, LI, LR, MH, MO, PO, RI, RO, ST, SU, TR, WA.

173a # *Cynoglossum virginianum* L. var. *boreale* (Fern.) Cooperrider
Northern Wild Comfrey
Cynoglossum boreale Fern.
Once rare, now presumably extirpated from Ohio; rich mesic and moist woods, thickets. AB: East Monroe Twp., Hicks, s.n., 12 Aug 1932 (OS 32718); along Grand River, Hicks s.n., 25 May 1930 (OS 70784). MH: Standing Rock, Berlin Twp., Vickers s.n., 19 Aug 1910 (Youngstown State University Herbarium 3687). PO: edge of woods, Shalersville Twp., Herrick s.n., 13 Jun 1959 (OS 64674).

173b *Cynoglossum virginianum* L. var. *virginianum*   Wild Comfrey
Frequent; rich mesic woods, wooded slopes. AB, CL, FA, GE, HL, LI, MD, PE, PO, RI, RO, SU, WA.

174 *Echium vulgare* L.   Blueweed, Blue Devil
Frequent, naturalized; roadsides, railroads, waste places. AB, CL, CU, GE, LK, LI, LR, MH, PO, RI, RO, ST, SU, TR, WA.

175 *Hackelia virginiana* (L.) I. M. Johnston   Stickseed
Common; railroads, thickets, roadsides, waste places. AS, AB, CL, CU, FA, GE, HL, KN, LK, LI, LR, PE, PC, PO, RO, ST, SU, WA.

176 *Heliotropium europaeum* L.   Heliotrope
Rare, naturalized; roadsides, waste places. MH: Youngstown, Vujevic s.n., 23 Sep 1980 (Youngstown State University Herbarium 21374).

177 *Lappula squarrosa* (Retz.) Dumort.   Prickly Stickseed
Lappula echinata Gilib.
Frequent, naturalized; railroads, waste places, roadsides. AB, CU, HL, LK, LI, LR, PC, PO, RI, ST, TR.

178 *Lithospermum arvense* L.   Corn Gromwell
Common, naturalized; thickets, roadsides, waste places. AB, CL, CU, FA, GE, LK, LI, LR, MO, PE, PC, PO, RI, RO, ST, SU, TR, WA.

179 *Lithospermum canescens* (Michx.) Lehm.   Hoary Puccoon
Infrequent; sandy fields, prairie remnants, open dry woods. AB, KN, PC, RO, SU.

180 *Lithospermum latifolium* Michx.   American Gromwell
Rare; woods, thickets. RO: Ross County, Bartley s.n., 13 May 1962 (OS 86163).

181 *Lithospermum officinale* L.   European Gromwell
Rare, naturalized; roadsides, waste places. RO: roadbank, Springfield Twp., Bryant 442, 11 Apr 1976 (Miami University Herbarium 121920).

182 *Lycopsis arvensis* L.   Small Bugloss
Rare, adventive; sandy fields, waste places. AB: escape near Austinburg, Hicks s.n., 17 Aug 1930 (OS 32907). ST: Congress Lake, Kellerman & Kellerman s.n., 28 Jun 1898 (OS 32905).

183 *Mertensia virginica* (L.) Persoon   Bluebells
Common; rich moist woods, wooded floodplains, stream terraces. ALL COUNTIES.

184 *Myosotis arvensis* (L.) Hill   Forget-me-not
Infrequent, naturalized; fields, roadsides, ditches. FA, LK, LR, MH, PC, PO.

185 *Myosotis discolor* Persoon
Myosotis versicolor (Persoon) Smith
Rare, adventive; waste places, lawns. PO: lawn, Franklin Twp., Cusick 4268, 24 May 1967 (KE 16781 & OS 26595).

186 *Myosotis laxa* Lehm.   Small Forget-me-not
Frequent; seepage banks, moist thickets, fens, marshes. AB, CL, CU, GE, HL, LK, LI, MH, PO, ST, SU, TR, WA.

187 **Myosotis macrosperma** Engelm. Large-seeded Forget-me-not
Rare; open woods, floodplains. PC: Washington Twp., Bartley & Pontius s.n., 7 May 1930 (OS 32868).

188 ***Myosotis scorpioides** L. True Forget-me-not
Common, naturalized; swamps, marshes, seeps, roadside ditches. AS, AB, CL, CU, FA, GE, HL, KN, LK, LI, LR, MH, MD, PO, RI, ST, SU, TR, WA.

189 ***Myosotis stricta** Link
Rare, adventive; roadsides, fields. MH: roadside park, Milton Twp., Cusick 23374, 16 May 1984 (KE 46571).

190 ***Myosotis sylvatica** Hoffm. Garden Forget-me-not
Infrequent, adventive; fields, roadsides. CU, GE, MD, PO, SU, TR.

191 **Myosotis verna** Nutt.
Infrequent; rocky woods, dry roadsides. AS, AB, CL, CU, FA, KN, LR, MH, PC, RO.

192 **Onosmodium molle** Michx. var. **hispidissimum** (Mackenz.) Cronq. False Gromwell
Onosmodium hispidissimum Mackenz.
Rare; dry hillsides, prairie remnants. LR: Lorain County, Day s.n., 30 Jul 1893 (OS 32843). RO: old canal, Union Twp., Bartley s.n., 3 Oct 1961 (OS s.n.); brushy hillside, Springfield Twp., Hall 434, 15 Jun 1954 (BHO 12590).

193 ***Symphytum asperum** Lepechin Prickly Comfrey
Infrequent, naturalized; roadsides, waste places. AS, HL, LK, SU, TR, WA.

194 ***Symphytum officinale** L. Common Comfrey
Infrequent, naturalized; damp roadsides, fields, waste places. AB, CU, GE, LK, RI, ST.

CACTACEAE (Cactus Family)

195 **Opuntia humifusa** Raf. Prickly Pear
Rare; dry sandy ground. FA: Pleasant Twp., Goslin s.n., Jan 1941 (OS 27559). RO: Paxton Twp., Bartley & Pontius s.n., 17 Sep 1933 (OS 27565).

CALLITRICHACEAE (Water-starwort Family)

196 **Callitriche heterophylla** Pursh Water Chickweed
Frequent; ponds, temporary pools, quiet streams. AB, CL, CU, FA, GE, HL, LK, LR, MH, PE, PO, ST, SU, TR, WA.

197 # **Callitriche palustris** L. (including **C. verna** L.) Water-starwort
Infrequent; mudflats, wet stream banks. AB, CU, GE, LR, PO, TR, WA.

198 **Callitriche terrestris** DC. Austin's Water-starwort
Callitriche deflexa A. Braun var. *austinii* (Engelm.) Hegelm.
Rare; mudflats, ditches. LK, LR, PE, PO, RO.

CAMPANULACEAE (Harebell Family)

199 **Campanula americana** L. Tall Bellflower
Common; rich mesic woods, roadside banks, wooded floodplains, swamps. All counties except MO, PC.

200a **Campanula aparinoides** Pursh var. **aparinoides** Marsh Bellflower
Rare; thickets, marshes. FA: Pleasant Run Swamp, Berne Twp., Goslin s.n., Jun 1934 (OS 101449). KN: marsh, Knox Lake, Keil, Roberts & McGill 9638, 17 Aug 1973 (OS 111824). RO: Blackwater Creek, Green Twp., Cusick 17413, 16 Aug 1977 (OS 122421); Liberty Twp., Pontius & Bartley s.n., 14 Jul 1929 (OS 37487).

200b **Campanula aparinoides** Pursh var. **grandiflora** Holz Marsh Bellflower
Campanula uliginosa Rydb.
Frequent; boggy thickets, meadows, fens. AS, CL, CU, FA, GE, HL, KN, LI, LR, PO, RI, RO, ST, SU, WA.

201 ***Campanula rapunculoides** L.
Infrequent, naturalized; roadsides, thickets, ditches, meadows. AS, AB, CU, GE, LK, LR, PO, ST, SU, TR, WA.

202 **Lobelia cardinalis** L. Cardinal Flower
Common; wooded floodplains, marshes, meadows, shaded roadside ditches. All counties except RO.

46

203 *Lobelia inflata* L.   Indian Tobacco
Common; fields, woodland borders, waste places. All counties except MO.

204 *Lobelia kalmii* L.   Kalm's Lobelia
Frequent; fens, wet meadows. CL, CU, FA, GE, HL, LR, PC, PO, RO, ST, SU.

205 *Lobelia siphilitica* L.   Great Lobelia
Common; roadside ditches, seeps, woodland borders, pastures. ALL COUNTIES.

206a *Lobelia spicata* Lam. var. *spicata*   Pale Spike Lobelia
Common; dry slopes, prairie remnants, railroads, open woods. AS, AB, CU, FA, GE, HL,
KN, LI, MH, PE, PO, RI, RO, ST, SU, WA.

206b *Lobelia spicata* Lam. var. *leptostachys* (DC.) Mackenz. & Bush
Rare; dry hillsides, woodland borders. FA: Kettle Hills, Thomas & Goslin s.n, 18 Jul 1933
(OS 37783); S of Lancaster, Chapman s.n., 10 Sep 1932 (OS 37678). LI: Toboso, Jones s.n.,
15 Jul 1891,(OC 81192). PC: Pickaway County, Bartley & Pontius s.n., 18 Sep 1937 (OS
37687).

207 *Triodanis perfoliata* (L.) Nieuwl.   Venus' Looking-glass
Specularia perfoliata (L.) A. DC.
Common; woodland borders, thickets, railroads, waste places. AS, AB, CL, CU, FA, GE,
HL, KN, LK, LI LR, PE, PO, RI, RO, ST, SU, TR, WA.

CANNABACEAE (Hemp Family)

208 **Cannabis sativa* L.   Marijuana, Hemp
Infrequent, adventive; waste places, roadsides, cultivated fields, railroads. CU, HL, KN, LK,
LI, PO, WA.

209 **Humulus japonicus* Sieb. & Zucc.   Japanese Hop
Infrequent, naturalized; roadsides, waste places, railroads. AB, CL, CU, PC, SU, TR, WA.

210 *Humulus lupulus* L.   Common Hop
Common; railroads, fence rows, waste places, roadsides. AS, CU, FA, GE, HL, KN, LK, LI,
LR, PE, PO, RI, RO, ST, SU, TR, WA.

CAPPARACEAE (Caper Family)

211 **Cleome hassleriana* Chodat   Spider-flower
Cleome spinosa Jacq.
Infrequent, adventive; roadsides, waste places, abandoned homesites. AB, CU, FA, LI, LR,
PE, ST.

212 **Cristatella jamesii* T. & G.
Rare, adventive; sandy soils, railroads. SU: railroad bed, Stow Twp., Cusick 14372, 10 Sep
1967 (KE 39244).

213 *Polanisia dodecandra* (L.) DC.   Clammyweed
Polanisia graveolens Raf.
Frequent; railroads, gravelly banks. AB, CU, FA, GE, KN, LK, LI, LR, PO, RO, ST, SU.

CAPRIFOLIACEAE (Honeysuckle Family)

214 # *Diervilla lonicera* Miller   Bush Honeysuckle
Common; shaded cliffs, dry woods, ravine summits. AS, AB, CL, CU, GE, HL, KN, LK,
LI, LR, MH, MD, PO, RI, RO, ST, SU, TR, WA.

215 # *Linnaea borealis* L. var. *americana* (Forbes) Rehder   Twinflower
Once rare, now presumably extirpated from Ohio; tamarack bogs. LK: Painesville, Beardslee
s.n., no date (OS 3796). SU: Akron, Wagner s.n., 17 Oct 1924 (OS 95195). Braun (1961)
indicates a record for PO.

216 **Lonicera* x *bella* Zabel (*L. morrowii* x *L. tatarica*)
Common, naturalized; thickets, roadsides, fields. AS, AB, CL, CU, GE, KN, LK, LI, LR,
MH, MD, PE, PO, ST, SU, WA.

217 # *Lonicera canadensis* Bart.   Fly Honeysuckle
Infrequent; hemlock slopes, cliffs, boggy woods. AS, AB, CL, CU, GE, LK, LR, MH, PO,
SU, TR.

218 *Lonicera dioica* L. var. *glaucescens* (Rydb.) Butters   Wild Honeysuckle
Common; boggy thickets, swamps, wooded slopes. All counties except FA, PE, PC.

219 *Lonicera japonica* Thunb.   Japanese Honeysuckle
   Common, naturalized; roadsides, thickets, woodland borders, woodlots. All counties except
   MO and TR.

220 *Lonicera maackii* (Rupr.) Maxim.
   Rare, naturalized; thickets, pastures. KN: railroad track, Liberty Twp., Pusey 854, 18 Sep
   1975 (KE 38143). ST: thicket, Jackson Twp., Amann 2089, 10 Oct 1960 (KE 3170); woods,
   Plain Twp., Amann 1540, 28 Aug 1960 (KE 3169).

221 *Lonicera morrowi* Gray   Morrow's Honeysuckle
   Frequent, naturalized; pastures, fields, waste places. AB, CL, CU, FA, GE, LK, LR, MD,
   PO, ST, SU, TR, WA.

222 # *Lonicera oblongifolia* (Goldie) Hooker   Swamp Fly Honeysuckle
   Once rare, now presumably extirpated from Ohio; swampy thickets, wet woods. AB: Pennline
   Bog, Hicks s.n., 17 Jun 1933 (OS 37203 & OS 37205). Braun (1961) indicates a record for
   CU.

223 *Lonicera prolifera* (Kirchner) Rehder   Grape Honeysuckle
   Rare; rocky woods, roadside banks. KN: Mohican River, Green s.n., 1893 (OS 95282). LI:
   Licking Reservoir, Jones s.n., 13 Jul 1892 (OC 81084). WA: Peewee Hollow, Selby & Duvel
   s.n., 15 Sep 1899 (OS 95284).

224 *Lonicera sempervirens* L.   Coral Honeysuckle, Trumpet Honeysuckle
   Rare; native in southern counties, adventive elsewhere (Braun, 1961); roadsides, thickets.
   AB, CU, FA, PO, ST.

225 *Lonicera tatarica* L.   Tartarian Honeysuckle
   Common, naturalized; pastures, fields, waste places. AS, AB, CU, FA, GE, KN, LK, LR,
   MH, MD, PE, PO, RI, SU, TR.

226 *Lonicera villosa* (Michx.) R. & S.   Mountain Fly Honeysuckle
   Braun (1961) indicates a record for AB.

227 *Lonicera xylosteum* L.   European Honeysuckle
   Infrequent, adventive; pastures, waste places. AB, FA, HL, LK, PO, TR.

228 *Sambucus canadensis* L.   Common Elderberry
   Common; woodlots, fields, roadsides, thickets, pond margins. All counties except PC.

229 # *Sambucus pubens* Michx.   Red-berried Elder
   Common; moist woods, hemlock slopes, cliffs, rocky moist woods. AS, AB, CL, CU, GE,
   HL, KN, LK, LR, MH, MD, PO, RI, ST, SU, TR, WA.

230a *Symphoricarpos albus* (L.) Blake var. *albus*   Snowberry
   Once rare, now presumably extirpated from Ohio; rocky banks, shores. LR: Elyria, Ricksecker
   s.n., Jun 1892 (OS 37152).

230b *Symphoricarpos albus* L. Blake var. *laevigatus* (Fern.) Blake   Snowberry
   Frequent, naturalized; mesic woods, roadside banks, waste places. AS, AB, CL, CU, FA,
   KN, LK, LI, MD, MO, PO, SU, WA.

231 *Symphoricarpos occidentalis* Moench   Wolfberry
   Rare, adventive; railroads, waste places. GE: weedy ground, Middlefield Twp., Hawver 730,
   27 Jun 1960 (KE 6233). HL: railroad, Washington Twp., Wilson 2201, 19 Jul 1971 (KE
   29327). KN: railroad, Miller Twp., Pusey 280, 11 Sep 1934 (KE 38136).

232 *Symphoricarpos orbiculatus* Moench   Coralberry
   Common; dry roadside banks, thickets, hillsides, waste places. AS, CL, CU, GE, HL, KN,
   LK, LI, LR, MD, PE, RO, ST, SU, TR, WA.

233 *Triosteum aurantiacum* Bickn.   Wild Coffee
   Common; rich woods, wooded lake margins, thickets. AB, CL, CU, GE, KN, LK, LI, MD,
   PE, PC, PO, RI, RO, ST, SU, WA.

234 *Triosteum perfoliatum* L.   Tinker's-weed
   Common; rocky open woods, thickets. AS, CU, FA, HL, KN, LK, LR, MD, MO, PE, PC,
   PO, ST, SU, TR, WA.

235 *Viburnum acerifolium* L.   Maple-leaf Viburnum
   Common; dry and mesic woods, rocky slopes, hemlock ravines, sandstone outcrops. All
   counties except PC.

236 # *Viburnum alnifolium* Marsh.   Hobble-bush
Infrequent; hemlock ravines, rich boggy woods. AB, CU, GE, LK, PO, SU.

237 *Viburnum cassinoides* L.   Witherod, Wild Raisin
Infrequent; boggy and moist woods, thickets. AS, AB, CU, GE, LR, MD, PO, ST, SU, TR, WA.

238 *Viburnum dentatum* L.   Southern Arrow-wood
Infrequent; moist thickets, woodland margins. CU, HL, KN, LI, RI, RO.

239 **Viburnum lantana* L.   Wayfaring-tree
Rare, adventive; railroads, fence rows, roadsides. GE: weedy roadside, Chester Twp., Hawver 433, 19 Sep 1959 (KE 3487). PO: railroad, Franklin Twp., Hobbs s.n., 23 Jul 1971 (KE 26022). RI: shaded slope, Worthington Twp., Brockett & Trelka 28, 6 May 1967 (KE 17050).

240 *Viburnum lentago* L.   Nannyberry, Sweet Viburnum
Common; stream banks, pond and lake margins, moist woods. All counties except MO, PC, RI.

241a **Viburnum opulus* L. var. *opulus*   European Highbush Cranberry
Frequent, naturalized; roadsides, marshes, boggy thickets, disturbed woodlots. AS, AB, CU, GE, LK, LR, MH, MD, PO, ST, SU.

241b # *Viburnum opulus* L. var. *americanum* Aiton   Highbush Cranberry
Viburnum trilobum Marsh.
Infrequent; marshes, boggy thickets, fens. AB, CU, GE, LK, MD, PO, ST, SU, TR.

242 *Viburnum prunifolium* L.   Black-haw
Common; marshes, moist roadsides, thickets. AS, CL, FA, HL, KN, LI, LR, MD, MO, PE, PC, PO, RI, RO, ST, TR, WA.

243 *Viburnum rafinesquianum* Schultes   Downy Arrow-wood
Rare; dry hillsides, dry open woods. LR: Oberlin, Hicks s.n., no date (CLM 11598); south woods, Dick s.n., 25 May 1895 (CLM 11597). SU: dry hillside, Boston Twp., Cooperrider & Herrick 7589, 27 Jul 1960 (KE 3950); hillside, Boston Twp., Herrick s.n., 28 Jul 1959 (OS 63387). TR: open woods, Southington Twp., Rood 2286, 18 Jun 1947 (KE 11478).

244 # *Viburnum recognitum* Fern.   Northern Arrow-wood
Common; damp thickets, boggy woods, floodplains, marshes. AS, AB, CL, CU, GE, HL, LK, LR, MH, MO, PO, RI, ST, SU, TR, WA.

CARYOPHYLLACEAE (Pink Family)

245 **Agrostemma githago* L.   Corn-cockle
Common, naturalized; cultivated fields, waste places. All counties except AS, MH, PE.

246 *Arenaria lateriflora* L.   Grove Sandwort
Infrequent; woodland borders, sandy fields, meadows. AB, LK, MO, PE, TR.

247 **Arenaria serpyllifolia* L.   Thyme-leaved Sandwort
Common, naturalized; dry sandy fields, roadsides, waste places. AB, CL, CU, FA, GE, HL, KN, LK, LI, LR, MH, PC, PO, RO, ST, SU.

248 *Arenaria stricta* Michx. var. *stricta*   Rock Sandwort
Rare; limestone ledges, rocks. RI: Mansfield, Wilkinson s.n., no date (OS 105239 & OC 209392).

249 *Cerastium arvense* L.   Field Chickweed
Infrequent; rocky soils, weedy lawns. CU, HL, KN, LK, LR, SU, TR.

250 **Cerastium fontanum* Baumg.   Common Mouse-eared Chickweed
Cerastium vulgatum L.
Common, naturalized; cultivated fields, lawns, meadows, open woods, gardens. All counties except RI.

251 *Cerastium nutans* Raf.   Nodding Chickweed
Common; rich moist woods, stream banks, wooded floodplains. AS, CL, CU, FA, GE, HL, KN, LK, LI, LR, MH, MO, PO, RO, ST, SU, TR, WA.

252 **Cerastium viscosum* L.
Infrequent, naturalized; fields, roadsides, waste places. CL, FA, LK, LI, RO, SU, TR.

253 **Dianthus armeria* L.   Deptford Pink
Common, naturalized; dry fields, roadsides, meadows. All counties except PC.

254 *Dianthus barbatus* L. Sweet William
Infrequent, adventive; fields, roadsides, abandoned homesites. AS, AB, GE, HL, LK, LI, PO, ST, SU, WA.

255 *Dianthus deltoides* L. Maiden Pink
Infrequent, adventive; roadsides, fields, open woods. CU, GE, KN, LK, PO.

256 *Holosteum umbellatum* L. Jagged Chickweed
Rare, naturalized; fields, roadsides, gardens. RO: railroad, Chillicothe, Bartley s.n., 14 May 1972 (OS 112947); railroad, Chillicothe, Bartley s.n., 11 Apr 1968 (OS 80415).

257 *Lychnis coronaria* (L.) Desr. Rose Campion
Infrequent, naturalized; roadsides, rocky fields, open woods, waste places. CL, CU, FA, HL, LK, MH, PO, ST.

258 *Lychnis dioica* L. Red Campion
Rare, naturalized; waste places. TR: yard, Braceville Twp., Rood 2965, 15 Jun 1956 (KE 11196).

259 *Lychnis flos-cuculi* L. Cuckoo-flower
Rare, adventive; fields, waste places. PO: low marshy ground, Close 548, 17 Aug 1933 (KE 480).

260 *Lychnis viscaria* L. German Catchfly
Rare, adventive; roadsides, weedy fields. PO: field, Brimfield Twp., Cooperrider 10830, 24 May 1969 (KE 21199).

261 *Myosoton aquaticum* (L.) Moench Giant Chickweed
Stellaria aquatica (L.) Scop.
Frequent, naturalized; wet ditches, marshes, pastures, meadows. AB, CL, CU, FA, HL, KN, LK, PO, RO, ST, SU, TR, WA.

262 *Paronychia canadensis* (L.) Wood Forked Chickweed
Common; sandy roadsides, fields, woodland borders. AS, AB, CL, CU, FA, HL, KN, LK, LI, LR, MH, PE, PO, RO, ST, SU, TR, WA.

263 *Paronychia fastigiata* (Raf.) Fern. Forked Chickweed
Infrequent; dry slopes, sandy openings. HL, LK, PE, RO, ST.

264 *Sagina decumbens* (Ell.) T. & G. Southern Pearlwort
Rare, adventive; sidewalks, fields. RO: stone walk, Chillicothe, Bartley s.n., 1 Jun 1952 (OS 45423).

265 *Sagina procumbens* L. Bird's-eye
Infrequent, naturalized; waste places, sidewalks. CU, LK, PC, RO, SU.

266 *Saponaria officinalis* L. Soapwort, Bouncing Bet
Common, naturalized; waste places, roadsides, fence rows. All counties except MO, PC.

267 *Scleranthus annuus* L. Knawel
Rare, naturalized; cultivated gardens, waste places. LK: sandy field, Mentor, Bissell 1980-170, 5 Sep 1980 (CLM 13809). MH: berry patch, Canfield, Knopp s.n., Jun 1960 (OS 63544). PO: flower bed, Franklin Twp., Cusick 14008, 7 Oct 1974 (KE 36577). SU: flower bed, Akron, Cusick 11614, 29 May 1971 (KE 26765).

268 *Silene antirrhina* L. Sleepy Catchfly
Common; sandy fields, waste places, railroads. All counties except CL, MO, RI.

269 *Silene armeria* L. Sweet William Catchfly
Infrequent, naturalized; lawns, waste places. AB, CU, HL, LK, LI, SU.

270 *Silene caroliniana* Walter var. *pensylvanica* (Michx.) Fern. Carolina Catchfly
Rare; dry sandy ridges. SU: sandy slope, Akron, Andreas 3285, 13 May 1979 (KE 42283); Gorge Park, Akron, "J. M." s.n., 16 May 1963 (University of Akron Herbarium s.n.); East Liberty, Hicks 1027, 9 Jun 1931 (OS 84319).

271 *Silene cserei* Baumg.
Rare, naturalized; railroads, waste places. TR: ballast, Phalanx, Rood 1343, 20 Jun 1935 (KE 11997).

272 *Silene dichotoma* Ehrh. Forking Catchfly
Rare, naturalized; fields, waste places. AB: south Wayne Twp., Chapman & Hicks s.n., 16 Jun 1933 (OS 17517). GE: Burton Twp., Chapman & Hicks s.n., 15 Jun 1933 (OS 17516); Burton, Dambach s.n., 13 Jun 1934 (OS 17414). PO: meadow, Windham Twp., Rood 2816,

18 Jun 1954 (KE 11989). RO: gravel pit, Union Twp., Cusick 14960, 21 Jun 1976 (KE 39439 & OS 122172).

273 *Silene noctiflora* L.   Night-flowering Catchfly
Frequent, naturalized; waste places, fields, roadsides. AS, AB, CU, FA, HL, KN, LK, LR, PC, PO, RO, SU, TR, WA.

274 *Silene pratensis* (Raf.) Godron & Gren.   White Campion
Lychnis alba Miller
Common, naturalized; shaded roadside banks, thickets, fields. All counties except LI, MO, RO.

275 *Silene stellata* (L.) Aiton f.   Starry Campion
Common; dry slopes, weedy thickets, shaded roadside banks. CL, CU, FA, GE, HL, KN, LI, LR, PE, PC, PO, RO, ST, SU, TR, WA.

276 *Silene virginica* L.   Fire-pink
Common; rich wooded slopes, steep roadside banks, rocky hillsides. AS, AB, CL, FA, HL, KN, LK, LI, MH, MD, PE, PC, PO, RI, RO, ST, SU, WA.

277 *Silene vulgaris* (Moench) Garcke   Bladder-campion
Silene cucubalus Wibel
Frequent, naturalized; railroads, waste places. AS, AB, CU, FA, GE, KN, LK, LR, MH, PC, PO, RO, ST, SU, TR.

278 *Spergula arvensis* L.   Corn Spurrey
Infrequent, naturalized; gardens, waste places, roadsides. AB, CU, LK, PC, PO, SU.

279 *Spergularia marina* (L.) Griseb.   Sand Spurrey
Infrequent, naturalized; saline marshes, roadsides. AB, CU, LK, LR, ST, SU, WA.

280 *Spergularia media* (L.) Griseb.   Sand Spurrey
Infrequent, adventive; saline gravels, roadsides. AB, CU, LK, LR, MD, MO, RI, ST, SU, WA.

281 *Spergularia rubra* (L.) J. & C. Presl
Rare, adventive; sandy soils, waste places. LK: highway, Carr 1413, 28 Jun 1979 (KE 45685). PO: sandy lot, Kent, Mazzer s.n., 6 Jun 1972 (KE 27725); weed lot, Herrick s.n., 20 Jun 1960 (KE 7368). SU: alkaline tailings, Franklin Twp., Andreas 6303, 19 Jul 1984 (KE 47805).

282 *Stellaria graminea* L.   Common Stitchwort
Frequent, naturalized; wet woods, fields, pastures. AS, AB, CL, CU, FA, GE, LK, LR, MD, MO, PO, ST, SU, TR, WA.

283 *Stellaria longifolia* Willd.
Common; wet meadows, fields, damp thickets. All counties except MO, RI, RO.

284 *Stellaria media* (L.) Vill.   Common Chickweed
Common, naturalized; cultivated fields, roadsides, waste places, gardens. All counties except RO.

285 *Stellaria pubera* Michx.   Great Chickweed
Infrequent; rich woods, rocky slopes, woodland borders. AB, FA, PE, RO, WA.

286 *Vaccaria hispanica* (Miller) Rauschert   Cow-cockle, Cowherb
Saponaria vaccaria L.
Infrequent, adventive; railroads, waste places. AB, LK, LR, PC, PO, RO, SU.

CELASTRACEAE (Staff-tree Family)

287 *Celastrus scandens* L.   Bittersweet
Common; fence rows, woodland borders, thickets. ALL COUNTIES.

288 *Euonymus alatus* (Thunb.) Sieb.   Winged Spindle-tree
Rare, adventive; woods, thickets. CU: secondary forest, Hunting Valley, Bissell 1975:109, 9 Jul 1975 (CLM 6210). FA: wooded area, Berne Twp., Goslin s.n., 7 Jun 1971 (BHO 31239). KN: woodlot, Clinton Twp., Pusey 139, 7 Aug 1974 (KE 38113). SU: mesophytic woods, Akron, Andreas 3393, 20 May 1979 (KE 46749).

289 *Euonymus atropurpureus* Jacq.   Wahoo, Burning-bush
Common; dry and mesic woods, wooded floodplains, roadside banks. ALL COUNTIES.

51

290 *Euonymus europaeus* L.
Rare, adventive; wet disturbed areas. LI: fence row, Licking County, Rypma s.n., 31 Oct 1982 (BHO 34110). RO: woods, Bartley s.n., 19 May 1963 (OS 73535).

291 *Euonymus fortunei* (Turcz.) Hand-Maz.   Wintercreeper
Rare, adventive; roadsides, woods. SU: stream terrace, Boston Twp., Cooperrider et al, 12511, 22 May 1980 (KE 42040).

292 *Euonymus obovatus* Nutt.   Running Strawberry-bush
Common; rich mesic and moist woods, thickets. ALL COUNTIES.

CERATOPHYLLACEAE (Hornwort Family)

293 *Ceratophyllum demersum* L.   Coontail
Common; ponds, lakes, quiet streams, rivers. All counties except PC.

294 *Ceratophyllum echinatum* Gray   Prickly Hornwort
Infrequent; quiet streams, ponds. AB, GE, LI, LR, PC, PO, RI, SU.

CHENOPODIACEAE (Goosefoot Family)

295 *Atriplex patula* L. (including *A. patula* var. *hastata* (L.) Gray)
Common; waste places, roadsides. All counties except LI.

296 *Chenopodium album* L.   Lamb's Quarters
Common, naturalized; roadsides, fields, cultivated soils, waste places. All counties except FA.

297 *Chenopodium ambrosioides* L.   Mexican Tea
Common, naturalized; fields, roadsides, waste places. AB, CU, FA, HL, KN, LK, LR, PE, PC, PO, RO, ST, SU, TR, WA.

298 *Chenopodium botrys* L.   Jerusalem-oak
Frequent, adventive; waste places, roadsides. AB, CU, GE, LK, LR, MH, PO, RI, ST, SU, TR, WA.

299 *Chenopodium capitatum* (L.) Ascherson   Strawberry-blite
Once rare, now presumably extirpated from Ohio; open dry soils, rock outcrops. GE: Thompson, Dodge s.n., 20 Sep 1920 (OS 18370). LK: Painesville, Beardslee s.n., (OC 78598). SU: Copley, Dillinga & Selby 1784, 30 Sep 1898 (OS 19369).

300 *Chenopodium glaucum* L.   Oak-leaved Goosefoot
Infrequent, adventive; waste places, cultivated fields, roadsides. AB, CU, LK, MH, PO, ST, SU, TR, WA.

301 *Chenopodium hybridum* L. var. *gigantospermum* (Aellen) Rouleau
Frequent; wet woods, rocky slopes, thickets, roadsides. AS, AB, CL, FA, HL, KN, LK, LI, LR, PO, RO, SU, TR.

302 *Chenopodium leptophyllum* Nutt.   Slender Goosefoot
Rare; waste places, garden, sandy fields. TR: dry hillside, Braceville Twp., Rood & Webb 2511, 21 Aug 1910 (KE 10964).

303 *Chenopodium murale* L.
Infrequent, adventive; cultivated fields, waste places, gardens. AB, CU, LK, PE, PC, PO, RO, SU.

304 *Chenopodium standleyanum* Aellen
Infrequent; dry woods, openings. FA, HL, KN, LK, RO.

305 *Chenopodium urbicum* L.
Rare, adventive; waste places. WA: dooryard, Wooster, Selby 553, 20 Sep 1899 (OS 93987).

306 *Chenopodium vulvaria* L.   Stinking Goosefoot
Rare, adventive; waste places. LR: Columbia Station, Thompson & Fullmer s.n., Summer 1911 (OS 18259).

307 *Corispermum nitidum* Kit.   Slender Bugseed
Rare, adventive; railroads, waste places. RO: railroad, Chillicothe, Bartley s.n., 18 Aug 1968 (OS 104882). WA: railroad, Milton Twp., Cusick 10574, 25 Sep 1969 (OS 102763).

308 *Cycloloma atriplicifolium* (Sprengel) Coulter   Winged Pigweed
Infrequent, naturalized; cultivated fields, sandy waste places, railroads. AB, CU, LK, LR, PO, RO, ST, TR.

309 ***Kochia scoparia** (L.) Schrader   Summer-cypress
Frequent, adventive; waste places, railroads. AB, CU, HL, LK, MH, PC, RO, SU, WA.

310 ***Salicornia europaea** L.   Glasswort
Rare, adventive; saline soils. WA: brine well margin, Milton Twp., Cusick 11518, 9 Oct 1970
(KE 35568).

311 ***Salsola kali** L. var. **tenuifolia** Tausch   Russian Thistle
Common, naturalized; railroads, roadsides, waste places. AB, CU, GE, HL, KN, LI, LK,
LR, MD, PE, PC, PO, RI, RO, SU, WA.

CISTACEAE (Rockrose Family)

312 **Helianthemum bicknellii** Fern.   Frostweed
Rare; sandy soils, steep hillsides. FA: field, Hocking Twp., Bartley s.n., 25 Jul 1950 (OS
43043). WA: Shreve, Duvel s.n., 7 Sep 1899 (OS 16441).

313 **Helianthemum canadense** (L.) Michx.   Canada Frostweed
Rare; sandy ridges, open woods, steep dry hillsides. PO: Brady Lake, Krebs s.n., 1891 (OS
16384). SU: prairie-like area, Akron, Andreas 3659, 7 Jun 1979 (KE 42230); Turkeyfoot
Lake, Hicks & Chapman s.n., 14 Jun 1933 (OS 16384); Gaylord's Grove, Laughlin s.n., 13
Aug 1913 (OS 77627).

314 # **Lechea intermedia** Leggett   Round-fruited Pinweed
Infrequent; dry eroding banks, wooded slopes, woodland borders. AS, LR, PO, RO, ST, SU,
WA.

315 # **Lechea leggettii** Britton & Hollick   Leggett's Pinweed
Infrequent; dry woods, fields, pond margins, woodland borders. MH, PO, RO, ST, SU, TR,
WA.

316 **Lechea minor** L.   Thyme-leaf Pinweed
Rare; dry sandy woods, woodland borders. HL: Holmes County, Selby s.n., 22 Jul 1893 (OS
94581).

317 **Lechea racemulosa** Michx.   Common Pinweed
Common; dry banks, rock outcrops, open woods. AS, AB, CL, CU, FA, HL, KN, LI, MH,
PE, ST, WA.

318 **Lechea tenuifolia** Michx.   Narrow-leaved Pinweed
Rare; dry woods, rocky slopes. AS: Mohican Forest, Bartley s.n,. 6 Oct 1956 (OS 56032).
HL: dry oak slope, Washington Twp., Wilson 2273, 30 Jul 1971 (KE 36887). RO: Liberty
Twp., Bartley s.n., 1932 (OS 16456).

319 **Lechea villosa** Ell.   Hairy Pinweed
Infrequent; dry open woods, eroding slopes, sandy fields. AS, AB, ST, SU, WA.

COMPOSITAE / ASTERACEAE (Sunflower Family)

320 ***Achillea lanulosa** Nutt.   Yarrow
Rare, adventive; sandy open ground. LR: ballast, Jones 67-6-18-547, 18 June 1967 (KE
30561).

321 ***Achillea millefolium** L.   Common Yarrow
Common, naturalized; fields, roadsides, meadows, pastures, prairie remnants, waste places.
ALL COUNTIES.

322 **Ambrosia artemisiifolia** L.   Common Ragweed
Common; roadsides, cultivated fields, waste places. All counties except PC.

323 ***Ambrosia psilostachya** DC.   Western Ragweed, Perennial Ragweed
Rare, adventive; dry fields, railroads, sandy fields. TR: dump, Braceville Twp., Rood 1794,
5 Oct 1941 (KE 11736).

324 **Ambrosia trifida** L.   Great Ragweed
Common; moist roadsides, waste places, cultivated fields. All counties except MO, PC.

325 **Anaphalis margaritacea** (L.) C. B. Clarke   Pearly Everlasting
Infrequent; dry hillsides, gravelly and sandy soils. AB, CU, FA, GE, LK, LI, LR, PO, TR.

326 **Antennaria neglecta** Greene   Pussytoes
Common; wooded slopes, roadside banks, woodland borders. AB, CL, CU, FA, GE, LI, LK,
LR, MH, MD, MO, PE, PC, PO, RI, ST, SU, WA.

53

327 *Antennaria neodioica* Greene   *(sensu lato)*
Common; roadsides, fields, pastures, woodland borders. AS, AB, CU, FA, GE, KN, LK, LI, LR, MO, PE, PC, RI, ST, SU, TR.

328 *Antennaria parlinii* Fern.   *(sensu lato)*
Common; roadsides, woodland borders, pastures. ALL COUNTIES.

329 *Antennaria solitaria* Rydb.
Rare; dry fields, rock exposures, roadsides. RO: Springfield Twp., Bartley & Pontius s.n., 23 Mar 1930 (OS 39341).

330 **Anthemis arvensis* L.   Corn Chamomile
Infrequent, naturalized; cultivated fields, roadsides. AS, CU, LK, LI, LR, PE, PC, PO, ST, SU.

331 **Anthemis cotula* L.   Mayweed, Stinking Chamomile
Common, naturalized; roadsides, railroads, fields, waste places. ALL COUNTIES.

332 **Anthemis nobilis* L.   Garden Chamomile
Rare, adventive; old gardens, abandoned homesites. LR: Lorain County, H.L. Jones 1898 (OC 76225).

333 **Anthemis tinctoria* L.   Yellow Chamomile
Rare, adventive; fields, roadsides, waste places. FA: Fairfield County, Bloomfield s.n., Sep 1937 (OS 48157). ST: open clearing, Canton Twp., Amann 503, 30 Jun 1960 (KE 2295). WA: Station farm, Wooster, Duvel s.n., 12 Oct 1898 (OS 90073).

334 **Arctium lappa* L.   Great Burdock
Frequent, naturalized; roadsides, moist woods. AB, CL, CU, HL, KN, LK, LR, MH, MD, PO, RI, SU, TR, WA.

335 *Arctium minus* Schkuhr   Common Burdock
Common; weedy fields, floodplains, waste places, roadsides, abandoned homesites, woodland blow-downs. ALL COUNTIES.

336 **Artemisia absinthium* L.   Wormwood
Rare, adventive; roadsides, pastures. LR: Elyria, no collector or date (OC 76326).

337 **Artemisia annua* L.   Sweet Wormwood
Rare, naturalized; waste places, roadsides, old gardens. PC: roadside, Walnut Twp., Weishaupt s.n., 29 Sep 1956 (OS 57221). RO: cliffs, Paint Twp., Roberts & Youger 3233, 23 Sep 1972 (OS 108215); Clarksburg, Dobbins s.n., 18 Jul 1937 (OS 49420).

338 **Artemisia biennis* Willd.
Rare, adventive; roadsides, waste places. AB: roadsides, Conneaut, Cusick 17639, 13 Sep 1977 (KE 45555); Ashtabula, South s.n., no date (OS 49797).

339 **Artemisia ludoviciana* Nutt.   Western Mugwort
Infrequent, adventive; railroads, waste places, roadsides. AB, CL, HL, LK, MH, MD, PE.

340 **Artemisia pontica* L.   Roman Wormwood
Rare, adventive; roadsides, waste places. AB: Rock Creek, Hicks s.n., 10 Jun 1931 (OS 49811). PO: Portage County, Krebs s.n., 1898 (OS 49805).

341 **Artemisia vulgaris* L.   Common Wormwood, Mugwort
Infrequent, naturalized; sandy soils, railroads, roadsides, waste places. AB, CL, CU, HL, LK, PO, RO, SU, WA.

342 # *Aster acuminatus* Michx.   Mountain Aster
Rare; dry and moist woods. AB: East Monroe Twp., Hicks s.n., 4 Sep 1931 (OS 40609). LK: rock outcrops, Concord Twp., Bradt 1938, 23 Aug 1985 (CLM 22915 and KE 49312).

343 *Aster azureus* Lindley
Infrequent; dry woods, clearings, thickets. AS, AB, LR, PE, PO, RO, WA.

344 **Aster brachyactis* Blake
Infrequent, naturalized; saline soil along roadsides, waste places. AS, LK, MD, MO, RI, WA.

345 *Aster cordifolius* L.   Heart-leaved Aster
Common; open woods, thickets, woodland clearings, shaded roadsides. ALL COUNTIES.

346 *Aster divaricatus* L.
Common; rich woods, wooded terraces, moist wooded slopes. All counties except CU, MO, PE.

347 *Aster drummondii* Lindley
Rare; dry fields, open woods. FA: Berne Twp., Goslin s.n., 24 Sep 1944 (OS 22243); Kettle Hills, Goslin s.n., 4 Oct 1936 (OS 98232). PO: oak barrens, Nelson Twp., Rood 2653, 26 Sep 1952 (KE 11668). WA: woods near Dalton, Duvel 1158, 4 Oct 1899 (OS 90118); roadside near Overton, Selby & Duvel 1156, 15 Sep 1899 (OS s.n.).

348 *Aster dumosus* L.   Bushy Aster
Rare; thickets, grassy fields, prairie remnants. LK: Painesville, Werner s.n., 5 Oct 1888 (OC 81390).

349 *Aster ericoides* L.
Common; weedy thickets, railroads, roadsides. AS, CU, FA, GE, HL, KN, LK, LR, MH, MD, PE, PO, RI, RO, ST, SU, TR.

350 *Aster infirmus* Michx.   Weak Aster
Rare; dry wooded slopes. PO: Pippin Lake, Franklin Twp., Webb 542, 6 Sep 1909 (KE 11679); Brady Lake, Krebs s.n., 1898 (OS 40873). SU: dry wooded slope, Boston Twp., Andreas 4828, 16 Sep 1979 (KE 42241); slope, Boston Twp., Herrick s.n., 29 Jul 1959 (OS 63481).

351 # *Aster junciformis* Rydb.   Swamp Aster
Frequent; sphagnous meadows, sedge meadows, fens, marshes. AB, CL, FA, GE, HL, KN, LI, PO, RO, ST, WA.

352 *Aster laevis* L.   Smooth Aster
Common; dry woods and slopes, eroding slopes, prairie remnants, sandy ridges. AS, AB, CL, CU, FA, GE, HL, KN, LK, LR, MH, PE, PO, RI, SU, WA.

353 *Aster lateriflorus* (L.) Britton
Common; roadsides, fields, clearings, thickets. ALL COUNTIES.

354 *Aster linariifolius* L.   Stiff-leaved Aster
Rare; dry open soils, rocky slopes. FA: Madison Twp., Rea s.n., 20 Sep 1942 (OS 40881). RO: Colerain Twp., Bartley & Pontius s.n., 7 Aug 1932 (OS 40876).

355 *Aster lowrieanus* Porter
Infrequent; open woods, clearings, thickets. KN, LK, LR, PE, WA.

356 *Aster macrophyllus* L.
Common; rich woods, shaded roadsides. All counties except LI, MO, RO.

357 *Aster novae-angliae* L.   New England Aster
Common; roadsides, fields, meadows, waste places. ALL COUNTIES.

358a *Aster patens* Aiton var. *patens*
Rare; dry woods. PO: oak woods, Franklin Twp., Cusick 11528, 17 Nov 1970 (KE 27095).

358b *Aster patens* Aiton var. *phlogifolius* (Muhl.) Nees
Frequent; dry woods, clearings, fields. AS, HL, KN, LK, LI, PE, PC, PO, RI, RO, WA.

359 *Aster paternus* Cronq.
Seriocarpus asteroides (L.) BSP.
Frequent; dry woods, woodland thickets. AS, CL, CU, FA, HL, KN, PE, PO, RO, ST, SU, TR, WA.

360a *Aster pilosus* Willd. var. *pilosus*
Common; roadsides, dry slopes, fields, railroads, thickets. ALL COUNTIES.

360b *Aster pilosus* Willd. var. *demotus* Blake
Rare; pastures, roadsides. AB: pasture, Speer 386, 5 Sep 1961 (OS 80986). SU: roadside, Twinsburg, Herrick s.n., 21 Sep 1957 (OS 640207).

360c *Aster pilosus* Willd. var. *pringlei* (Gray) Blake
Rare; fields, roadsides. GE: Thompson Ledges, Hopkins s.n., 19 Aug 1911 (OS 42550). LK: fields and woods, Painesville Twp., Tandy 886, 30 Sep 1972 (KE 40274).

361 *Aster praealtus* Poiret
Frequent; bogs, prairie remnants, roadsides. AB, CL, FA, LK, LI, LR, MD, PO, RO, TR, WA.

362 *Aster prenanthoides* Muhl.
Common; roadsides, moist woods, damp thickets, marsh margins. ALL COUNTIES.

363 *Aster puniceus* L.   Fen Aster
Common; fens, sedge meadows, marshes, wet thickets, meadows. All counties except PC.

364 *Aster sagittifolius* Wedemeyer
Common; dry open woods, roadsides, thickets. All counties except MD, MO, PE.

365 *Aster schreberi* Nees
Common; dry woods, thickets. ALL COUNTIES.

366 *Aster shortii* Lindley
Frequent; dry or moist woods, wooded slopes, thickets. CU, KN, LK, LI, LR, MD, PC, PO, RI, RO, ST, SU, WA.

367 *Aster simplex* Willd.
Common; wet fields, roadsides, waste places. All counties except LI, MO.

368 **Aster subulatus* Michx.
Rare, naturalized; saline soils. MD: ditch, Westfield Twp., Cusick 24021, 22 Sep 1984 (KE 47739).

369 *Aster umbellatus* Miller
Common; fields, thickets, railroads, open woods. All counties except LI, MO, RI.

370 *Aster undulatus* L.
Common; roadsides, dry woods, open slopes. AS, CL, CU, FA, GE, HL, KN, LK, LI, LR, PE, PO, RO, ST, SU, WA.

371 *Aster vimineus* Lam.
Frequent; moist fields, clearings, roadsides. AB, CL, CU, HL, LK, LR, MH, MD, PE, RO, TR, WA.

372 **Bellis perennis* L.   English Daisy
Infrequent, naturalized; lawns, roadsides, abandoned homesites. AB, CU, LK, LR, PO, RO, TR, WA.

373 *Bidens aristosa* (Michx.) Britton   Tickseed Sunflower
Infrequent; roadside ditches, swamps, wet meadows. CL, LI, LR, MH, MD, RI, ST, SU, TR.

374 *Bidens bipinnata* L.   Spanish Needles
Frequent; rocky woods, roadsides, waste places. CL, CU, FA, HL, KN, LI, LI, LR, PE, PC, RO, ST, WA.

375 *Bidens cernua* L.   Stick-tight
Common; marshes, stream and pond margins, roadsides, wet meadows, pastures. ALL COUNTIES.

376 *Bidens coronata* (L.) Britton   Tickseed Sunflower
Common; roadside ditches, stream and pond margins, fields. All counties except MD, MO.

377 # *Bidens discoidea* (T. & G.) Britton
Infrequent; pond margins, marshes, sphagnous thickets, bog margins. FA, GE, KN, LR, PC, PO, ST, SU, TR.

378 *Bidens frondosa* L.   Beggar Ticks
Common; wet fields, roadsides, waste places. All counties except MO.

379 *Bidens polylepis* Blake
Infrequent; wet sandy fields, roadsides, prairie remnants. CU, LK, LR, MD, PO, RI, RO, SU, TR.

380 *Bidens tripartita* L. (including *B. connata* L. and *B. comosa* (Gray) Weig.)
Common; marshes, swamps, wet ditches, thickets, stream and pond margins. All counties except AB, PE, RI.

381 *Bidens vulgata* Greene   Beggar Ticks
Frequent; ditches, roadsides, waste places. AB, CL, CU, GE, HL, KN, LR, MH, PC, PO, RI, ST, TR.

382 *Cacalia atriplicifolia* L.   Pale Indian-plantain
Common; dry wooded slopes, roadside banks, prairie remnants. AB, CL, CU, FA, GE, HL, LK, LI, LR, MH, PE, PC, PO, RI, RO, ST, SU, TR, WA.

383 *Cacalia muhlenbergii* (Schultz-Bip.) Fern.   Great Indian-plantain
Rare; rich woods, wooded slopes. RO: Tar Hollow, Colerain Twp., Goslin s.n., 30 Jul 1967 (BHO 29479).

384 *Cacalia suaveolens* L.
Infrequent; wet meadows, thickets, fens. AB, CU, FA, LK, LR, PC, RO, ST.

385 *Cacalia tuberosa* Nutt.   Tuberous Indian-plantain
Rare; fens. FA: Pleasant Run Swamp, Berne Twp., Goslin s.n., 7 Jul 1935 (OS 50005). RO:
Spring Bank Bog, Union Twp., Bartley s.n., 3 Oct 1961 (OS 66763).

386 **Carduus acanthoides* L.   Plumeless Thistle
Rare, naturalized; waste places, roadsides. LK: farm, Mentor, Werner s.n., Sep 1925 (OS
46160). PC: Pickaway Twp., Bartley & Pontius s.n., 17 Jul 1931 (OS 50056).

387 **Carduus nutans* L.   Musk Thistle
Rare, naturalized; waste places, roadsides. LI: roadside, Forsyth s.n., 13 Jun 1972 (OS
111293). LR: pasture, Jones & Long 79-7-11-291, 11 Jul 1979 (OC 221597). PC: roadbank,
Pickaway Twp., Burns 3876, 25 Jun 1985 (KE 47538).

388 **Centaurea cyanus* L.   Bachelor's Buttons
Rare, adventive; roadsides, abandoned gardens, waste places. AB: Andover, Hicks s.n., 17
Jun 1932 (OS 50337).

389 **Centaurea dubia* Suter.   Short-fringed Knapweed
Infrequent, adventive; roadsides, fields. AB, CU, LR, PC, TR.

390 **Centaurea jacea* L.   Brown Knapweed
Infrequent, naturalized; roadsides, fields, waste places. AB, CU, GE, LK, PC, RI, SU, TR.

391 **Centaurea maculosa* Lam.   Spotted Knapweed
Frequent, naturalized; roadsides, fields, waste places. AS, CU, HL, KN, LK, LR, PE, PC,
PO, RI, ST, SU.

392 **Centaurea nigra* L.   Black Knapweed, Spanish Buttons
Infrequent, naturalized; dry fields, roadsides, waste places. AB, CU, GE, LK, PC, RI, ST,
WA.

393 **Centaurea solstitialis* L.   Yellow Star-thistle
Rare, adventive; fields, waste places. RO: alfalfa fields, Frankfort, Roseboom s.n., 20 Sep
1904 (OS 95832). TR: field, Braceville Twp., Rood 975, 19 Jul 19008 (KE 11852).

394 **Chrysanthemum balsamita* L.   Costmary
Infrequent, adventive; waste places, roadsides. CU, PC, PO, TR, WA.

395 **Chrysanthemum leucanthemum* L.   Ox-eye Daisy, White Daisy
Common, naturalized; pastures, fields, roadsides, waste places. ALL COUNTIES.

396 **Chrysanthemum maximum* L.   Shasta Daisy
Rare, adventive; roadsides. AB: roadside, Windsor Twp., Tandy 1648, 4 Aug 1973 (KE
40582).

397 **Chrysanthemum parthenium* (L.) Bernh.   Feverfew
Infrequent, adventive; abandoned homesites, roadsides. AB, CU, FA, KN, LK, MD, ST.

398 *Chrysogonum virginianum* L.   Golden-knees
Rare; open woods, fields. FA: grassy open, Hocking Twp., Cusick, Burns, & Howes 22457,
23 May 1983 (OS 164749). SU: Peninsula, Conger s.n., 17 May 1921 (OS 95823) [This
specimen is probably an escape from cultivation].

399 **Chrysopsis camporum* Greene   Prairie Golden Aster
Heterotheca camporum (Greene) Shinners
Rare, adventive; dry roadsides, sandy soils. WA: field along railroad, Milton Twp., Andreas
1622, 19 Sep 1977 (KE 42418).

400 **Cichorium intybus* L.   Chicory
Common, naturalized; roadsides, fields, waste places. All counties except RO.

401 *Cirsium altissimum* (L.) Sprengel
Common; thickets, weedy floodplains, moist ditches. CL, CU, FA, GE, HL, LK, LI, LR,
MD, PE, PO, RO, ST, SU, WA.

402 **Cirsium arvense* (L.) Scop.   Canada Thistle
Common, naturalized; roadsides, pastures, cultivated fields. All counties except CL, PC.

403 *Cirsium discolor* (Muhl.) Sprengel   Prairie Thistle
Frequent; floodplains, riverbanks, prairie remnants. AS, FA, GE, HL, KN, LI, MH, MO,
PE, PC, PO, WA.

404 **Cirsium muticum** Michx. Swamp Thistle
Common; wet ditches, swamps, calcareous fens, wet thickets. AS, AB, CL, CU, FA, GE, HL, LN, LK, LI, LR, MD, PO, RI, RO, ST, SU, TR, WA.

405 ***Cirsium plattense** (Rydb.) Cockerell
Rare, adventive; waste places. TR: railroad ballast, Braceville Twp., Rood 1764, 1 Aug 1926 (KE 11858).

406 **Cirsium pumilum** (Nutt.) Sprengel Pasture Thistle
Infrequent; dry fields, pastures. AS, AB, CL, HL, KN, PO, ST, TR.

407 ***Cirsium vulgare** (Savi) Tenore Common Thistle
Common, naturalized; roadsides, pastures, fields. AS, AB, CL, CU, GE, HL, KN, LK, LI, LR, MD, PE, PO, RI, RO, ST, SU, TR, WA.

408 **Conyza canadensis** (L.) Cronq. Horseweed
Erigeron canadensis L.
Common; roadsides, fields, waste places. All counties except MO.

409 **Conyza ramosissima** Cronq.
Erigeron divaricatus Michx.
Rare; waste places, fields. RO: Hopetown, Bartley s.n., 24 Aug 1958 (OS 67941); pasture, Hopetown, Bartley s.n., 18 Aug 1949 (OS 44087).

410 ***Coreopsis grandiflora** Hogg Large-flowered Tickseed
Infrequent, adventive; rocky fields, roadsides. CU, GE, LK, PO, ST, SU.

411 ***Coreopsis lanceolata** L.
Infrequent, adventive; roadsides, waste places. HL, KN, LK, PO, ST.

412 ***Coreopsis tinctoria** Nutt.
Infrequent, adventive; roadsides, fields. AB, CU, FA, LK, PC.

413 **Coreopsis tripteris** L. Tall Coreopsis
Common; dry roadside banks, open dry woods, fields, prairie remnants. AS, CL, CU, FA, GE, HL, LK, LI, LR, MH, MD, PC, PO, RO, ST, SU, TR.

414 ***Cosmos bipinnatus** Cav. Cosmos
Rare, adventive; abandoned homesites, roadsides. AB: Wayne Twp., Hicks s.n., 14 Sep 1928 (OS 38616). FA: railroad, Berne Twp., Goslin s.n., 6 Sep 1948 (BHO 22285). RO: Stoney Creek, Crowl s.n., 30 Aug 1937 (OS 53861).

415 ***Crepis capillaris** (L.) Wallr. Hawk's-beard
Infrequent, adventive; fields, lawns, waste places. AB, CU, KN, LK, LR, RI, SU.

416 ***Crepis pulchra** L.
Rare, adventive; roadsides, waste places. KN: weedy cemetery, College Twp., Pusey 687, 23 Jul 1975 (KE 38311). RO: Mt. Logan, Bartley s.n., 3 Jul 1973 (OS 115565).

417 ***Crepis tectorum** L.
Rare, adventive; roadsides, cultivated fields. LR: lawn, Oberlin, Grover s.n., 18 Aug 1912 (CLM 12074).

418 ***Dyssodia papposa** (Vent.) Hitchc. Fetid Marigold
Infrequent, adventive; roadsides, waste places. AS, LK, MD, MO, RI, WA.

419 **Echinacea papposa** (Vent.) Moench Purple Coneflower
Rare; dry soils, prairie remnants, HL: moist soil, Plimpton, Stair s.n., 1 Aug 1898 (CLM s.n.); swampy ground, Lakeville, Wilkinson s.n., 20 Aug 1896 (OC 81517); Holmes County, Lageman s.n., 1899 (OS 38248). PC: dry bluff, Crownover Bridge, Bartley s.n., 3 Jul 1959 (OS 65813). RO: Union Twp., Bartley & Pontius s.n., 14 Sep 1932 (OS 38243).

420 ***Eclipta alba** (L.) Hassk. Yerba-de-Tago
Infrequent, adventive; moist soils, ditches, waste places. CU, FA, HL, LI, PC, RO, WA.

421 **Elephantopus carolinianus** Raeuschel Elephant's-foot
Rare; open woods, thickets. RO: Mt. Logan, Hall 842, 5 Sep 1954 (BHO 15634); Paxton, Bartley & Pontius s.n., 31 Aug 1930 (OS 41324).

422 **Erechtites hieracifolia** (L.) Raf. Pilewort, Fireweed
Common; railroads, damp roadsides, open woodland borders, fields, waste places. ALL COUNTIES.

423 **Erigeron annuus** (L.) Persoon Daisy Fleabane, White-top
Common; fields, waste places, roadsides. ALL COUNTIES.

424 *Erigeron philadelphicus* L.   Philadelphia Fleabane
Common; fields, roadsides, woodland margins, thickets. ALL COUNTIES.

425 *Erigeron pulchellus* Michx.   Robin's Plantain
Common; dry wooded slopes, open woods, moist fields. All counties except MO, PC, RO.

426 *Erigeron strigosus* Muhl. ex Willd.   Daisy Fleabane
Common; fields, meadows, roadsides, railroads. All counties except PC, RO.

427 *Eupatorium altissimum* L.
Infrequent; railroads, roadsides, clearings. AB, CU, GE, KN, PC, RO, SU.

428 *Eupatorium coelestinum* L.   Blue Boneset, Ageratum
Infrequent; damp thickets, clearings, woodland terraces. AB, FA, LI, LR, PC, RO, SU, WA.

429 *Eupatorium fistulosum* Barratt   Joe-Pye Weed
Common; moist ditches, woodland margins, marshes, thickets, damp meadows. AS, AB, CL, CU, GE, HL, KN, LK, LR, MH, MD, PE, PO, RO, ST, SU, TR, WA.

430 *Eupatorium maculatum* L.   Mottled Joe-Pye Weed
Common; boggy thickets, marshes, roadside ditches, fens. AS, AB, CL, CU, FA, GE, HL, KN, LK, LR, MH, MD, PO, RI, RO, ST, SU, WA.

431 *Eupatorium perfoliatum* L.   Thoroughwort, Boneset
Common; moist woods, stream and pond margins, thickets, wet meadows. All counties except MO.

432 *Eupatorium purpureum* L.   Green-stemmed Joe-Pye Weed
Common; moist woods, shaded clearings. All counties except MH, PC.

433 *Eupatorium rugosum* Houtt.   White Snakeroot
Common; rich woods, wooded slopes, shaded roadside banks. All counties except MO.

434 *Eupatorium serotinum* Michx.
Rare; dry thickets, woodland borders, clearings, fields. MH: Canfield, Beardslee s.n., 2 Oct 1925 (OS 95888).

435 *Eupatorium sessilifolium* L.   Upland Boneset
Frequent; dry woods, wooded slopes, thickets. CL, CU, FA, HL, KN, LK, LI, PC, PO, RO, ST, SU, WA.

436 *Euthamia graminifolia* (L.) Nutt.   Flat-topped Goldenrod
Solidago graminifolia (L.) Salisb.
Common; moist fields, thickets, roadside banks. ALL COUNTIES.

437 **Gaillardia pulchella* Foug.   Gaillardia, Blanket-flower
Rare, adventive; waste places, dry roadsides. AB: Jefferson, Hicks s.n., 22 Jun 1932 (OS 39107). FA: Bloom Twp., Goslin 44-128-2, 11 Sep 1971 (BHO 31163).

438 **Galinsoga parviflora* Cav.
Frequent, adventive; waste places, cultivated fields. AB, CL, CU, LK, LI, LR, MO, SU, TR.

439 **Galinsoga quadriradiata* Ruiz & Pavon
Galinsoga ciliata (Raf.) Blake
Common, adventive; roadsides, fields, waste places. All counties except LI.

440 *Gnaphalium obtusifolium* L.   Catfoot
Common; dry fields, roadside banks, woodland broders. ALL COUNTIES.

441 *Gnaphalium purpureum* L.   Purple Cudweed
Infrequent; dry fields, woodland borders. AB, CU, FA, LK, LI, LR, ST, SU, TR, WA.

442 *Gnaphalium uliginosum* L.   Low Cudweed
Frequent; sandy soils, clearings, waste places. AB, CL, CU, GE, HL, LK, LR, PC, PO, ST, SU, TR, WA.

443 # *Gnaphalium viscosum* HBK   Macoun's Everlasting
Gnaphalium macounii Greene
Once rare, now presumably extirpated from Ohio; clearings, open woods, woodland borders, meadows. LR: Oberlin, Steele s.n., 1881 (OC 76661).

444 **Grindelia squarrosa* (Pursh) Dunal   Curlycup Gumweed
Rare, adventive; dry fields, roadsides, disturbed places. MH: ballast, Berlin Twp., Vickers 1423, 11 Jul 1915 (KE 16157). PC: Pickaway Twp., Bartley & Pontius s.n., 27 Jul 1933 (OS

39373). RO: roadside, Delano Station, Bartley s.n., 16 Sep 1950 (OS 43038). TR: dooryard, Southington Twp., Rood 1423, 27 Jul 1925 (KE 16164).

445 *Helenium autumnale* L. Sneezeweed
Common; wet roadside ditches, meadows, fens, thickets. All counties except PE, PC.

446 *Helenium flexuosum* Raf. Brown Sneezeweed
Helenium nudiflorum Nutt.
Frequent; roadsides, meadows, fields. AB, CU, FA, GE, HL, LK, PC, PO, RO, ST, SU, TR.

447 **Helianthus annuus* L. Common Sunflower
Frequent, adventive; roadsides, waste places. CL, CU, FA, GE, HL, KN, LR, PE, PO, RO, ST, SU.

448 *Helianthus decapetalus* L.
Common; dry fields, thickets, woodland borders, roadsides. AS, AB, CL, CU, GE, HL, KN, LK, LI, LR, MH, MD, PE, PC, PO, RI, ST, SU, TR, WA.

449 *Helianthus divaricatus* L.
Common; dry roadside banks, dry woods. AS, CL, CU, FA, GE, HL, KN, LK, LI, LR, MH, PE, PO, RI, RO, ST, SU, TR, WA.

450 *Helianthus giganteus* L.
Common; wet meadows, sedge meadows, fens, thickets, wet prairies. AS, CU, FA, GE, HL, KN, LR, MH, PE, PC, PO, RI, RO, ST, SU, TR, WA.

451 *Helianthus grosseserratus* Martens
Frequent; roadsides, thickets, prairie remnants. AS, AB, CL, CU, FA, GE, HL, KN, LK, PC, PO, RO, ST, SU, WA.

452 *Helianthus hirsutus* Raf.
Frequent; dry woods, woodland borders, thickets. AS, CL, FA, HL, LR, KN, MD, PE, PC, RO, ST, SU, WA.

453 *Helianthus x laetiflorus* Persoon (*H. rigidus* x *H. tuberosus*)
Rare; dry roadside banks, thickets, open woods. RO: Ross County, Bartley & Pontius s.n. 6 Sep 1932 (OS 38519). TR: railroad embankment, Braceville Twp., Rood s.n., 14 Sep 1944 (KE 11766). WA: dry field, Chippewa Twp., Andreas 1611, 18 Sep 1977 (KE 42614).

454 **Helianthus maximilianii* Schrader Maximilian's Sunflower
Rare, adventive; roadsides, waste places. AB: Grand River, Harpersfield, Hicks s.n., 4 Sep 1931 (OS 38357); Dorset, Hicks s.n., 6 Sep 1932 (OS 85666). HL: dry ground, Plimpton, Stair s.n., 1 Aug 1898 (OS 38342).

455 *Helianthus mollis* Lam. Ashy Sunflower
Rare; dry woods, sandy fields, railroads. RO: Liberty Twp., Bartley & Pontius s.n., Nov 1931 (OS 38397).

456 *Helianthus occidentalis* Riddell Western Sunflower
Rare; dry woods, rocky banks, wooded slopes. HL: dry roadbank, Washington Twp., Wilson 2489, 23 Aug 1971 (KE 29393).

457 **Helianthus petiolaris* Nutt.
Infrequent, adventive; dry roadside banks, fields, railroads. CU, FA, LK, RO, TR.

458 *Helianthus strumosus* L.
Common; dry woods, roadsides, clearings, fields. AS, AB, CL, CU, FA, GE, HL, KN, LK, LI, LR, PC, PO, RI, RO, ST, SU, TR, WA.

459 *Helianthus tuberosus* L. Jerusalem Artichoke
Common; roadside ditches, woodland borders, thickets, wooded floodplains, waste places. ALL COUNTIES.

460 *Heliopsis helianthoides* (L.) Sweet Ox-eye
Common; roadside ditches, shaded floodplains, woodland borders. All counties except MO.

461 **Hieracium aurantiacum* L. Devil's Paint-brush
Common, naturalized; fields, roadsides, waste places. AS, AB, CL, CU, GE, HL, KN, LK, LI, LR, MH, MD, MO, PO, RI, ST, SU, TR, WA.

462 **Hieracium caespitosum* Dumort. Meadow King Devil
Hieracium pratense Tausch
Common, naturalized; fields, roadsides, waste places. All counties except CL, MO, RO.

463 *Hieracium florentinum* All. King Devil
Frequent, naturalized; roadsides, fields, pastures, waste places, lawns. AS, CL, FA, HL, KN, LK, LI, MD, PE, PO, ST, SU, WA.

464 *Hieracium floribundum* Wimm. & Grab. King Devil
Frequent, naturalized; fields, roadsides, waste places. CU, GE, HL, KN, LK, MD, PE, PO, SU, TR, WA.

465 *Hieracium gronovii* L. Hawkweed
Frequent; dry banks, open woods, wooded slopes. AS, AB, CL, CU, FA, HL, KN, LK, LI, LR, PE, PC, PO, RO, ST, SU, WA.

466 *Hieracium paniculatum* L.
Common; wooded slopes, dry woods, hemlock slopes. All counties except FA, MO, PC.

467 *Hieracium pilosella* L. Mouse-ear Hawkweed
Frequent, naturalized; dry fields, roadsides, waste places. AB, CU, GE, HL, LK, LR, MD, PO, ST, SU, TR, WA.

468 *Hieracium scabrum* Michx.
Common; dry fields, wooded slopes, railroads, open woods. All counties except MD, MO.

469 *Hieracium trailii* Greene
Rare; dry slopes, oak woods. FA: Berne Twp., Goslin s.n, 24 Jul 1934 (OS 50557). WA: Peewee Hollow, Selby 2536, 15 Sep 1900 (OS 89903).

470 *Hieracium venosum* L. Rattlesnake Weed
Common; dry woods, hogback ridges, wooded slopes, bluffs. All counties except MO, PC.

471 *Hieracium venosum* L. x *H. gronovii* L.
Rare; woods. SU: Stow Corners, Bartley & Pontius s.n., 18 Jun 1933 (OS 50526).

472 *Hypochoeris radicata* L. Cat's Ear
Hypochaeris radicata L.
Frequent, naturalized; cultivated fields, waste places, dry thickets. AB, CU, GE, LK, LR, MD, PE, PO, RO, ST, SU, TR, WA.

473 *Inula helenium* L. Elecampane
Common, naturalized; fields, pastures, woodland borders. All counties except CL, PC, RO.

474 *Iva xanthifolia* Nutt. Marsh-elder
Rare, adventive; waste places. GE: farm, Parkman Twp., Webb 1646, 30 Aug 1924 (KE 11781). PC: stockyard, Circleville Twp., Bartley s.n., 10 Jul 1949 (OS 68376).

475 *Krigia biflora* (Walter) Blake Dwarf Dandelion
Common; dry woods, woodland borders, dry roadsides, fens. All counties except CL, MO.

476 *Krigia virginica* (L.) Willd.
Rare; dry soil. LI: railyard, Carr 1922, 9 Aug 1979 (OS 139029).

477 *Kuhnia eupatorioides* L. False Boneset
Infrequent; roadside banks, open woods, thickets, rocky slopes. CU, HL, PC, RO, ST.

478 *Lactuca biennis* (Moench) Fern. Tall Blue Lettuce
Common; damp woods, fields, waste places. AB, CL, CU, FA, GE, HL, KN, LK, LI, LR, MH, MD, PE, PO, RO, SU, TR, WA.

479 *Lactuca canadensis* L. Wild Lettuce
Common; fields, woods, wet thickets, roadsides. All counties except CL, RO, TR.

480a *Lactuca floridana* (L.) Gaertner var. *floridana*
Frequent; wet thickets, roadsides, woodland borders, shaded floodplains. AB, CU, HL, KN, LK, LI, LR, MO, PE, PO, RO, SU, WA.

480b *Lactuca floridana* (L.) Gaertner var. *villosa* (Jacq.) Cronq.
Rare; wet thickets, floodplains. AB: Phelps Creek, Hicks s.n., 2 Sep 1931 (OS 50818). WA: Little Killbuck, Selby & Duvel 1304, 31 Jul 1899 (OS 89968).

481 *Lactuca pulchella* (Pursh) DC. Blue Lettuce
Rare, adventive; waste places. TR: dump, Braceville Twp., Rood 1830, 27 Aug 1939 (KE 11890).

482 *Lactuca saligna* L. Willow-leaf Lettuce
Infrequent, adventive; roadsides, waste places. AS, CU, FA, MD, MO, PC.

483 *Lactuca serriola* L.   Prickly Lettuce
Lactuca scariola L.
Common, naturalized; roadsides, railroads, waste places. AS, CL, CU, FA, GE, HL, KN, LK, LI, LR, MH, MD, MO, PO, RI, RO, ST, TR, WA.

484 *Lapsana communis* L.   Common Nipplewort
Rare, naturalized; waste places, roadsides. LR: fill, Oberlin, Jones 67-6-10-734, 10 Jul 1960 (KE 30238); yard, Oberlin, Grover s.n., 32 Jul 1920 (CLM 11823). ST: woodland edge, Canton Twp., Amann 487, 29 Jun 1960 (KE 2504). SU: shaded hedge row, Hudson Twp., Cusick 14334, 29 Jun 1975 (KE 37385).

485 *Leontodon hastilis* L.
Leontodon hispidus L.
Rare, adventive; waste places. AB: lawn weed, Rock Creek, Herrick s.n., 29 Jul 1960 (KE 7679). CU: Orange village, Pinkava 457, 11 Sep 1959 (OS 69651). GE: lawn weed, Chardon Twp., Herrick s.n., 29 Jul 1960 (OS 79674). LK: Painesville, Hacker s.n., 9 Jul 1900 (OS 50346).

486 *Leontodon taraxacoides* (Vill.) Merat
Leontodon leysseri (Wallr.) G. Beck
Rare, adventive; fields, gardens. PO: door yard, Windham Twp., Rood 2530, 23 Aug 1951 (KE 11892). ST: field, Plain Twp., Amann 1361, 19 Aug 1960 (KE 2496).

487 *Liatris aspera* Michx.   Rough Blazing Star
Rare; dry wooded slopes, sandy banks. RO: fire tower, Crowl s.n., 31 Aug 1937 (OS s.n.); Jefferson Twp., Bartley & Pontius s.n., 6 Sep 1931 (OS s.n.). SU: slump, Boston Twp., Andreas 4825, 16 Sep 1979 (KE 42613).

488 *Liatris pycnostachya* Michx.
Infrequent, naturalized; wet fields, roadside ditches. CU, GE, LK, MD, RI, SU.

489 *Liatris scariosa* (L.) Willd.
Rare; wet meadows, dry woodland borders, meadows. ST: damp meadows, Canton, Case s.n., 20 Aug 1900 (OS 1814).

490 *Liatris spicata* (L.) Willd.   Spicate Blazing Star
Frequent; fens, damp meadows, open seeps. CU, FA, LK, LR, MH, PE, PC, RO, SU.

491 *Liatris squarrosa* (L.) Michx.   Scaly Blazing Star
Rare; dry open woods, prairie remnants. SU: prairie-like area, Akron, Andreas 4414, 2 Aug 1979 (KE 42240).

492 *Matricaria maritima* L. var. *agrestris* (Knaf.) Wilmott
Rare, adventive; waste places. PO: railroad ballast, Windham Twp., Rood 2793, 28 Jun 1954 (KE 11826).

493 *Matricaria matricarioides* (Less.) Porter   Pineapple-weed
Common, naturalized; roadsides, cultivated fields, waste places. All counties except FA.

494 # *Megalodonta beckii* (Torrey) Greene   Water-marigold
Once infrequent, now presumably extirpated from Ohio; ponds, lakes, slow-moving rivers. AB, LK, MD, PO, ST.

495 *Petasites hybridus* (L.) Gaertner, Meyer & Scherber   Sweet Coltsfoot
Rare, adventive; floodplains, waste places. CU: floodplain, Gates Mills, Feher s.n., 21 Apr 1978 (CLM 16493); willow bar, Hunting Valley, Bissell 1975:139, 23 Jul 1975 (CLM 6407).

496 *Picris echioides* L.   Ox-tongue
Rare, adventive; waste places, fields, roadsides. TR: field, Braceville Twp., Rood 976, 19 Jul 1908 (KE 11895).

497 *Picris hieracioides* L.   Bitterweed
Rare, adventive; railroads, waste places. HL: railroad tracks, Washington Twp., Wilson 2394, 9 Aug 1971 (KE 29457).

498 *Polymnia canadensis* L.   Leafcup
Frequent; boggy thickets, moist woods. AS, CU, FA, GE, KN, LI, LR, PO, RO, ST, SU, WA.

499 *Polymnia uvedalia* (L.) L.   Uvedale's Leafcup
Infrequent; moist woods, thickets. CU, FA, LK, RO, TR.

500 **Prenanthes alba** L. White Lettuce, Rattlesnake-root
Frequent; dry woods, roadside embankments. AB, CU, GE, HL, LK, LR, PC, PO, ST, SU, WA.

501 **Prenanthes altissima** L. Tall Rattlesnake-root
Common; moist woods, shaded roadsides. All counties except MO.

502 **Prenanthes crepidinea** Michx. Nodding Rattlesnake-root
Infrequent; dry woods, thickets, open slopes. CU, FA, LK, LI, LR, MD, PC, RO.

503 **Prenanthes racemosa** Michx. Prairie Rattlesnake-root
Rare; prairie remnants, fens, thickets. PO: fen, Streetsboro Twp., Andreas 6205, 15 Sep 1983 (KE 46463); Mantua Bog, Rood 2637, 27 Sep 1952 (KE 11896).

504 **Prenanthes serpentaria** Pursh Lion's-foot
Rare; dry barrens, thickets, open woods. FA: Madison Twp., Goslin s.n., 4 Sep 1938 (OS 50718). RO: Harrison Twp., Bartley & Pontius s.n., 28 Aug 1932 (OS 50715).

505 **Ratibida pinnata** (Vent.) Barnhart Prairie Coneflower
Infrequent; wet and dry fields, dry slopes, roadsides. HL, KN, LK, PC, RO, ST, SU, WA.

506 **Rudbeckia fulgida** Aiton Coneflower
Rare; dry and moist wooded slopes, thickets, fens. HL: roadside, Washington Twp., Burns 3367, 13 Sep 1983 (KE 46916). RO: Immell Bog, Green Twp., Bartley & Pontius s.n., 27 Jul 1937 (OS 38137).

507 **Rudbeckia hirta** L. Black-eyed Susan
Common; dry fields, pastures, roadsides, fens, prairie remnants. ALL COUNTIES.

508 **Rudbeckia laciniata** L. Cut-leaf Coneflower
Common; fields, floodplains, fens, thickets. ALL COUNTIES.

509 **Rudbeckia triloba** L. Three-lobed Coneflower
Common; roadsides, meadows, thickets, fields. ALL COUNTIES.

510 **Senecio anonymus** A. Wood Small's Squaw-weed
Senecio smallii Britton
Rare; thickets, waste places, fields. CU: cinders, Bedford Twp., Herrick s.n., 14 Jun 1955 (OS 563411). MH: railroad embankment, Berlin Twp., Cooperrider 6785, 15 Jul 1960 (KE 5117). SU: cinders, Twinsburg Twp., Herrick s.n., 14 Jun 1955 (OS 54057).

511 **Senecio aureus** L. Golden Ragwort, Squaw-weed
Common; boggy thickets, moist woods, meadows, fens, woodland seeps. ALL COUNTIES.

512 **Senecio obovatus** Muhl. ex Willd. Obovate-leaved Squaw-weed
Common; wooded hillsides, dry slopes, old fields. All counties except PE.

513 **Senecio pauperculus** Michx. Balsam Squaw-weed
Rare; calcareous soils, fens. GE: cultivated field, Bainbridge Twp., Herrick s.n., 10 Jun 1957 (OS 58322). RO: marl bog, Colerain Twp., Bartley & Pontius s.n., 3 Jun 1947 (OS s.n.).

514 ***Senecio vulgaris** L. Common Groundsel
Infrequent, naturalized; cultivated areas, roadsides, waste places. AB, CU, GE, LK, LR, MD, PO, RI, ST, SU.

515 **Silphium laciniatum** L. Compass Plant
Rare; prairies, fields. SU: Ira, Hine s.n., 13 Oct 1938 (OS 38998 and OS 38999); Ira, Hine s.n., 2 Apr 1938 (OS 38997).

516 **Silphium perfoliatum** L. Cup-plant
Common; weedy floodplains, thickets, stream terraces. AS, AB, CU, FA, HL, KN, LK, LI, LR, MD, MO, PE, PC, RI, RO, SU, TR, WA.

517 **Silphium terebinthinaceum** Jacq. Prairie-dock
Infrequent; dry fields, prairie remnants. CU, HL, PC, RO, ST, WA.

518 **Silphium trifoliatum** L. Whorled Rosinweed
Common; woodland borders, dry slopes, fens, roadside banks. All counties except MO.

519 ***Silybum marianum** (L.) Gaertner Milk-thistle
Rare, adventive; abandoned gardens, waste places. AB: Unionville, Floyd 2421, 21 Sep 1898 (OS 90395). LR: yard, Oberlin, Grover s.n., 29 Sep 1902 (OS 80887). WA: plains, Mellinger 2534, 29 Sep 1900 (OS 90303).

520 *Solidago bicolor* L. Silverrod, White Goldenrod
Common; dry woodland borders, rocky slopes, roadside banks. All counties except MO, PC, RO.

521 *Solidago caesia* L. Blue-stem Goldenrod
Common; rich woods, woodland borders, shaded roadside banks, thickets. ALL COUNTIES.

522 *Solidago canadensis* L. (including *S. altissima* L.) Field Goldenrod
Common; dry fields, roadsides, railroads, thickets. ALL COUNTIES.

523 *Solidago erecta* Pursh
Rare; dry woods, open slopes. FA: loop road, Wolfe, Wareham & Scofield s.n., 2 Aug 1941 (OS 43381); Hocking Twp., Goslin s.n., 22 Sep 1935 (OS 39535); Berne Twp., Goslin s.n., 15 Sep 1935 (OS 39537). RO: Colerain Twp., Bartley & Pontius s.n., 18 Sep 1929 (OS 39543).

524 *Solidago flexicaulis* L. Zigzag Goldenrod
Common; mesic and moist woods, hemlock slopes. ALL COUNTIES.

525 *Solidago gigantea* Aiton Field Goldenrod
Common; moist thickets, roadside ditches, fens. All counties except RI.

526 *Solidago hispida* Muhl.
Infrequent; weedy thickets, dry rocky woods, wooded slopes. AS, FA, GE, LK, LR, PC, PO, RO, SU, TR.

527 *Solidago juncea* Aiton
Common; dry fields, open woods, roadsides. All counties except PC, RO.

528 *Solidago nemoralis* Aiton Early Goldenrod
Common; dry roadside banks, dry fields, open woods, eroded slopes. ALL COUNTIES.

529 *Solidago ohioensis* Riddell Ohio Goldenrod
Infrequent; fens, wet prairie remnants. FA, PC, PO, RO, ST, SU.

530 *Solidago patula* Muhl. Rough-leaved Goldenrod
Common; boggy thickets, fens, moist woodland borders, marshes. All counties except MO, PE, PC.

531 *Solidago riddellii* Frank Riddell's Goldenrod
Infrequent; fens, wet prairie remnants. FA, HL, PC, RO, ST.

532 *Solidago rigida* L. Stiff Goldenrod
Rare; calcareous soils, rocky slopes, prairie remnants, thickets. PC: Pickaway County, Bartley s.n., 1 Sep 1943 (OS 61653). SU: railroad, Hudson, Beardlsee s.n., 1932 (OC 98670).

533 *Solidago rugosa* Miller Wrinkled-leaf Goldenrod
Common; fields, swamps, marshes, thickets. All counties except AS, MO, PC.

534 **Solidago sempervirens* L. Seaside Goldenrod
Rare, naturalized; saline roadside soils. SU: gravel, Richfield Twp., Bissell 1982:328, 18 Sep 1982 (KE 46356). WA: saline gravel, Congress Twp., Bissell 1981:167, 29 Sep 1981 (CLM 15562).

535 *Solidago speciosa* Nutt. Showy Goldenrod
Infrequent; dry open woods, thickets, wooded roadsides. LK, LR, PC, SU, WA.

536 *Solidago squarrosa* Muhl. Leafy Goldenrod
Infrequent; rocky woods, dry thickets, clearings, wooded slopes. AB, CU, GE, HL, LK, LI, LR, RO, SU.

537 *Solidago uliginosa* Nutt. Marsh Goldenrod
Frequent; fens, marshes, moist peaty fields. CU, FA, GE, HL, LK, LI, LR, MH, PE, PO, RO, ST, SU, WA.

538 *Solidago ulmifolia* Muhl. Elm-leaved Goldenrod
Common; woods, thickets, woodland borders. All counties except RI, ST.

539 **Sonchus arvensis* L. (including *S. uliginosus* Bieb.) Field Sow-thistle
Common; naturalized; roadsides, waste places. AS, AB, CL, CU, GE, HL KN,LK, LR, MH, MD, PC, PO, RO, ST, SU, TR, WA.

540 **Sonchus asper* (L.) Hill Prickly Sow-thistle
Common, naturalized; roadsides, fields, waste places. All counties except PC, RO.

541 *Sonchus oleraceus* L.   Common Sow-thistle
Frequent, naturalized; railroads, roadsides, waste places. AS, AB, CU, GE, HL, LK, LI, LR, MD, PO, RI, RO, ST, SU, WA.

542 *Tanacetum vulgare* L.   Common Tansy
Common, naturalized; roadsides, woodland borders, waste places. AS, AB, CL, CU, GE, HL, KN, LK, LI, LR, MH, MD, PO, RI, ST, SU, WA.

543 *Taraxacum laevigatum* (Willd.) DC.   Red-seeded Dandelion
Taraxacum erythrospermum Andrz.
Rare, adventive; waste places. LR: dooryard, Oberlin, Grover s.n., 12 Jul 1910 (OC 51048). PO: Randolph, Kellerman s.n., 8 Dec 1899 (OS 17622).

544 *Taraxacum officinale* Weber   Common Dandelion
Common, naturalized; roadsides, woodland borders, fields, lawns, waste places. ALL COUNTIES.

545 *Tragopogon dubius* Scop.
Tragopogon major Jacq.
Common, naturalized; roadsides, railroads, waste places. AS, AB, CL, CU, GE, HL, KN, LK, MH, MD, MO, PO, ST, SU, TR, WA.

546 *Tragopogon porrifolius* L.   Salsify, Oyster-plant
Common, naturalized; railroads, roadsides, waste places. AS, AB, CU, FA, GE, HL, KN, LK, LR, MH, MD, MO, PO, RI, ST, SU, TR, WA.

547 *Tragopogon pratensis* L.   Goat's Beard
Frequent, naturalized; railroads, roadsides, waste places. AB, CL, CU, FA, GE, KN, LK, LR, MH, PO, RO, ST, TR, WA.

548 *Tussilago farfara* L.   Coltsfoot
Common, naturalized; roadsides, eroding slopes, stream banks, railroads, waste places. All counties except LI, PC, RO.

549 *Verbesina alternifolia* (L.) Britton   Wingstem
Actinomeris alternifolia (L.) DC.
Common; thickets, wooded floodplains, woodland borders. ALL COUNTIES.

550 *Verbesina virginica* L.   Tickseed
Rare, adventive; thickets, waste places. LI: Buckeye Lake, Craig s.n., 25 May 1909 (OS 38558).

551 *Vernonia fasciculata* Michx.   Prairie Ironweed
Rare; prairie remnants, weedy fields. PC: Pickaway Twp., Bartley & Pontius s.n., 16 Jul 1937 (OS 95948).

552 *Vernonia gigantea* (Walter) Branner & Coville   Ironweed
Vernonia altissima Nutt.
Common; fields, pastures, pond margins, fens, moist woodland borders. ALL COUNTIES.

553 *Vernonia noveboracensis* (L.) Michx. x *Vernonia gigantea* (Walter) Branner & Coville
Rare; dry woods, ridges. LK: open area, Concord Twp., Bradt 1855, 14 Sep 1985 (KE 48288).

554 *Xanthium strumarium* L.   Common Cocklebur
Common; waste places, cultivated fields, roadsides. ALL COUNTIES.

CONVOLVULACEAE (Morning Glory Family)

555 *Calystegia hederacea* Wallich
Rare, adventive; weedy fields. LR: fill dirt, Oberlin, Jones 78-6-21-38, 21 Jun 1978 (OC 220631).

556 *Calystegia pubescens* Lindley
Convolvulus pellitus Ledeb.
Rare, adventive; weedy fields, roadsides, waste places. MD: Medina Twp., Nettleton s.n., 1897 (OS s.n.). ST: roadside, Sugar Creek Twp., Amann 664, 16 Jul 1960 (KE 4021).

557 *Calystegia sepium* (L.) R. Br.   Hedge Bindweed
Convolvulus sepium L.
Common; fields, roadsides, moist thickets. All counties except PC.

558 *Calystegia spithamaea* (L.) Pursh   Low Bindweed
Convolvulus spithamaeus L.

Frequent; dry sandy fields, open dry woods, lightly wooded slopes. CL, CU, GE, HL, LK, LI, MH, PC, PE, PO, ST, SU, TR, WA.

559 *Convolvulus arvensis* L. Field Bindweed
Common, naturalized; fields, roadsides, waste places. AS, CU, FA, GE, HL, KN, LK, LR, MH, MD, MO, PC, PO, RI, RO, ST, SU, TR, WA.

560 *Cuscuta campestris* Yuncker Prairie Dodder
Infrequent; river banks, railroad embankments, waste places. GE, LK, LR, PO, RI, SU.

561 *Cuscuta coryli* Engelm. Hazel Dodder
Rare; sandy soils, pond margins. CU: swamp, Cleveland, Claassen s.n., 1891 (OS 29572). PO: dry sandy island, Suffield Twp., Rood & Webb 1417, 24 Aug 1919 (KE 11247).

562 *Cuscuta epilinum* Weihe Flax Dodder
Rare, adventive; parasitic on flax. WA: Wooster, Taggart s.n., 1895 (OS 29616 & OS 94642).

563 *Cuscuta gronovii* Willd. Common Dodder
Common; floodplains, thickets, marshes, roadsides. ALL COUNTIES.

564 *Cuscuta pentagona* Engelm. Five-angled Dodder
Rare; floodplains, roadsides. AB: weed, Pymatuning River, Hicks s.n., 7 Sep 1931 (OS 29552). MH: weed in clover, Beloit, Hunt s.n., 10 Sep 1960 (OS 64460). WA: Blachleyville, Selby & Duvel 957, 6 Jul 1899 (OS 94645).

565 *Cuscuta polygonorum* Engelm.
Rare; floodplains, roadsides. CU: disturbed floodplain, Rocky River, Bissell 1974: 167 (CLM 18975). LR: Elyria, Watson s.n., 19 Aug 1895 (OS s.n.). PE: Reservoir Park, Kellerman s.n., 18 Jul 1895 (OS 29554). SU: roadside, Twinsburg Twp., Herrick s.n., 31 Jul 1955 (OS 56400).

566 *Ipomoea coccinea* L. Red Morning Glory
Rare, naturalized; thickets, roadsides, waste places. LR: Oberlin, Ricksecker s.n., 31 Aug 1894 (CLM 10623). MD: weedy lot, Wadsworth, Cusick 12342, 21 Oct 1971 (KE 36071).

567 *Ipomoea hederacea* (L.) Jacq. Ivy-leaved Morning Glory
Infrequent, naturalized; waste places, roadsides, weedy fields. FA, KN, LK, LR, PC, PO, RO, WA.

568 *Ipomoea lacunosa* L.
Rare; weedy fields, waste places, roadsides. RO: Scioto Twp., Bartley & Pontius s.n., 26 Sep 1930 (OS 29418).

569 *Ipomoea pandurata* (L.) G. F. W. Meyer Wild Potato-vine
Infrequent; roadsides, weedy fields, fence rows. AS, FA, HL, KN, LK, LI, PE, RO, ST.

570 *Ipomoea purpurea* (L.) Roth Common Morning Glory
Infrequent, naturalized; roadsides, waste places, abandoned gardens. AS, CU, FA, HL, KN, LI, LR, MD, PC, RO, ST.

CORNACEAE (Dogwood Family)

571 *Cornus alternifolia* L. f. Alternate-leaf Dogwood, Pagoda Dogwood
Common; thickets, moist woods, woodland borders. ALL COUNTIES.

572 *Cornus amomum* Miller (including *C. obliqua* Raf.) Red Willow Dogwood
Common; marshes, fens, swamps, damp thickets. ALL COUNTIES.

573 # *Cornus canadensis* L. Bunchberry
Infrequent; damp boggy woods, sphagnous thickets. AB, CU, GE, LK, LI, PO, ST, SU.

574 *Cornus drummondii* Meyer Roughleaf Dogwood
Infrequent; dry slopes, fields, woodland borders. CU, KN, LI, LR, PC, RO, SU.

575 *Cornus florida* L. Flowering Dogwood
Common; rich woods, old fields, woodland borders. All counties except PC.

576 *Cornus racemosa* Lam. Gray Dogwood
Common; wet fields, weedy floodplains, fens, marshes. ALL COUNTIES.

577 # *Cornus rugosa* Lam. Roundleaf Dogwood
Infrequent; dry eroding slopes, river bluffs, rocky slopes. AB, CU, GE, LK, LR, PC, RO, SU.

578 *Cornus stolonifera* Michx. Red Osier
Common; wet thickets, fens, damp fields, marshes. AS, AB, CL, CU, FA, GE, LK, LI, LR, MH, MD, PO, ST, SU, TR, WA.

CRASSULACEAE (Orpine Family)

579 *Sedum acre* L. Mossy Stonecrop
Rare, naturalized; rocky banks, waste places. MH: cemetery, Green Twp., Cusick 3188, 15 Sep 1966 (KE 27820). PO: Cuyahoga River, Kent, Ogg & Campbell 226, 1 Jul 1931 (KE 683).

580 *Sedum album* L.
Rare, adventive; fields, woodland borders. PO: woods, Streetsboro Twp., Bliss s.n., no date (KE 682).

581 *Sedum sarmentosum* Bunge
Rare, adventive; rock ledges, woodland borders. CL: woodlot, Hanover Twp., Cusick s.n., 17 Jun 1962 (OS 70062). KN: roadbank, Union Twp., Pusey 793, 4 Oct 1975 (KE 38340). TR: yard, Southington Twp., Rood 2964, 29 Jun 1956 (KE 12232).

582 *Sedum telephium* L. (including *S. purpureum* L.) Garden Orpine
Infrequent, adventive; roadbanks, disturbed woods, waste places. AS, AB, CU, GE, HL, KN, LR, PO, ST, SU.

583 *Sedum ternatum* Michx.
Common; mesic and moist woods, wooded floodplains, stream margins, hemlock ravines. ALL COUNTIES.

CRUCIFERAE / BRASSICACEAE (Mustard Family)

584 *Alliaria petiolata* (Bieb.) Cavara & Grande Garlic-mustard
Alliaria officinalis Andrz.
Common, naturalized; moist roadside banks, wet ditches, disturbed woods. AS, AB CL, CU, FA, HL, KN, LK, LR, PO, RI, RO, ST, SU, TR, WA.

585 *Alyssum alyssoides* (L.) L. Alyssum
Rare, adventive; roadsides, waste places. PC: along railroad, Circleville Twp., Bartley s.n., 6 May 1955 (OS 52972).

586 *Arabidopsis thaliana* (L.) Heynh. Mouse-ear Cress
Common, naturalized; disturbed fields, waste places, roadsides. AB, CL, CU, FA, HL, KN, LK, LI, LR, MH, MD, PE, PC, PO, RO, ST.

587 *Arabis canadensis* L. Sickle-pod
Frequent; dry woods, eroding slopes, wooded roadside banks. AB, CL, CU, FA, GE, HL, PE, PO, RO, SU, WA.

588 *Arabis drummondii* Gray Drummond's Rock-cress
Rare; limestone ledges, gravelly banks. LI: Newark, Jones s.n., 19 Jun 1891 (OC 78998). LR: river bank, Day s.n., 20 May 1896 (OC 99416).

589 *Arabis glabra* (L.) Bernh. Tower Mustard
Common; rocky woods, ledges, fields. AS, AB, CL, CU, FA, GE, HL, KN, LK, LI, LR, MO, PE, PO, RI, RO, ST, SU, WA.

590 *Arabis hirsuta* (L.) Scop. var. *pycnocarpa* (Hopkins) Rollins
Infrequent; cliffs, ledges, gravelly banks. CU, FA, LI, PC, SU.

591 *Arabis laevigata* (Muhl.) Poiret
Common; mesic and dry woods, steep ravines. All counties except PC.

592 *Arabis lyrata* L. Lyre-leaf Rock-cress
Infrequent; sandy banks, dry ledges, eroding slopes. AB, FA, LK, LI, SU.

593 *Arabis perstellata* E. L. Braun
Infrequent; rich woods, ledges. AS, KN, LR, PC, RO.

594 *Armoracia aquatica* (Eaton) Wieg. Lake-cress
Once rare, now presumably extirpated from Ohio; lakes, quiet creeks. LI: margin, Buckeye Lake, Detmers s.n., 11 Jul 1910 (OS 13444); reservoir, Craig s.n., Jun 1889 (OS 91823). LR: edge of golf course, Oberlin, Grover s.n., 18 Jun 1913 (OC 73237); Oberlin, McCormick s.n., 20 Jun 1888 (OC 73239). PE: reservoir, Kellerman s.n., 20 Jul 1895 (OS s.n.) PC: Hitler Pond, Pickaway Twp., Bartley & Pontius s.n., 10 Jun 1936 (OS 13442).

595 *Armoracia rusticana* Gaertner, Meyer & Schreber Horseradish
Armoracia lapathifolia Gilib.
Frequent, adventive; roadside ditches, abandoned homesites, fields. AS, CU, FA, GE, KN, LI, LR, PE, PO, ST, TR, WA.

596 ***Barbarea verna** (Miller) Aschers. Early Winter-cress
Infrequent, naturalized; pastures, fields, woodland margins. AS, AB, CU, FA, HL, LK, PO, ST, TR.

597 ***Barbarea vulgaris** R. Br. Wild Mustard, Yellow Rocket
Common, naturalized; roadsides, cultivated fields, pastures, waste places. ALL COUNTIES.

598 ***Berteroa incana** (L.) DC. Hoary Alyssum
Infrequent, adventive; waste places, roadsides. AB, CU, FA, GE, KN, LK, PC, PO.

599 ***Brassica alba** (L.) Rabenh. White Mustard
Brassica hirta Moench
Rare, adventive; waste places, fields. LR: railroad, Oberlin, French s.n., 6 Jun 1892 (OC 732663).

600 ***Brassica juncea** (L.) Czern. Chinese Mustard, Brown Mustard
Frequent, adventive; cultivated fields, waste places. CL, FA, HL, KN, LR, MD, PE, PO, RO, TR, WA.

601 ***Brassica kaber** (DC.) Wheeler Charlock
Common, adventive; fields, railroads, roadsides, waste places. AS, CL, CU, FA, GE, HL, LK, LI, LR, MD, MO, PO, RI, ST, SU, TR, WA.

602 ***Brassica napus** L. Rape
Rare, adventive; waste places. FA: Lancaster Twp., Goslin s.n., 3 May 1935 (OS 14538). RO: Ross County, Bartley & Pontius s.n., 30 Apr 1933 (OS 14532). WA: Creston, Bunnels s.n., 25 Apr 1945 (OS s.n.).

603 ***Brassica nigra** (L.) Koch
Common, adventive; waste places, cultivated fields. AB, CU, GE, HL, KN, LK, LI, LR, MD, PC, PO, RO, ST, SU, TR, WA.

604 ***Brassica oleracea** L. Cabbage
Rare, adventive; gardens, waste places. SU: fallow field, Northampton Twp., Tomci 653, 22 May 1976 (KE 41470).

605 ***Brassica rapa** L. (including **B. campestris** L.) Field Mustard
Infrequent, adventive; waste places, roadsides, lawns. AB, CU, PO, RO, SU.

606 **Cakile edentula** (Bigelow) Hooker Sea-rocket
Rare; beaches. A dubious record occurs from RI: Wilkinson s.n., 21 Aug 1895 (KE 18886).

607 ***Camelina microcarpa** DC. False Flax
Frequent, naturalized; waste places, railroads, roadsides. AB, CU, FA, GE, HL, LR, PC, RO, ST, SU, TR, WA.

608 ***Camelina sativa** (L.) Crantz Gold-of-pleasure
Infrequent, adventive; roadsides, waste places, fields. CU, FA, LK, LR, RO.

609 ***Capsella bursa-pastoris** (L.) Medicus Shepherd's-purse
Common, adventive; roadsides, fields, waste places, railroads. ALL COUNTIES.

610 **Cardamine bulbosa** (Muhl.) BSP. Spring-cress
Common; woodland seeps, wet meadows, wooded stream terraces. ALL COUNTIES.

611 **Cardamine douglassii** Britton Purple Spring-cress
Common; moist rich woods, seeps, wooded floodplains. All counties except CL, RO.

612 ***Cardamine hirsuta** L.
Frequent, naturalized; roadsides, waste places, fields, lawns. AS, AB, CU, FA, HL, KN, LK, PE, PO, RO, TR.

613 ***Cardamine impatiens** L.
Rare, adventive; waste places, wet fields. AB: seepage ditch, Richmond Twp., Andreas 5968, 2 Jul 1982 (KE 46998). PC: field, Pickaway Twp., Bartley s.n., 6 Jun 1961 (OS s.n.).

614 **Cardamine parviflora** L. var. **arenicola** (Britton) Schulz
Infrequent; woodland margins, sandy soils, roadsides. AB, FA, LK, LI, MO, PC, PO, RO.

615 **Cardamine pensylvanica** Muhl.
Common; open wet woods, ponds, marshes, stream margins, thickets. ALL COUNTIES.

616 ***Cardamine pratensis** L. Cuckoo-flower
Rare, adventive; lawns, waste places, fields. LR: lawns, Oberlin, Grover s.n., 8 May 1913 (OC 3313). PO: railroad, Nelson Twp., Barker 69, 6 May 1977 (KE 41621); Ravenna, Webb

s.n., 24 May 1909 (OS 14280). ST: railroad, Massillon, Amann 160, 21 May 1960 (KE 8562). SU: ditch, Stow Twp., Cusick 14615, 23 Apr 1976 (KE 39495).

617 *Cardamine rotundifolia* Michx.   Mountain Watercress
Rare; wet banks, creek margins. RO: Jefferson Twp., Bartley & Pontius s.n., 18 May 1930 (OS 14268).

618 **Cardaria draba* (L.) Desv.   Hoary Cress
Infrequent, naturalized; waste places, fields, railroads, roadsides. AB, CU, LK, PC, PO, SU.

619 **Chorispora tenella* (Willd.) DC.
Rare, adventive; fields, roadsides. PC: Thatcher, Pickaway Twp., Bartley s.n., 5 May 1961 (OS 67130).

620 **Conringia orientalis* (L.) Dumort.   Hare's Ear Mustard
Infrequent, adventive; waste places, fields, gardens. CU, FA, GE, PO, WA.

621 *Dentaria diphylla* Michx.   Two-leaved Toothwort
Common; rich mesic and moist woods, wooded floodplains. All counties except MO, PC, RO.

622 *Dentaria heterophylla* Nutt.   Slender Toothwort
Rare; rich moist woods, stream terraces. KN: beech woods, Liberty Twp., Stuckey 8759, 8 May 1970 (OS 104172). PO: woods, Windham, Webb 465, 16 May 1920 (KE 11102).

623 *Dentaria laciniata* Muhl.   Cut-leaf Toothwort
Common; rich mesic and moist woods, pastures, moist clearings, stream terraces. ALL COUNTIES.

624 *Descurainia pinnata* (Walter) Britton   Tansy-mustard
Rare; dry sandy fields, railroads, waste places. FA: Kettle Hills, Jones s.n., 25 May 1942 (OS 45507); railroad, Berne Twp., Goslin s.n., 3 May 1935 (OS 13740). PC: Pickaway Twp., Bartley & Pontius s.n., 8 May 1932 (OS 13728). PO: railroad, Garrettsville, Webb s.n., 25 Jul 1913 (OS 13741).

625 **Descurainia sophia* (L.) Prantl
Rare, adventive; waste places, railroads, roadsides. RO: Green Twp., Bartley & Pontius s.n., 26 Apr 1931 (OS 13748).

626 **Diplotaxis tenuifolia* (L.) DC.   Narrow-leaved Wall-rocket
Rare, adventive; waste places, roadsides. WA: Wayne County, Thomas s.n., 1 Jun 1945 (OS 87838 & OS 87839).

627 *Draba reptans* (Lam.) Fern.
Rare; dry fields, roadside banks. RO: Green Twp., Bartley s.n., 19 Apr 1933 (BHO 18650).

628 **Erophila verna* (L.) Besser   Whitlow-grass
Draba verna L.
Frequent, adventive; waste places, roadsides, fields. AS, AB, FA, HL, KN, LK, LI, MO, PE, PC, PO, RO, SU.

629 **Erucastrum gallicum* (Willd.) O. E. Schulz
Infrequent, adventive; waste places, roadsides, pastures, railroads. FA, HL, LK, PC, PO, ST, WA.

630 **Erysimum cheiranthoides* L.   Wormseed-mustard
Infrequent, adventive; waste places, fields, meadows. AB, FA, LK, LR, PC, PO, RO, ST, TR.

631 **Erysimum inconspicuum* (S. Watson) MacM.
Rare, adventive; waste places. TR: dump, Rood 1343, 20 Jul 1938 (KE 11098 & KE 11099).

632 **Erysimum repandum* L.   Treacle Mustard
Infrequent, adventive; waste places, fields. AB, FA, LR, MO, PC, PO, RO.

633 **Hesperis matronalis* L.   Dame's-rocket
Common, naturalized; roadside banks, fields, floodplains, waste places, disturbed woods. AS, AB, CL, CU, FA, GE, HL, KN, LK, LR, MH, MD, PE, PO, RI, ST, SU, TR, WA.

634 **Iberis umbellata* L.   Candytuft
Rare, adventive; abandoned gardens, waste places, roadsides. LR: roadsides, Kelsey s.n., 1895 (OC 73209).

635 *Iodanthus pinnatifidus* (Michx.) Steud. Purple Rocket
Infrequent; moist roadside banks, floodplains, moist slopes. AS, CU, FA, LI, LR, MO, PE, PC, RO.

636 *Lepidium campestre* (L.) R. Br. Cow-cress
Common, naturalized; waste places, fields, railroads, roadsides. All counties except RO.

637 *Lepidium densiflorum* Schrader
Frequent, naturalized; waste places, fields, roadsides. AS, AB, CU, LI, LR, PO, RO, ST, SU, TR, WA.

638 *Lepidium perfoliatum* L.
Rare, adventive; waste places, roadsides. PO: lawns, Kent, Hopkins s.n., 2 Jun 1914 (OS 62871). PC: Pickaway Twp., Bartley s.n., 30 May 1968 (BHO 33644).

639 *Lepidium sativum* L. Garden-cress
Rare, adventive; waste places. LR: Oberlin, Koford s.n., 1 Sep 1890 (OC 73226).

640 *Lepidium virginicum* L. Poor-man's Pepper
Common; roadsides, railroads, fields, waste places. ALL COUNTIES.

641 *Lunaria annua* L. Money-plant, Honesty
Rare, adventive; waste places, roadsides. LR: river bank, Columbia Twp., Herrick s.n., 22 Jun 1960 (OS 79853). RO: Green Twp., Bartley & Pontius s.n., 20 May 1932 (OS 13326). ST: Alliance, Lanum s.n., 19 Jun 1922 (OS 87787).

642 *Lunaria rediviva* L. Honesty
Rare, adventive; roadsides, waste places. GE: woods, Parkman, Herrick s.n., 15 May 1967 (KE 16575). PO: woods, Kent, Kunst s.n., 21 May 1971 (KE 26015). ST: roadside, Plain Twp., Amann 518, 13 Jul 1960 (KE 8674).

643 *Nasturtium officinale* R. Br. Watercress
Common, naturalized; springs, roadside ditches, fens, quiet streams. All counties except MH, PE, PC.

644 *Raphanus raphanistrum* L. Wild Radish
Frequent, adventive; fields, waste places, roadsides. AS, AB, CU, GE, KN, LK, MD, PC, PO, ST, TR, WA.

645 *Raphanus sativus* L. Radish
Infrequent, adventive; fields, waste places, abandoned gardens. LI, LR, MO, PC, RO, ST, SU, WA.

646 *Rorippa palustris* (L.) Besser Yellow-cress
Rorippa islandica (Oeder) Borbas
Common; wet ditches, pond and lake margins, disturbed wetlands. ALL COUNTIES.

647 *Rorippa sessiliflora* (Nutt.) Hitchc. Sessile Yellow-cress
Rare; sandy creek margins, wet pastures. RO: river, Chillicothe, Scioto Twp., Roberts 2225, 26 Sep 1971 (OS 114789); wet ground, Green Twp., Bartley s.n., 29 Aug 1943 (OS 43888).

648 *Rorippa sylvestris* (L.) Besser Creeping Yellow-cress
Frequent, naturalized; waste places, roadsides, wet pastures. AS, AB, CL, CU, HL, KN, LK, MH, PO, RO, ST, SU, WA.

649 *Sisymbrium altissimum* L. Tumble-mustard
Frequent, naturalized; waste places, barnyards, pastures, fields, roadsides, railroads. AB, CL, CU, FA, GE, HL, LK, PC, PO, RO, ST, SU, TR, WA.

650 *Sisymbrium officinale* (L.) Scop.
Common, naturalized; waste places, fields. AS, AB, CU, FA, GE, HL, KN, LI, LR, MO, PO, RI, RO, ST, SU, TR, WA.

651 *Thlaspi arvense* L. Field Penny-cress
Common, naturalized; roadsides, fields, waste places. AS, AB, CU, FA, GE, HL, KN, LK, LI, LR, MO, PC, PO, RO, ST, SU, WA.

652 *Thlaspi perfoliatum* L. Fanweed
Rare, naturalized; fields, roadsides. RO: roadside, Paint Twp., Cusick 19074, 4 Apr 1974 (OS 131443); Chillicothe, Bartley s.n., April 1965 (OS 80079).

CUCURBITACEAE (Gourd Family)

653 *Citrullus lanatus* (Thunb.) Matsumura & Nakai   Watermelon
Rare, adventive; waste places. PO: waste ground, Franklin Twp., Cusick 6932, 7 Oct 1967 (KE 27842). RO: barnyard, Green Twp., Cusick 17427, 16 Aug 1977 (KE 44954).

654 *Cucumis melo* L.   Muskmelon
Rare, adventive; waste places. KN: landfill, Clinton Twp., Pusey 1579, 20 Aug 1975 (KE 38367).

655 *Cucurbita pepo* L.   Pumpkin
Rare, adventive; waste places. KN: landfill, Morris Twp., Pusey 721, 12 Sep 1975 (KE 38368).

656 *Echinocystis lobata* (Michx.) T. & G.   Prickly Cucumber
Common; moist woodland borders, wooded floodplains, wet thickets, roadsides. All counties except MO, MH, PE.

657 *Sicyos angulatus* L.   Bur-cucumber
Common; river banks, fence rows, woodland borders. AS, AB, CU, FA, GE, HL, KN, LK, LI, LR, PC, PO, RI, RO, ST, SU, WA.

DIPSACACEAE (Teasel Family)

658 *Dipsacus fullonum* L.   Wild Teasel
Dipsacus sylvestris Hudson
Common, naturalized; fields, roadsides, railroads, waste places. ALL COUNTIES.

659 *Dipsacus laciniatus* L.
Rare, adventive; fields, waste places. TR: field, Bazetta Twp., Burns 160, 2 Oct 1976 (KE 46067).

660 *Dipsacus sativus* (L.) Honck.
Rare, adventive; waste places. RO: Green Twp., Bartley s.n., Oct 1957 (OS 153047).

DROSERACEAE (Sundew Family)

661 # *Drosera intermedia* Hayne   Spathulate-leaved Sundew
Rare; peaty bog margins, organic floating mats. GE: *Sphagnum*, Burton Twp., Woodrich s.n., 19 Jun 1967 (KE 18805). SU: bog, Akron, Karlo s.n., 5 Aug 1962 (KE 42181); Copley Swamp, Foltz s.n., 25 Aug 1889 (CLM s.n.); Copley Swamp, Claypole s.n., 7 Aug 1889 (University of Akron Herbarium s.n.). WA: bog, Fox Lake, Selby & Duvel 646, 27 Jul 1899 (OS 87931).

662 *Drosera rotundifolia* L.   Round-leaved Sundew
Frequent; organic hummocks in fens, bogs, damp sands. AB, CU, GE, HL, LK, LI, PE, PO, RI, ST, SU, WA.

EBENACEAE (Ebony Family)

663 *Diospyros virginiana* L.   Common Persimmon
Infrequent; dry woods, fields, clearings. FA, LK, LI, MH, MD, PE, PC, PO, RO, TR.

ELAEAGNACEAE (Oleaster Family)

664 *Elaeagnus angustifolia* L.   Russian Olive
Rare, adventive; fields, thickets. CU: barrens and shrubby thicket, Bissell 1983: 70, 18 Jun 1983 (CLM 19955).

665 *Elaeagnus umbellata* Thunb.   Oleaster
Frequent, naturalized; fields, roadsides, waste places. AB, CL, GE, KN, LK, LR, MH, PO, SU, TR.

666 # *Shepherdia canadensis* (L.) Nutt.   Buffalo-berry
Infrequent; dry calcareous eroding slopes, lacustrine soils, sandy banks. AB, CU, GE, LK, LR, SU.

ERICACEAE (Heath Family)

667 # *Andromeda glaucophylla* Link   Bog-rosemary
Once infrequent, now presumably extirpated from Ohio; sphagnous bogs. AB, GE, ST, SU, WA. Dachnowski (1912) indicates a record for PO.

668 **# Chamaedaphne calyculata** (L.) Moench   Leatherleaf
Infrequent; acidic bogs, buttonbush swamps. AB, GE, LI, LR, PO, ST, SU, WA.

669 **Chimaphila maculata** (L.) Pursh   Spotted Wintergreen
Common; dry woods, hemlock slopes. AS, AB, CL, CU, FA, GE, HL, KN, LK, LI, LR, PE, PC, PO, RI, RO, ST, SU, TR, WA.

670 **Chimaphila umbellata** (L.) Bart. var. **cisatlantica** Blake   Pipsissewa
Infrequent; dry woods, hemlock slopes, pine plantations. AB, FA, GE, KN, LK, MH, PO, SU, TR. Braun (1961) indicates a record for CU.

671 **Epigaea repens** L.   Trailing Arbutus
Common; dry woods, hemlock slopes. AS, AB, CL, CU, FA, GE, HL, KN, LK, LI, LR, MH, MD, PO, RO, ST, SU, TR, WA.

672 ***Erica tetralix** L.   Cross-leaved Heath
Rare, adventive; waste places. MH: under ornamental shrub, Mill Creek Park, Vickers 2620, 1 Jul 1952 (KE 12911).

673 **# Gaultheria hispidula** (L.) Bigelow   Creeping Snowberry
Once rare, now presumably extirpated from Ohio; sphagnous bogs, tamarack swamps. AB: Pymatuning Bog, Hicks s.n., 1933 (OS s.n.). LK: Painesville, Beardslee 1871 (OC 80266). ST: tamarack bog, Meyer's Lake, Vickers 1284, 4 Jul 1893 (KE 12889 & OC 80267). SU: Turkey Foot Lake, Akron, Selby s.n., 4 Jul 1899 (OS 28906).

674 **Gaultheria procumbens** L.   Wintergreen, Mountain Tea
Common; rich woods, dry wooded slopes, sphagnous thickets. AS, AB, CL, CU, FA, GE, HL, KN, LK, LI, LR, MH, MD, PO, RO, ST, SU, TR, WA.

675 **Gaylussacia baccata** (Wang.) K. Koch   Huckleberry
Common; dry woods, eroding slopes, sphagnous bogs. All counties except MO, PE, PC.

676 **Kalmia latifolia** L.   Mountain Laurel
Infrequent, almost entirely limited to unglaciated Ohio; dry woods, hemlock ravines. FA: Jacob's Ladder, Hocking Twp., Barden s.n., 15 May 1963 (KE 6861). KN: Pipesville, Moldenke 12186, 11 Apr 1942 (OS 48734). RO: Colerain Twp., Bartley & Pontius s.n., 16 Jun 1929 (OS 28542).

677 **# Ledum groenlandicum** Oeder   Labrador-tea
Rare; sphagnous thickets, tamarack bogs. AB: east Andover Twp., Hicks s.n., 12 Jul 1928 (OS 65271). LI: Cranberry Island [probably introduced (Braun, 1961)], Hicks s.n., 25 May 1933 (OS 28463 & OS 28464). PO: bog, Roostown Twp., Andreas & Cooperrider 2172, 30 Jun 1978 (KE 41933); Cooperrider 6403, 11 Jun 1960 (KE 1952); peat bog, Shalersville Twp., Webb 661, 16 Jun 1899 (KE 12897); peat bog, Shalersville Twp., Webb & Robinson s.n., 24 Jun 1909 (OS 28460); Way Swamp, Shalersville Twp., Webb s.n., 29 Jun 1913 (OS 42433).

678 **Lyonia ligustrina** (L.) DC.   Maleberry
Once rare, now presumably extirpated from Ohio; thickets. RO: Liberty Twp., Bartley & Pontius s.n., 20 Sep 1931 (OS 28613 & OS 28614).

679 **Monotropa hypopithys** L.   Pinesap
Common; rich mesic woodlands, pine plantations. AS, AB, CL, CU, FA, GE, HL, LK, LR, MD, PE, PO, RI, RO, ST, SU, TR, WA.

680 **Monotropa uniflora** L.   Indian-pipe
Common; rich dry and mesic woodlands. ALL COUNTIES.

681 **Pyrola elliptica** Nutt.   Shinleaf
Common; rich mesic woods, wooded stream terraces. All counties except KN, MO, PC.

682 **Pyrola rotundifolia** L. var. **americana** (Sweet) Fern.   Round-leaf Shinleaf
Frequent; rich mesic woods. AB, CU, FA, GE, HL, LK, LI, PO, RO, ST, SU, TR, WA.

683 **# Pyrola secunda** L.   One-sided Wintergreen
Once infrequent, now presumably extirpated from Ohio; dry woods, sandstone ledges. AB, CU, GE, LK, LR, MH, PO, SU.

684 **Rhododendron nudiflorum** (L.) Torrey (including **R. roseum** (Loisel.) Rehd.)
       Northern Rose Azalea
Infrequent; boggy woods, sphagnous thickets, dry woods. AB, GE, LK, PO, RO, ST, TR.

685 **#** *Vaccinium angustifolium* Aiton   Low Sugarberry
Frequent; dry sandy slopes, sandstone ledges. AB, CL, CU, GE, HL, KN, LK, MH, PO, ST, SU, TR.

686 *Vaccinium corymbosum* L.   Northern Highbush Blueberry
Common; sphagnous bogs, sandy fields, kettlehole depressions, wet thickets, dry woods. AS, AB, CL, CU, GE, HL, KN, LK, LR, MH, MD, PO, RI, ST, SU, TR, WA.

687 *Vaccinium macrocarpon* Aiton   Large Cranberry
Frequent; moist sands, sphagnous bogs, sphagnous hummocks in fens. AB, CU, GE, HL, LI, LR, PO, RI, ST, SU, WA.

688 **#** *Vaccinium myrtilloides* Michx.   Velvet-leaf Blueberry
Rare; moist woods, swamp woods. AB: wet swamp woods, Rome Twp., Bissell 1982:230, 24 Jul 1982 (KE 43789); east Andover Twp., Hicks s.n., 14 Jun 1931 (OS 28897); Pymatuning swamp, Davis s.n., 11 Aug 1910 (OS 1078040). GE: wet woods, Montville Twp., Bissell 1983:108, 26 Jul 1983 (CLM 19719).

689 **#** *Vaccinium oxycoccos* L.   Small Cranberry
Rare; tamarack swamps, sphagnous hummocks. GE: Burton Twp., Klein s.n., 12 Oct 1957 (OS 59236). PO: in tamaracks, Brimfield Twp., Cooperrider and Zavortink 8548, 30 Jul 1961 (KE 8932), Andreas 1999, 20 Jun 1978 (KE 41977); hummocks, Mantua Twp., Andreas 3736, 12 Jun 1979 (KE 42344); Mantua Swamp, Mantua Twp., Rood 2629, 15 Jun 1952 (KE 12876). ST: near Beach City, Sugar Creek Twp., Camp s.n., 11 Sep 1933 (OS 28976). Cooperrider (1985a) indicates a record for LR.

690 *Vaccinium pallidum* Aiton   Early Sweet Blueberry
Vaccinium vacillans Torrey
Common; sandstone ledges, dry woods, fields. All counties except MO, PC.

691 *Vaccinium stamineum* L.   Deerberry
Frequent; dry fields, dry woods, openings. AB, CL, CU, FA, LK, LI, LR, PE, PO, RI, RO, ST, SU, TR.

EUPHORBIACEAE (Spurge Family)

692 **Acalypha ostryaefolia* Riddell   Hornbeam-leaved Copperleaf
Rare, naturalized; waste places, cultivated fields. RO: truck patch, Chillicothe, Bartley s.n., 26 Jul 1970 (OS 10559).

693 *Acalypha virginica* L. (including *A. rhomboidea* Raf.)   Three-seeded Mercury
Common; roadsides, cultivated fields, waste places. All counties except LI.

694 **Croton glandulosus* L. var. *septentrionalis* Muell. Arg.   Northern Croton
Rare, adventive; fields, waste places. RO: cinders, Chillicothe, Carr 2047, 20 Aug 1979 (OS 15041 & KE 44995); railroad yard, Chillicothe, Bartley s.n., 26 Jul 1964 (OS 76291).

695 **Croton monanthogynus* Michx.   Prairie-tea
Rare, naturalized; calcareous soils. RO: Green Twp., Bartley & Pontius s.n., 11 Sep 1937 (OS 49452).

696 *Euphorbia commutata* Engelm.   Wood Spurge
Infrequent; moist woods, stream margins, eroding slopes. LI, LR, MD, PC, RO, ST.

697 *Euphorbia corollata* L.   Flowering Spurge
Common; dry fields, roadsides, eroding slopes, dry hillsides, prairie remnants. All counties except MO.

698 **Euphorbia cyathophora* Murray   Painted-leaf Spurge
*Euphorbia heterophylla of authors, not L. (Cooperrider, 1985a).
Rare, adventive; waste places, railroad. PC: waste place, Thatcher, Bartley s.n., 28 Sep 1963 (OS 73928).

699 **Euphorbia cyparissias* L.   Cypress Spurge
Frequent; naturalized; waste places, roadsides, fields, abandoned homesites. AS, AB, CL, CU, KN, LK, LI, LR, MH, PO, RI, TR, WA.

700 **Euphorbia dentata* Michx.
Common, naturalized; railroads, waste places. AS, AB, CL, CU, FA, GE, HL, KN, LK, LI, LR, PE, PC, PO, RO, ST, SU, TR.

701 **Euphorbia esula* L.   Wolf's-milk
Rare, naturalized; roadsides, waste places. RI: railroad right-of-way, Butler, Chuey & Shaffer 2919, 24 Aug 1979 (Youngstown State University Herbarium 17118).

702 **Euphorbia falcata** L.
Rare, adventive; waste places, railroads. RO: Mt. Logan, Bartley s.n., 13 Jun 1933 (OS 115455).

703 **Euphorbia humistrata** Engelm.
Rare; sandy fields, moist creek banks. RO: Colerain Twp., Bartley & Pontius s.n., 17 Aug 1933 (OS 15704).

704 **Euphorbia lathyris** L. Caper-spurge, Mole-plant
Rare, adventive; waste places. PO: Atwater, Woods s.n., 19 Sep 1924 (OS 38963).

705 **Euphorbia maculata** L. (including **E. supina** Raf.)
Common; roadsides, waste places, railroads. ALL COUNTIES.

706 **Euphorbia marginata** Pursh Snow-on-the-mountain
Infrequent, adventive; roadsides, fields. AB, CU, FA, HL, KN, LK, LR, PO, ST.

707 **Euphorbia nutans** Lag.
Common; railroads, roadsides, waste places. ALL COUNTIES.

708 **Euphorbia obtusata** Pursh
Rare; damp woods, thickets, fields. RO: cultivated field, Franklin Twp., Cusick 16693, 9 Jun 1971 (KE 44921); Harrison Twp., Bartley & Pontius s.n., 25 May 1933 (OS 15451). ST: boggy thicket, Lawrence Twp., Andreas 3823, 18 Jun 1979 (KE 44300).

709 **Euphorbia peplus** L. Petty Spurge
Frequent, adventive; cultivated fields, waste places. AB, CU, GE, LK, LR, MD, PO, SU.

710 **Euphorbia serpens** HBK. Roundleaf Spurge
Rare; weedy fields, waste places. PC: in lowland, Pickaway Twp., Hicks & Bartley s.n., 24 Oct 1955 (OS 52973). RO: abandoned road, Union Twp., Bartley s.n., 16 Oct 1964 (OS 77361); Scioto Twp., Bartley s.n., 7 Oct 1945 (OS 44725).

711 **Euphorbia vermiculata** Raf.
Frequent; waste places, fields, railroads, roadsides. CL, CU, GE, LK, LR MD, MO, PO, SU, TR, WA.

712 **Phyllanthus caroliniensis** Walter Carolina Leaf-flower
Rare; gravelly banks, sandy soils. RO: Springfield Twp., Bartley & Pontius s.n., 13 Aug 1933 (OS 15353).

FAGACEAE (Beech Family)

713 **Castanea dentata** (Marsh.) Borkh. Chestnut
Common, as basal sprouts; rich dry and mesic woods. All counties except PC.

714 **Fagus grandifolia** Ehrh. American Beech
Common; rich mesic and moist woods, ravines. ALL COUNTIES.

715 **Quercus alba** L. White Oak
Common; dry woods, eroding slopes, ridges, river bluffs. ALL COUNTIES.

716 **Quercus bicolor** Willd. Swamp White Oak
Common; wooded floodplains, swamps, stream margins. ALL COUNTIES. Braun (1961) indicates a record for RO.

717 **Quercus coccinea** Muenchh. Scarlet Oak
Common; dry woods. AS AB, CL, CU, FA, GE, HL, KN, LK, LI, LR, MH, PO, RI, RO, ST, SU, TR, WA.

718 **Quercus x exacta** Trel. **(Q. imbricaria x Q. palustris)**
Rare; sandy soil. AS: Ashland Co., Harwood s.n., 17 Oct 1954 (OS 51891). PO: sandy soil, Kent, Cooperrider 11146, 2 Aug 1970 (KE 22323).

719 **Quercus x hawkinsiae** Sudw. **(Q. rubra x Q. velutina)**
Rare; sandy woods. LR: roadside, Herrick s.n., 21 Jul 1960 (KE 808). PO: campus, Kent State University, Cooperrider & Cooperrider 12407, 20 Aug 1975 (KE 36925).

720 **Quercus imbricaria** Michx. Shingle Oak
Common; dry woods, fields, sandy knolls. All counties except AB.

721 **Quercus x leana** Nutt. **(Q. imbricaria x Q. velutina)**
Infrequent; dry woods, open slopes. CL, FA, KN, LI, PC, PO, SU, TR. Braun (1961) indicates a record for ST.

722 *Quercus macrocarpa* Michx.   Bur Oak
Common; moist woods, meadows, swamps. ALL COUNTIES.

723 *Quercus muehlenbergii* Engelm.   Chinquapin Oak
Common; dry woods, eroding slopes, wooded ridges. AS, CU, FA, GE, HL, KN, LI, LR,
MD, MO, PE, PC, PO, RO, ST, SU, WA.

724 *Quercus palustris* Muenchh.   Pin Oak
Common; wet floodplains, swamps, bog margins, thickets. ALL COUNTIES.

725 *Quercus prinus* L. (including *Q. montana* Willd.)   Rock Chestnut Oak
Common; dry rocky woods. AB, CL, CU, FA, GE, HL, KN, LK, MD, PE, PC, PO, RO,
ST, SU, TR, WA.

726 *Quercus rubra* L.   Red Oak
Common; rich woods, wooded ridges. All counties except FA. Braun (1961) indicates a record
for RI.

727 *Quercus stellata* Wang.   Post Oak
Rare; dry sterile soils, open woods. FA: campground, Greenfield Twp., Goslin s.n., 1 Jul
1959 (BHO 21468); Fairfield County, Kellerman s.n., Oct 1892 (OS 25759). RO: Green Twp.,
Bartley & Pontius s.n., 2 Sep 1935 (OS 25764); Mt. Logan, Chillicothe, Kellerman s.n., 12
Aug 1898 (OS s.n.).

728 *Quercus velutina* Lam.   Black Oak
Common; dry woods, ridges. ALL COUNTIES.

FUMARIACEAE (Fumitory Family)

729 *Adlumia fungosa* (Aiton) BSP.   Climbing Fumitory
Infrequent; cliffs, rocky woods, clearings. AS, CL, CU, GE, KN, LK, LR, PO, RI, SU.

730 *Corydalis flavula* (Raf.) DC.   Yellow Harlequin
Infrequent; rocky slopes, open sandy woods. AB, CL, FA, KN, PC, RO.

731 *Corydalis sempervirens* (L.) Persoon   Rock Harlequin
Frequent; rocky woods, cliffs, ledges. AB, CL, CU, FA, GE, KN, LK, LI, LR, PO, SU.

732 *Dicentra canadensis* (Goldie) Walp.   Squirrel-corn
Common; rich woods, wooded floodplains, stream terraces. All counties except CL, MH, PC.

733 *Dicentra cucullaria* (L.) Bernh.   Dutchman's Breeches
Common; rich woods, wooded floodplains, stream terraces. ALL COUNTIES.

734 *Fumaria officinalis* L.   Common Fumitory
Infrequent, adventive; waste places, fields. AB, CL, GE, LK, LR, PC.

GENTIANACEAE (Gentian Family)

735 *Bartonia virginica* (L.) BSP.   Yellow Bartonia
Frequent; sphagnous thickets, dry acidic fields. AB, CU, GE, LK, LI, LR, PO, RO, ST, SU,
TR.

736 *Centaurium pulchellum* (Swartz) Druce   Centaury
Rare, adventive; fields, roadsides, waste places. PO: roadside gravel, Windham Twp., Carr
1932, 14 Aug 1979 (KE 41776).

737 *Gentiana andrewsii* Griseb.   Bottle Gentian
Common; shaded roadside ditches, pond and stream margins, thickets. CL, LK, CU, FA,
GE, HL, KN, LK, LI, LR, MH, PO, RI, ST, SU, TR, WA.

738 # *Gentiana clausa* Raf.   Closed Gentian, Blind Gentian
Infrequent; moist openings, pond margins, thickets. AB, CL, CU, GE, LK, MH, ST, SU,
TR. Andreas & Cooperrider (1981) and Pringle (1967) indicate a record for PO.

739 *Gentiana crinita* Froel.   Fringed Gentian, Eastern Gentian
Rare; calcareous eroding slopes. AB: seepage area, Harpersfield Twp., Cusick 17596, 12 Sep
1977 (KE 39203). CU: moist slope, Valley View, Bissell 79:169, 19 Oct 1975 (CLM s.n.);
clay slope, Hunting Valley, Bissell 1982:17, 4 Oct 1982 (CLM 18422). RI: Lucas, Stair s.n.,
Sep 1898 (CLM 10009). SU: field, Boston Twp., Holeski s.n., 6 Sep 1967 (University of
Akron Herbarium s.n.).

740 *Gentiana procera* Holm   Great Plains Gentian
Infrequent; fens, calcareous thickets. AB, CU, FA, GE, LK, LR, PC, PO, RO, ST, SU.

75

741 *Gentiana quinquefolia* L. var. *occidentalis* (Gray) Hitchc. Stiff Gentian
Frequent; moist slopes, woodland openings, roadside banks. CU, FA, GE, KN, LK, LR, MD, PC, PO, RO, SU, WA.

742 *Obolaria virginica* L. Pennywort
Infrequent; rich mesic and moist woods. AB, CU, FA, HL, LK, LR, PO, RO, SU.

743 *Sabatia angularis* (L.) Pursh Marsh-pink, Bitter-bloom
Frequent; dry to moist fields, eroding clay slopes, open woods, roadsides. AS, AB, CU, FA, HL, KN, LI, MH, PE, PC, PO, RO, SU, TR, WA.

744 *Swertia caroliniensis* (Walter) Kuntze American Columbo
Infrequent; dry woods, fields. AB, GE, HL, LI, MD, PO, RO, SU.

GERANIACEAE (Geranium Family)

745 **Erodium cicutarium* (L.) L'Her. Pin-clover, Stork's-bill
Rare, adventive; waste places, roadsides. LI: Granville, Jones s.n., 27 May 1890 (OC 79668). PO: cinders, Kent, Cooperrider 5431, 26 May 1960 (KE 1851).

746 *Geranium bicknellii* Britton Bicknell's Geranium
Rare; open woods, clearings, roadsides, waste places. LR: railroad, Oberlin, Koford s.n., 23 Jun 1891 (OC 74021).

747 *Geranium carolinianum* L.
Infrequent; dry rocky woods, fields, waste places. AB, CU, FA, HL, LK, LI, LR, PC, RO, WA.

748 **Geranium dissectum* L.
Rare, adventive; waste places, roadsides. FA: Fairfield County, Bartley s.n., 25 Jul 1950 (OS 114753).

749 *Geranium maculatum* L. Wild Geranium
Common; rich woods, wooded floodplains, pastures. ALL COUNTIES.

750 **Geranium molle* L. Dove's-foot Cranesbill
Infrequent, adventive; lawns, waste places. AB, CU, LK, LR, PO.

751 **Geranium pusillum* L.
Infrequent, naturalized; lawns, waste places. AB, CU, GE, FA, LK, LR, MH, PO, RO, SU, TR.

752 *Geranium robertianum* L. Herb-Robert
Frequent; rich woods, roadside banks, rocky slopes. AB, CL, CU, GE, HL, LK, LR, MD, PO, RI, SU, TR, WA.

753 **Geranium sanguineum* L. Blood-red Cranesbill
Rare, naturalized; waste places, roadsides. SU: cemetery, Twinsburg Twp., Herrick s.n., 15 Jun 1955 (OS 54001).

GUTTIFERAE / CLUSIACEAE (St. John's-wort Family)

754 # *Hypericum boreale* (Britton) Bickn. Northern St. John's-wort
Once rare, now presumably extirpated from Ohio; damp sands, sphagnous mats. GE: Geauga Lake, Jennings s.n., 22 Aug 1903 (OS 16251). WA: Browns Lake, Hopkins 111, 20 Jul 1911 (OS 42445); Fox Lake, Selby & Duvel 798, 27 Jul 1899 (OS 49570); Doner's Lake, Selby s.n., 20 Sep 1899 (OS 16247).

755 *Hypericum canadense* L. Canadian St. John's-wort
Rare; pond margins, slopes. AB: pond, Orwell Twp., Cusick 18850, 21 Sep 1978 (OS s.n.).

756 *Hypericum drummondii* (Grev. & Hooker) T. & G. Nits-and-lice
Infrequent; dry rocky slopes, sandy soils. AB, CU, FA, LK, LR, PC, PO, RO, ST, SU.

757 # *Hypericum ellipticum* Hooker Few-flowered St. John's-wort
Rare; stream margins, damp sandy soil. AB: muddy shore, Harpersfield Twp., Bissell 1985:170, 28 Jun 1985 (CLM 21768).

758 *Hypericum gentianoides* (L.) BSP. Orange-grass, Pineweed
Infrequent; dry sandy soils. AS, CL, FA, LK, PO, RO.

759 *Hypericum hypericoides* (L.) Crantz   St. Andrew's Cross
Ascyrum hypericoides L.
Rare; dry sandy soils. FA: Amanda, Kellerman s.n., Aug 1892 (OS 16380). PE: dry field,
Pleasant Twp., Cusick 11068, 20 Jul 1970 (KE 27855). RO: Colerain Twp., Bartley & Pontius
s.n., 29 Oct 1929 (OS 16370).

760 *Hypercium kalmianum* L.   Kalm's St. John's-wort
Rare; rocky calcareous soils. SU: Silver Lake, Krebs & Claassen s.n., 1891 (OS 16072).

761 *Hypericum majus* (Gray) Britton   Tall St. John's-wort
Rare; open woods, waste places. AS: weedy field, Emmitt 2109, 16 Jul 1980 (KE 45252).
LR: Camden Bog, Kipton, Metcalf s.n., 2 Aug 1887 (OS 16266). PO: boggy woods, Shalersville
Twp., Andreas 2694, 18 Aug 1978 (KE 41956). SU: cinders, Akron, Carr 1976, 15 Aug 1979
(KE 45053).

762 *Hypericum mutilum* L.
Common; wet pastures, pond margins, waste places. ALL COUNTIES.

763 **Hypericum perforatum* L.   Common St. John's-wort
Common, naturalized; fields, roadsides, waste places. All counties except PC.

764 *Hypericum prolificum* L.   Shrubby St. John's-wort
Hypericum spathulatum (Spach) Steud.
Common; dry and moist roadside banks, thickets, pastures. AS, AB, CL, CU, FA, GE, HL,
KN, LK, LI, MH, MD, PE, PO, RO, ST, SU, TR, WA.

765 *Hypericum punctatum* L.   Dotted St. John's-wort
Common; weedy thickets, roadsides, woodland borders. ALL COUNTIES.

766 *Hypericum pyramidatum* Aiton   Great St. John's-wort
Common; weedy floodplains, sedge meadows, pastures, thickets. AS, AB, CU, GE, HL, KN,
LK, LI, LR, MD, PO, RI, ST, SU, TR, WA.

767 *Hypericum sphaerocarpum* Michx.
Rare; rocky woods. PC: Pickaway Twp., Bartley & Pontius s.n., 23 Aug 1930 (OS 16144).
RO: north Union Twp., Bartley & Pontius s.n., 4 Aug 1929 (OS 16142); Colerain Twp.,
Bartley & Pontius s.n., 23 Aug 1929 (OS 16143).

768a *# Hypericum virginicum* L. var. *virginicum*   Marsh St. John's-wort
Frequent; sphagnous bogs, buttonbush swamps, thickets. AB, CL, CU, GE, HL, LI, LR,
PO, RI, ST, SU, WA.

768b *Hypericum virginicum* L. var. *fraseri* (Spach) Fern.   Marsh St. John's-wort
Infrequent; peaty soils, swampy woods. AS, AB, CL, CU, FA, LK, MH, PO, ST, WA.

HALORAGACEAE (Water-milfoil Family)

769 *Myriophyllum heterophyllum* Michx.   Two-leaved Water-milfoil
Rare; ponds, streams. TR: Mosquito Creek, Howland Twp., Rood & Shanks 1059, 4 Jul
1935 (KE 12742).

770a **Myriophyllum spicatum* L. var. *spicatum*   Water-milfoil
Common, naturalized; ponds, lakes, slow-moving rivers. All counties except PC, RO.

770b *Myriophyllum spicatum* L. var. *squamosum* Laestad   American Water-milfoil
Myriophyllum exalbescens Fern.
Rare; neutral to calcareous water in natural lakes. HL: shallow water, Washington Twp.,
Wilson 2402, 11 Aug 1971 (KE 29594). PO: shallow water, Suffield Twp., Andreas 35, 13
Aug 1968 (KE 20266); Brady Lake, Franklin Twp., Webb 1249, 29 July 1913 (KE 12741).
ST: Congress Lake, Lake Twp., Roberts 1972, 8 Sep 1971 (OS 106691).

771 *Proserpinaca palustris* L. var. *crebra* Fern. & Grisc.   Mermaid-weed
Frequent; mud flats, marshes. AB, CU, GE, LK, MH, PO, ST, SU, TR, WA.

HAMAMELIDACEAE (Witch-hazel Family)

772 *Hamamelis virginiana* L.   Witch-hazel
Common; dry and moist woods, thickets, pastures. ALL COUNTIES.

HIPPOCASTANACEAE (Buckeye Family)

773 *Aesculus glabra* Willd.   Ohio Buckeye
Common; rich floodplains, stream margins, wet ravines. All counties except AB, GE.

774 *Aesculus hippocastanum* L.   Horse-chestnut
Infrequent, adventive; abandoned homesites, waste places. FA, MD, PC, PO, RO.

775 *Aesculus octandra* Marsh.   Yellow Buckeye
Rare; rich woods. RO: low terrace, Liberty Twp., Beatley 19831, 6 May 1958 (OS 59876).

HYDROPHYLLACEAE (Waterleaf Family)

776 *Hydrophyllum appendiculatum* Michx.   Waterleaf
Common; rich woods, ravines. AS, CL, CU, GE, HL, KN, LK, LI, LR, MD, MO, PE, PC, PO, RI, RO, SU, WA.

777 *Hydrophyllum canadense* L.   Canadian Waterleaf
Common; rich woods. All counties except FA, PE, PC.

778 *Hydrophyllum macrophyllum* Nutt.
Infrequent; rich damp woods. FA, LI, PE, RO, WA.

779 *Hydrophyllum virginianum* L.   John's Cabbage
Common; rich woods, wooded floodplains. ALL COUNTIES.

780 *Phacelia dubia* (L.) Trel.   Small-flowered Scorpion-weed
Once rare, now presumably extirpated from Ohio; rich woods, thickets. FA: Clark's Crossing, Griggs s.n., 8 May 1904 (OS 45518).

781 *Phacelia purshii* Buckl.   Miami-mist
Common; floodplains, moist woods, wet meadows. AS, AB, CL, CU, FA, GE, HL, KN, LI, LR, MD, PE, PC, PO, RI, RO, ST, SU, WA.

JUGLANDACEAE (Walnut Family)

782 *Carya cordiformis* (Wang.) K. Koch   Bitternut Hickory
Common; rich dry and mesic woods, hemlock slopes, pastures. ALL COUNTIES.

783 *Carya glabra* (Miller) Sweet   Pignut
Common; dry woods, wooded slopes, pastures. All counties except MO, PC.

784 *Carya laciniosa* (Michx. f.) G. Don   Big Shellbark Hickory
Frequent; rich mesic and moist woods. AS, AB, CL, CU, FA, HL, KN, LK, LI, LR, MO, PE, PO, RI, TR.

785 *Carya ovalis* (Wang.) Sarg.   Sweet Pignut
Common; rich mesic and dry woods. CL, FA, GE, HL, KN, LK, LR, MD, PE, PO, ST, SU, TR, WA.

786 *Carya ovata* (Miller) K. Koch   Shagbark Hickory
Common; dry and mesic woods. ALL COUNTIES.

787 *Carya tomentosa* Nutt.   Mockernut Hickory
Common; dry and moist woods. CL, CU, FA, GE, HL, KN, LK, LR, MH, MD, PE, PC, PO, RI, RO, ST, SU, TR, WA.

788 *Juglans cinerea* L.   Butternut
Common; wooded floodplains, roadsides, thickets. All counties except PC.

789 *Juglans nigra* L.   Black Walnut
Common; moist thickets, rich woods. ALL COUNTIES.

LABIATAE / LAMIACEAE (Mint Family)

790 *Agastache nepetoides* (L.) Kuntze   Yellow Giant Hyssop
Infrequent; thickets, woodland borders. AS, CU, FA, HL, KN, LK, LI, LR, PC, RO.

791 *Agastache scrophulariaefolia* (Willd.) Kuntze   Purple Giant Hyssop
Infrequent; rich woods, woodland borders, thickets. AB, CL, FA, KN, LI, MD, RO, ST, SU, TR.

792 *Ajuga reptans* L.   Creeping Bugleweed
Frequent, naturalized; lawns, abandoned homesites, roadsides, fields. AB, CL, CU, GE, LK, LI, PO, RO, ST, SU, TR.

793 *Blephilia ciliata* (L.) Benth.
Common; thickets, woodland borders. AS, AB, CL, CU, FA, HL, KN, LK, LI, LR, MH, MD, PE, PC, ST, TR, WA.

78

794 *Blephilia hirsuta* (Pursh) Benth.   Wood-mint
Common; moist roadsides, moist woods, wooded floodplains, woodland borders. All counties
except AB, PC, TR.

795 *Clinopodium vulgare* L.   Basil
Satureja vulgaris (L.) Fritsch
Common; roadsides, fields, woodland borders. All counties except MO, PC, RO.

796 *Collinsonia canadensis* L.   Richweed
Common; rich woods, wooded floodplains, shaded roadside banks. ALL COUNTIES.

797 ***Dracocephalum parviflorum*** Nutt.   Dragonhead
Rare, adventive; rocky slopes, waste places. PC: railroad, Circleville Twp., Bartley & Hicks
s.n., 14 May 1955 (OS 52969). TR: railroad ballast, Braceville Twp., Rood 1628, 4 Jul 1924
(KE 11315).

798 ***Galeopsis tetrahit*** L.   Hemp-nettle
Rare, adventive; roadsides, waste places, cultivated fields. AB: near Plymouth, Hicks s.n.,
9 Aug 1932 (OS 34457). KN: shady road, Brown Twp., Pusey 559, 25 Jul 1975 (KE 38494).
PO: Garrettsville, Webb 1112, 31 Jul 1924 (KE 11392).

799 ***Glecoma hederacea*** L.   Gill-over-the-ground
Common, naturalized; waste places, roadsides, rich woods, lawns. ALL COUNTIES.

800 *Hedeoma hispida* Pursh   Rough Pennyroyal
Rare; railroads, sandy soils. LR: Lorain County, Kelsey s.n., 5 Jun 1896 (OC 34081 & OC
75363). RO: Green Twp., Bartley & Pontius s.n., 18 May 1932 (OS 46312).

801 *Hedeoma pulegioides* (L.) Persoon   Pennyroyal
Common; dry wooded hillsides, ditches, waste places. All counties except RI, MO.

802 *Isanthus brachiatus* (L.) BSP.   False Pennyroyal
Infrequent; dry eroding banks, railroads, fields. CL, FA, LK, RO, ST.

803 ***Lamium amplexicaule*** L.   Henbit
Common, naturalized; roadsides, waste places. AS, CL, CU, FA, GE, HL, KN, LK, LI, MH,
PE, PC, PO, RO, ST, SU, TR, WA.

804 ***Lamium maculatum*** L.   Spotted Henbit
Rare, adventive; roadsides, waste places. AB, KN, LI, LR, MH, PO.

805 ***Lamium purpureum*** L.   Purple Henbit
Common, naturalized; fields, roadsides, waste places. All counties except PC.

806 ***Leonurus cardiaca*** L.   Common Motherwort
Common, naturalized; roadsides, thickets, woodland borders. ALL COUNTIES.

807 *Lycopus americanus* Muhl.   American Bugleweed
Common; pond and lake margins, fens, sphagnous thickets, ditches. All counties except MH.

808 *Lycopus rubellus* Moench
Infrequent; marshes, stream margins. AS, AB, FA, LK, LI, LR, PO, TR, SU.

809 *Lycopus virginicus* L. (including *L. uniflorus* Michx.)   Water-horehound
Common; marshes, stream banks, fens, wet meadows, sphagnous thickets. ALL COUNTIES.

810 ***Marrubium vulgare*** L.   Common Horehound
Infrequent, adventive; waste places. AB, FA, HL, LK, LI, LR, PC.

811 *Meehania cordata* (Nutt.) Britton   Meehania
Rare; rich woods, shaded roadside banks. MH: wet woods, Berlin Center, Vickers 1187, 20
May 1915 (KE 11309). PO: fen thicket, Streetsboro Twp., Steward 220, 15 Sep 1985 (Miami
University Herbarium s.n.).

812 ***Melissa officinalis*** L.   Common Balm
Infrequent, adventive; fields, roadsides, woods. AB, CU, FA, GE, LK, LI, LR, PC, ST, WA.

813 *Mentha arvensis* L. var. *glabrata* (Benth.) Fern.
Common; stream and river margins, swamps, wet fields. All counties except MO, PC.

814 ***Mentha longifolia*** (L.) Hudson   Horse-mint
Rare, adventive; thickets, roadsides. KN: open field, Pike Twp., Pusey 1007, 4 Sep 1974
(KE 38483). PC: marshy area, Pickaway Twp., Cooperrider, Sabo & Bartley 9412, 16 Sep
1964 (KE 8476).

815a ***Mentha** x **piperita** L. var. **piperita** **(M. aquatica** x **M. spicata)** Peppermint
Mentha piperita L.
Common, naturalized; wet ditches, stream and pond margins, marshes. All counties except
MO, PE, RO.

815b ***Mentha** x **piperita** L. var. **citrata** (Ehrh.) Briq. **(M. aquatica** x **M. spicata)**
Mentha citrata Ehrh.
Rare, naturalized; pond margins, wet ditches. LR: pastures, Oberlin, Grover s.n., 15 Sep
1902 (CLM 10884). PO: Lake Brady, Brown 1814, 10 Oct 1939 (OS 33583).

816 ***Mentha rotundifolia** (L.) Hudson
Rare, adventive; roadsides, fields, thickets. CL: swamp border, Perry Twp., Carbone 11, 12
Aug 1975 (Youngstown State University Herbarium 11661). PO: near campus, Franklin
Twp., Uguoue 7, 13 Aug 1975 (Youngstown State University Herbarium 11364). RO: Paxton
Twp., Bryant 1093, 31 Aug 1980 (Miami University Herbarium 129000).

817 ***Mentha spicata** L. Spearmint
Common, naturalized; wet ditches, roadside banks, woodland margins. All counties except
AB, MO, PE.

818 ***Mentha** x **villosa** Hudson var. **alopecuroides** (Hull) Harley
Mentha alopecuroides Hull
Rare, adventive; roadsides, waste places, pond margins. GE: gravel talus, Thompson Twp.,
Bissell 1983:154, 3 Sep 1983 (CLM 20906). HL: pond margin, Washington Twp., Wilson
2664, 22 Sep 1971 (KE 29564).

819 **Monarda didyma** L. Bee-balm, Oswego-tea
Frequent; wooded floodplains, roadsides, thickets, rich woods. AB, CL, CU, FA, GE, HL,
LK, MH, MD, PO, ST, SU, TR.

820 **Monarda fistulosa** L. (including **M. clinopodia** L.)
Common; dry hillsides, open woods, thickets, roadsides. ALL COUNTIES.

821 **Monarda** x **media** Willd. **(M. didyma** x **M. fistulosa)**
Monarda media Willd.
Infrequent; dry woods, thickets. AB, GE, MH, PO, ST.

822 ***Nepeta cataria** L. Catnip
Common, naturalized; rich woods, roadside thickets, railroads. All counties except AB.

823 ***Origanum vulgare** L. Wild Marjoram
Rare, naturalized; dry fields, roadsides. PO: roadside, Kent, Suster 6, 3 Sep 1984 (KE 46526).
WA: dry hillside, Chester Twp., Andreas 1286, 19 Jul 1977 (KE 42412).

824 ***Perilla frutescens** (L.) Britton
Rare, adventive; waste places. AB: Hartsgrove, Hicks s.n., 20 Aug 1928 (OS 34355). MH:
roadside, Canfield, Moor s.n., 4 Sep 1910 (Youngstown State University Herbarium 4123).

825 **Physostegia virginiana** (L.) Benth. False Dragonhead
Common, native at some sites, naturalized elsewhere (Cooperrider, 1985a); roadside ditches,
stream banks, thickets, prairie remnants. AB, CU, FA, GE, HL, KN, LK, LI, LR, MD, PE,
PC, PO, RI, RO, ST, SU, WA.

826 **Prunella vulgaris** L. Self-heal, Heal-all
Common; roadsides, waste places, lawns, fields. ALL COUNTIES.

827 **Pycnanthemum incanum** (L.) Michx.
Infrequent; dry woods, thickets. AS, AB, CU, FA, HL, LK, LI, PE, PO, RO, ST, SU, WA.

828 # **Pycnanthemum muticum** (Michx.) Persoon Blunt Mountain-mint
Infrequent; moist roadside ditches, fields, thickets, fens. AB, CU, GE, LK, MD, PO, SU.

829 **Pycnanthemum tenuifolium** Schrader Slender-leaved Mountain-mint
Common; fields, thickets, fens. AB, CL, CU, FA, GE, HL, KN, LK, LI, LR, MD, MO, PE,
PC, PO, ST, SU, TR.

830 **Pycnanthemum verticillatum** (Michx.) Persoon (including **P. pilosum** Nutt.)
Whorled Mountain-mint
Frequent; dry thickets, clearings. AB, CU, FA, GE, HL, LI, PE, PO, ST, SU, TR.

831 **Pycnanthemum virginianum** (L.) Durand & Jackson
Common; moist fields, thickets, fens. CL, CU, FA, GE, HL, KN, LI, LR, MH, PC, PO,
RO, ST, SU, TR, WA.

80

832 **Salvia lyrata** L.   Lyre-leaved Sage, Cancer Weed
Rare, adventive; sandy woods, thickets. PO: university woods, Kent, Ogg 538, 16 Aug 1933 (KE 953).

833 **Salvia officinalis** L.   Scarlet Salvia
Rare, adventive; waste places. AB: near Colebrook, Hicks 1875, 4 Sep 1931 (OS 70707).

834 **Salvia pratensis** L.
Rare, adventive; waste places, fields. PO: grassy area, Brimfield Twp., Cooperrider 11201, 27 May 1973 (KE 34110). PC: pasture field, Circleville Twp., Bartley s.n., 5 Jun 1948 (BHO 19993).

835 **Salvia reflexa** Hornem.
Rare, naturalized; dry banks, woodland borders. PC: cornfield, Pickaway Twp., Hicks & Bartley s.n., 30 Oct 1955 (OS 53031). RO: north Union Twp., Bartley & Pontius s.n., no date (OS 34721).

836 **Scutellaria epilobiifolia** A. Hamilton   Common Skullcap
Common; sphagnous thickets, marshes, sedge meadows, pond and lake margins. AS, AB, CL, CU, FA, GE, HL, KN, LK, LI, LR, MH, MO, PE, PC, PO, ST, SU, WA.

837 **Scutellaria incana** Biehler   Hoary Skullcap
Common; rich woods, shaded roadside banks. All counties except MH, MO.

838 **Scutellaria lateriflora** L.   Mad-dog Skullcap
Common; wet thickets, fens, marshes, ditches. ALL COUNTIES.

839 **Scutellaria nervosa** Pursh (including **S. nervosa** var. **calvifolia** Fern.)
Infrequent; moist woods, thickets. AS, CL, LI, LR, MD, PE, PC, RO, TR, WA.

840 **Scutellaria ovata** Hill var. **versicolor** (Nutt.) Fern.
Rare; rocky woods, creek banks. LR: Lorain, Day s.n., 29 Jun 1892 (OS 33375). RO: Mt. Logan, Chillicothe, Kellerman s.n., 12 Aug 1898 (OS 3391); Green Twp., Bartley & Pontius s.n., 6 Jun 1935 (OS 51090).

841a **Scutellaria parvula** Michx. var. **parvula**
Rare; sandy soils, thickets. RO: north of Hallsville, Colerain Twp., Bartley s.n., 30 May 1948 (OS 44102).

841b **Scutellaria parvula** Michx. var. **leonardii** (Epling) Fern.
Infrequent; dry soils, sandy cliffs, dry woods. FA, HL, KN, PC, WA.

842 **Stachys aspera** Michx.
Rare, naturalized; dry woods, roadsides. SU: woods, Boston Twp., Herrick s.n., 28 Jul 1958 (KE 2676).

843 **Stachys germanica** L.
Rare, adventive; waste places, fields, roadsides. RO: parking lot, Colerain Twp., Bartley & Hicks s.n., 29 May 1955 (OS 52982). ST: Congress Lake, Kellerman s.n., 28 Jun 1898 (OS 34729).

844 **Stachys nuttallii** Shuttlew.   Hedge-nettle
Rare; moist thickets, floodplains. LR: Oberlin, Kelsey s.n., 29 Jun 1896 (OS 34343).

845 **Stachys palustris** L.   Woundwort
Infrequent; ditches, wet roadsides, meadows. AB, FA, KN, LK, PO.

846 **Stachys tenuifolia** Willd. (including **S. hispida** Pursh)   Hedge-nettle
Common; marshes, fields, roadsides, weedy floodplains, railroads. All counties except MO, RI.

847 **Synandra hispidula** (Michx.) Britton   Synandra
Rare; moist woods, floodplains, stream terraces. RO: bluff, Green Twp., Bartley s.n., 10 May 1946 (OS 44051).

848a **Teucrium canadense** L. var. **canadense**   American Germander
Common; fields, roadsides, thickets. All counties except MO, PC.

848b **Teucrium canadense** L. var. **occidentale** (Gray) McClintock & Epling
Infrequent; fields, roadsides, thickets. AB, LK, LI, PE, PC.

849 **Thymus serpyllum** L.   Creeping Thyme
Rare, naturalized; fields, lawns, waste places. GE, LK, RO, ST, TR.

850a **_Trichostema dichotomum_** L. var. **_dichotomum_**   Bluecurls
Infrequent; dry sandy soils, railroads. AB, FA, KN, LI, PE.

850b **_Trichostema dichotomum_** L. var. **_lineare_** (Walter) Pursh
Rare; dry sandy soils. FA: Hocking Twp., Goslin s.n., 2 Sep 1935 (OS 62320); Kettle Hills, Goslin s.n., 23 Aug 1931 (OS 33191).

LAURACEAE (Laurel Family)

851 **_Lindera benzoin_** (L.) Blume   Spicebush
Common; rich woods, wooded floodplains, thickets. ALL COUNTIES.

852 **_Sassafras albidum_** (Nutt.) Nees   Sassafras
Common; woodland borders, mesic woods, thickets, old fields. ALL COUNTIES.

LEGUMINOSAE / FABACEAE (Pea Family)

853 **_Amorpha fruticosa_** L.   False Indigo, Indigo-bush
Infrequent; dry and moist roadside banks, fields, shrubby floodplains. AS, CU, GE, LK, LR, PO, SU, WA.

854 **_Amphicarpaea bracteata_** (L.) Fern.   Hog-peanut
Common; open woods, woodland borders, fens. All counties except PC.

855 **_Apios americana_** Medicus   Ground-nut, Wild Bean
Common; moist thickets, fens, roadsides. All counties except RI.

856 **_Astragalus canadensis_** L.   Canada Milk-vetch
Rare; thickets, shores, railroads. RO: railroad yard, Chillicothe, Chapman & Hicks s.n., 11 Jun 1932 (OS 22429).

857 **_Baptisia lactea_** (Raf.) Thieret   Prairie False Indigo
Baptisia leucantha T. & G.
Rare; woods, bluffs, prairie remnants. MD: wooded ravine, Hickley Twp., Mutz 647, 20 Aug 1965 (KE 19324). RO: Paxton Twp., Bartley & Pontius·s.n., 21 Nov 1930 (OS 21908).

858 **_Baptisia tinctoria_** (L.) R. Br.   Wild Indigo, Rattleweed
Infrequent; dry sandy woods, woodland borders, barren sandy fields, eroding slopes. AB, CU, FA, GE, LK, LR, PO, ST, SU, TR.

859 **_Cassia chamaecrista_** L.   Partridge-pea
Cassia fasciculata Michx.
Infrequent; sandy soils, railroads, roadsides. AB, CU, FA, LK, PC, PO, RO, ST, SU.

860 **_Cassia hebecarpa_** Fern.   Wild Senna
Infrequent; gravelly soils, roadsides, thickets, floodplains. AS, FA, HL, LK, LI, LR, PO, ST, SU, WA.

861 **_Cassia marilandica_** L.   Wild Senna
Infrequent; dry roadsides, thickets, railroads. CU, FA, LI, RO, ST, TR, WA.

862 **_Cassia nictitans_** L.   Wild Sensitive Plant
Infrequent; roadsides, fields, eroding slopes. FA, HL, KN, LI, PE, PC, RO.

863 **_Cercis canadensis_** L.   Redbud
Frequent, native in southern counties, adventive elsewhere; roadsides, floodplains, fields, woodland borders. FA, GE, HL, KN, LK, LI, PE, PC, RO, ST, SU.

864 **_Clitoria mariana_** L.   Butterfly-pea
Rare; dry slopes. FA: Kettle Hills, Thomas & Goslin s.n., 18 Jul 1933 (OS 23216 & BHO 21737).

865 ***_Coronilla varia_** L.   Crown-vetch
Frequent, naturalized; eroding roadside banks, dry slopes. AS, CL, CU, GE, HL, KN, LK, LR, MD, PO, ST, SU, WA.

866 **_Desmanthus illinoensis_** (Michx.) MacM.   Prairie-mimosa, Prickle-weed
Rare; prairies. AB: Ashtabula County, South s.n., 1901 (OS 21696).

867 **_Desmodium canadense_** (L.) DC.   Canada Tick-trefoil
Frequent; dry fields, open woods. AB, CU, GE, LK, LI, LR, MH, MD, PO, ST, SU, TR.

868 **_Desmodium canescens_** (L.) DC.   Hoary Tick-trefoil
Frequent; dry woods, fields. AS, AB, CU, FA, HL, KN, LK, LI, LR, PC, RI, RO, SU, WA.

869 *Desmodium ciliare* (Willd.) DC.
Infrequent; sandy woods, clearings. AS, FA, HL, LI, LR, PC, PO, RO, SU.

870 *Desmodium cuspidatum* (Muhl.) Loudon
Infrequent; dry woods, sandy openings. AB, CU, HL, LK, LR, PO, RO, SU.

871 *Desmodium glutinosum* (Muhl.) Wood
Common; dry and mesic woods, woodland borders. AS, AB, CU, GE, HL, KN, LK, LI, LR, MD, MO, PE, PC, PO, RI, ST, SU, TR, WA.

872 *Desmodium illinoense* Gray   Prairie Tick-trefoil
Rare; dry sandy fields, prairie remnants. CL: weedy ground, Salem Twp., Cooperrider 7783, 5 Aug 1960 (KE 4198). LI: near Buckeye Lake, Moore 1275, Aug 1939 (BHO 8238). RO: Colerain Twp., Bartley s.n., 18 Aug 1932 (BHO 18937).

873 *Desmodium laevigatum* (Nutt.) DC.   Smooth Tick-trefoil
Infrequent; dry slopes, sandy woods. CU, FA, LI, RO, SU, WA.

874 *Desmodium marilandicum* (L.) DC.
Rare; dry woods, clearings. FA: Hocking Twp., Goslin s.n., Aug 1934 (OS 22696). HL: open woods, Washington Twp., Cooperrider 8123, 19 Aug 1960 (KE 4191). RO: Colerain Twp., Bartley & Pontius s.n., 18 Aug 1932 (OS 22690).

875 *Desmodium nudiflorum* (L.) DC.
Common; moist, mesic and dry woods, stream terraces. AS, AB, CU, FA, GE, HL, KN, LK, LI, LR, MH, MD, PC, PO, RI, ST, SU, TR, WA.

876 *Desmodium obtusum* (Willd.) DC.
Desmodium rigidum (Ell.) DC.
Rare; sandy woods, roadside banks. LI: Toboso, Jones s.n., 19 Aug 1890 (OC 79533). RO: Colerain Twp., Bartley & Pontius s.n., 18 Aug 1932 (OS 22682).

877 *Desmodium paniculatum* (L.) DC.
Common; dry fields, woodland borders, clearings, railroads, pastures. All counties except MO.

878 *Desmodium pauciflorum* (Nutt.) DC.   Few-flowered Tick-trefoil
Rare; mesic woods, wooded slopes. FA: Greenfield Twp., Goslin s.n., 25 Aug 1935 (OS 22727).

879 *Desmodium perplexum* Schub.
Frequent; railroads, disturbed areas, sandy roadside banks. AS, AB, CU, FA, GE, LK, LI, LR, MD, PO, RI, RO, ST, SU, WA.

880 *Desmodium rotundifolium* DC.   Round-leaved Tick-trefoil
Common; dry roadside banks, woodland borders, eroding slopes, open woods. AS, AB, CL, CU, FA, HL, KN, LK, LI, LR, PE, PC, RO, ST, SU, WA.

881 *Desmodium sessilifolium* (Torrey) T. & G.   Sessile Tick-trefoil
Rare; dry sandy soils. PC: Deercreek Twp., Bartley & Pontius s.n., 18 Sep 1937 (OS 22714).

882 *Desmodium viridiflorum* (L.) DC.   Velvety Tick-trefoil
Infrequent; dry woods, roadsides, moist thickets. FA, LR, MD, PE, PC, PO, RO, ST.

883 *Gleditsia triacanthos* L.   Honey-locust
Common; roadside thickets, fields, pastures. ALL COUNTIES.

884 *Glycine max* (L.) Mer.   Soybean
Rare, adventive; abandoned fields, roadsides, waste places. FA: railroad, Hocking Twp., Goslin s.n., 26 Aug 1945 (BHO 21745). KN: landfill, Morris Twp., Pusey 706, 12 Sep 1975 (KE 38451). MD: pasture, Harrisville Twp., Mutz 198, 16 Jul 1964 (KE 15330).

885 *Gymnocladus dioica* (L.) K. Koch   Kentucky Coffee-tree
Infrequent, native in southern counties, adventive elsewhere; rich woods. AS, KN, LI, LR, MD, RI, RO, ST.

886 *Lathyrus latifolius* L.   Perennial Pea
Frequent, adventive; roadsides, railroads, waste places. AS, AB, CU, GE, HL, KN, LK, LI, MH, MD, PE, PO, ST, SU, TR.

887 *Lathyrus ochroleucus* Hooker   Yellow Vetchling
Infrequent; dry roadside banks, eroding slopes. AS, AB, CU, LK, LR, TR.

888 *Lathyrus odoratus* L.   Sweet Pea
Rare, adventive; waste places. WA: roadside ditch, Franklin Twp., Andreas 1181, 8 Jul 1977 (KE 49682).

83

889 *Lathyrus palustris* L.   Marsh Vetchling
Frequent; boggy thickets, fens, marshes. AB, CU, GE, HL, LK, LI, LR, PC, PO, RI, RO, ST, SU, WA.

890 **Lathyrus pratensis* L.   Yellow Vetchling
Rare, adventive; wet slopes, meadows, roadsides. WA: Blachleysville, Duvel 835, 13 Oct 1899 (OS 98049).

891 **Lathyrus tuberosus* L.   Tuberous Vetchling
Rare, adventive; fields, roadsides. LR: cultivated field, Henrietta Twp., Jones s.n., 12 Jul 1964 (OC 207130). TR: Canfield, Longnecker s.n., Summer 1930 (OS 23200).

892 *Lespedeza capitata* Michx.   Bush-clover
Frequent; dry sandy fields, woodland borders, roadsides. AS, CU, FA, HL, KN, LK, PC, RO, ST, SU, WA.

893 **Lespedeza cuneata* (Dumont.) G. Don
Rare, naturalized; fields, roadsides. PE: lake bank, Cline 433, 23 Jul 1974 (KE 43546). PO: dry field, Kent, Theiss s.n., 10 Sep 1985 (KE 48258). ST: field, Plain Twp., Amann 2046, 8 Oct 1960 (KE 3750).

894 *Lespedeza hirta* (L.) Hornem.   Pubescent Bush-clover
Common; dry banks, open woods, slopes, fields. AS, AB, CL, CU, FA, GE, HL, KN, LK, LI, LR, PE, PO, RI, RO, ST, SU, TR, WA.

895 *Lespedeza intermedia* (S. Watson) Britton
Common; dry open woods, roadsides, thickets. All counties except MD, MO.

896 *Lespedeza nuttallii* Darl.   Nuttall's Bush-clover
Infrequent; dry woods, openings. CL, CU, FA, LK, RO.

897 *Lespedeza procumbens* Michx.   Trailing Bush-clover
Frequent; dry sandy fields, woodland borders. AB, CL, FA, KN, LI, LR, PE, PC, PO, RO, ST, WA.

898 *Lespedeza repens* (L.) Bart.   Creeping Bush-clover
Infrequent; sandy woods, thickets, rocky slopes. AS, AB, CL, CU, FA, HL, KN, LK, PE, RO.

899 **Lespedeza striata* (Thunb.) H. & A.   Japanese Clover
Rare, adventive; roadsides, dry fields. FA: Madison Twp., Rea s.n., 28 Aug 1938 (OS 23044). PE: open upland woods, Reading Twp., Cline 1246, 6 Sep 1975 (KE 43542). PC: Pickaway, Schwinn s.n., 22 Sep 1941 (OS 23036). RO: Scioto River, Bartley & Pontius s.n., 2 Sep 1928 (OS 23032); Stoney Creek, Crowl s.n., 30 Aug 1937 (OS 23033).

900 *Lespedeza violacea* (L.) Persoon
Infrequent; dry open woods, sandy slopes. AB, CU, FA, HL, KN, LK, LR, PC, PO, WA.

901 *Lespedeza virginica* (L.) Britton
Frequent; dry woods, thickets. AS, AB, CL, FA, HL, KN, LK, LI, LR, PE, PC, RO.

902 **Lotus corniculata* L.   Birdsfoot Trefoil
Common, naturalized; roadsides, lawns, fields, waste places. All counties except RI.

903 *Lupinus perennis* L.   Wild Lupine
Rare; dry sandy soils, railroads, prairie remnants. PO: Portage County, Hopkins 485, 25 May 1913 (KE 16298); Dorgen swamp, Ravenna, Krebs s.n., 1891 (OS 21946). SU: prairie-like opening, Akron, Andreas & Karlo 3288, 13 May 1979 (KE 42266); dry open woods, Akron, Dean s.n., 1 Jun 1961 (University of Akron Herbarium s.n.); Akron, Clark s.n., 1 Jun 1915 (OS 68846).

904 **Medicago lupulina* L.   Black Medic
Common, naturalized; roadsides, waste places, fields. ALL COUNTIES.

905 **Medicago sativa* L.   Alfalfa
Common, naturalized; roadsides, fields, waste places. All counties except MO.

906 **Melilotus alba* Medicus   White Sweet Clover
Common, naturalized; roadsides, woodland borders, fields, weedy thickets. ALL COUNTIES.

907 **Melilotus officinalis* (L.) Pallas   Yellow Sweet Clover
Common, naturalized; fields, roadsides, thickets, waste places. All counties except MO, PC.

908 **Phaseolus polystachios** (L.) BSP. Wild Bean
Rare; dry woods, sandy thickets. RO: Colerain Twp., Bartley & Pontius s.n., 7 Aug 1932 (OS 23311). SU: Cuyahoga River, Akron, Foltz s.n., 18 Aug 1889 (OS 23315). WA: station farm, Duvel s.n., 1898 (OS 88262).

909 ***Phaseolus vulgaris** L. Kidney Bean
Rare, adventive; abandoned gardens, waste places. KN: landfill, Clinton Twp., Pusey 1598, 20 Aug 1975 (KE 38436).

910 **Psoralea onobrychis** Nutt. Scurf-pea
Rare; rich woods, thickets, clearings. FA: Fairfield, Goslin s.n., Jul 1934 (OS 97269). LI: Licking Reservoir, Jones s.n., 13 Jul 1892 (OC 79601); Newark, Jones s.n., 18 Jul 1890 (OC 79602). RO: Adelphi, Bartley & Thomas s.n., 21 Oct 1928 (OS 22456).

911 **Psoralea psoralioides** (Walter) Cory False Scurf-pea
Rare; stream banks, clearings, fields. RO: stream, Miller s.n., 11 Jun 1923 (OS 22479).

912 ***Robinia hispida** L. Bristly Locust, Rose-acacia
Infrequent, adventive; highway rights-of-way, strip-mined areas. CL, CU, FA, MH, PO, ST, SU.

913 **Robinia pseudo-acacia** L. Black Locust
Common, native to southern counties, naturalized elsewhere (Braun, 1961); woodland borders, pastures, thickets. ALL COUNTIES.

914 ***Robinia viscosa** Vent Clammy Locust
Infrequent, adventive; dry woods, woodland borders. AB, CU, FA, LK, LR.

915 **Strophostyles helvula** (L.) Ell. Wild Bean
Infrequent; dry roadsides, thickets, sandy banks, railroads. AB, CU, LK, LI, LR, ST.

916 **Stylosanthes biflora** (L.) BSP. Pencil-flower
Rare; dry woods, thickets, prairie remnants. FA: dry prairie, Madison Twp., Rae s.n., 7 Jul 1940 (OS 23058); Kettle Hills, Berne Twp., Thomas & Wickert s.n., 19 Jun 1938 (OS 23059). RO: Liberty Twp., Bartley & Pontius s.n., 21 Sep 1930 (OS 23049); Colerain Twp., Bartley & Pontius s.n., 10 Sep 1931 (BHO 19016).

917 **Tephrosia virginiana** (L.) Persoon Goat's-rue, Rabbit's-pea
Infrequent; dry sandy woods, dry slopes. AB, FA, HL, KN, LI, PC, PO, RO.

918 ***Trifolium arvense** L. Rabbitfoot Clover
Infrequent, naturalized; dry roadsides, fields. CU, HL, KN, LK, LI, LR, PO, RO, ST.

919 ***Trifolium aureum** Poll Yellow Clover, Hop Clover
Trifolium agrarium L.
Common, naturalized; roadsides, fields, clearings, waste places. All counties except MO, PC.

920 ***Trifolium campestre** Schreber Low Hop Clover
Trifolium procumbens L.
Common, naturalized; roadsides, lawns, waste places. AS, CU, FA, GE, HL, KN, LK, LI, LR, MD, PE, PO, RO, ST, SU, WA.

921 ***Trifolium dubium** Sibth. Little Hop Clover
Frequent, naturalized; fields, roadsides, lawns. CU, FA, GE, LK, MD, PE, PO, ST, SU, TR, WA.

922 ***Trifolium hybridum** L. Alsike Clover
Common, naturalized; roadsides, fields, waste places. All counties except PE, PC, RI.

923 ***Trifolium incarnatum** L. Crimson Clover
Infrequent, adventive; waste places, roadsides, fields. CU, LK, LI, LR, PO, RO, SU, WA.

924 ***Trifolium pratense** L. Red Clover
Common, naturalized; roadsides, fields, waste places. ALL COUNTIES.

925 **Trifolium reflexum** L. Buffalo Clover
Once rare, now presumably extirpated from Ohio; native in southern counties, possibly adventive elsewhere. AB: near Austinburg, Hicks s.n., 18 Jun 1932 (OS 22248). PC: Devil's Backbone, Washington Twp., Bartley s.n., 15 Jun 1932 (OS 22244 & BHO 19026).

926 ***Trifolium repens** L. White Clover
Common, naturalized; roadsides, pastures, fields, waste places. All counties except FA, RO.

927 *Vicia americana* Willd.
Infrequent; gravelly shores, thickets, eroding slopes, dry roadside banks. GE, LK, LR, SU, TR.

928 **Vicia angustifolia* Reichard   Common Vetch
Infrequent, adventive; railroads, roadsides. HL, KN, LK, PE, PO, SU.

929 *Vicia caroliniana* Walter   Wood Vetch
Infrequent; open eroding slopes, rich woods, thickets. AS, AB, GE, LK, LR, PE, PO, RO, SU.

930 **Vicia cracca* L.   Tufted Vetch, Bird Vetch
Frequent, naturalized; railroads, roadsides. AS, AB, CL, CU, FA, LK, PC, PO, RO, ST, TR, WA.

931 **Vicia hirsuta* (L.) S. F. Gray
Rare, adventive; roadsides, waste places. KN: Mt. Vernon, Kellerman s.n., 23 Jun 1898 (OS 23138). LR: College campus, Oberlin, Cowles s.n., 10 Oct 1892 (OC 73969).

932 **Vicia sativa* L.   Spring Vetch, Common Vetch
Rare, adventive; waste places. FA: Kettle Cave, Chapman s.n., 3 Jul 1932 (OS 23147).

933 **Vicia villosa* Roth   Hairy Vetch
Frequent, naturalized; fields, thickets, railroads, roadsides. AS, AB, CU, FA, GE, HL, KN, LR, MH, MD, PO, ST, SU, TR, WA.

LENTIBULARIACEAE (Bladderwort Family)

934 *Utricularia cornuta* Michx.   Horned Bladderwort
Rare; marly areas, wet sandy soils. PO: Muzzy Lake, "R.H.J" s.n., 12 Jul 1888 (College of Wooster Herbarium s.n.). SU: Long's Lake, Claassen s.n., 1890 (OS 32513).

935 *Utricularia gibba* L.   Humped Bladderwort
Infrequent; roadside ditches, bog lakes, ponds. AB, CU, FA, GE, HL, PO, ST, SU, TR, WA.

936 # *Utricularia intermedia* Hayne   Flat-leaved Bladderwort
Infrequent; ponds, fen pools, shallow streams. LK, PO, ST, SU, WA.

937 # *Utricularia minor* L.   Lesser Bladderwort
Infrequent; ponds, ditches, bog lakes. AB, CL, GE, KN, LI, PO, SU.

938 *Utricularia vulgaris* L.   Common Bladderwort
Frequent; ponds, lakes, ditches, oxbows, bog lakes. AS, AB, CU, GE, HL, LI, LR, PO, RI, ST, SU, TR, WA.

LIMNANTHACEAE (False Mermaid Family)

939 *Floerkea proserpinacoides* Willd.   False Mermaid
Common; rich moist woods, wooded floodplains. ALL COUNTIES.

LINACEAE (Flax Family)

940 *Linum medium* (Planch.) Britton var. *texanum* (Planch.) Fern.   Flax
Frequent; dry woods, open soils. AB, CL, CU, LK, LR, MD, PE, PO, RO, WA.

941 *Linum striatum* Walter   Furrowed Flax
Rare; moist peaty soils, boggy woods. AB: near Andover, Hicks s.n., 20 Aug 1930 (OS 14976 & OS 842323). CU: peaty ground, Olmsted Twp., Jones 73-8-8, 8 Aug 1973 (OC 217349). RO: Liberty Twp., Bartley & Pontius s.n., 14 Jul 1924 (OS 14970).

942 **Linum usitatissimum* L.   Common Flax
Infrequent, adventive; waste places, roadsides. AB, CU, HL, LK, LI, LR, PC, RI, SU.

943 *Linum virginianum* L.   Virginia Flax
Common; dry woods, eroding slopes, thickets. AB, CL, CU, FA, HL, KN, LK, LI, LR, MH, PE, PO, RO, ST, SU, TR, WA.

LYTHRACEAE (Loosestrife Family)

944 *Ammannia robusta* Heer & Regel
Rare; muddy shores. PC: mudflat, Circleville Twp., Stuckey 9464, 6 Aug 1973 (OS 116698); Stages Pond, Walnut Twp., Bourdo & Roberts 4307, 11 Oct 1973 (OS 113926). RO: Green Twp., Bartley & Pontius s.n., 23 Jul 1933 (OS 23718).

945 *Cuphea viscosissima* Jacq.   Clammy Cuphea
Cuphea petiolata (L.) Koehne
Frequent; fens, prairie remnants, stream banks. CL, CU, FA, LK, LR, MH, PE, PC, RO, TR, WA.

946 *Decodon verticillatus* (L.) Ell.   Swamp Loosestrife
Common; swamps, marshes, buttonbush swamps, bogs, lake margins. AS, AB, CU, FA, GE, HL, LK, LI, LR, MH, MD, PE, PO, RI, RO, ST, SU, TR, WA.

947 *Lythrum alatum* Pursh   Winged Loosestrife
Frequent; ditches, wet pastures, marshes. AB, CU, FA, GE, LK, LI, LR, MD, PC, PO, RI, RO, ST, SU, WA.

948 **Lythrum hyssopifolia* L.
Rare, adventive; marshes. PC: wet place, Thatcher, Bartley & Pontius s.n., 21 Jul 1955 (OS 78992).

949 **Lythrum salicaria* L.   Purple Loosestrife
Common, naturalized; roadside ditches, swamps, marshes, fens. AB, CL, CU, FA, GE, HL, KN, LK, LR, MH, MD, PO, RO, ST, SU, TR.

950 *Rotala ramosior* (L.) Koehne   Tooth-cup
Rare; sandy shores, mudflats, fields. FA: Clearcreek Twp., Bartley & Pontius s.n., 5 Sep 1943 (OS 43891). LK: reservoir, Stockberger s.n., 22 Aug 1890 (OS 23726). PC: mudflat, Circleville Twp., Stuckey 9461, 6 Aug 1973 (OS 116701); Stages Pond, Walnut Twp., Bourdo & Roberts 4293, 11 Oct 1973 (OS 112545); cultivated field, Circleville Twp., Carr 2387, 17 Sep 1979 (KE 45050).

MAGNOLIACEAE (Magnolia Family)

951 *Liriodendron tulipifera* L.   Tulip-tree, Tulip-popular
Common; rich mesic woods, woodlots. ALL COUNTIES.

952 *Magnolia acuminata* L.   Cucumber-tree, Cucumber-magnolia
Common; rich mesic woods. ALL COUNTIES.

MALVACEAE (Mallow Family)

953 **Abutilon theophrasti* Medicus   Velvet-leaf
Common, naturalized; grain fields, waste places. AS, AB, CL, CU, HL, KN, LK, LI, LR, MH, MD, MO, PO, RI, RO, ST, SU, TR, WA.

954 **Alcea rosea* L.   Hollyhock
Althaea rosea (L.) Cav.
Infrequent, adventive; roadsides, abandoned homesites. GE, HL, KN, LR, MD, ST.

955 **Althaea officinalis* L.   Marsh Mallow
Rare, adventive; wet meadows, marshes. GE: Fowler's Mills, Dudley 2025, 17 Aug 1900 (OS 95663). HL: roadside ditch, Salt Creek Twp., Wilson 2332, 4 Aug 1971 (KE 29660); pasture, Paint Twp., Wilson 2166, 17 Jul 1971 (OS s.n.). WA: marsh, Sugar Creek Twp., Andreas 1115, 27 Jun 1977 (KE 49683).

956 **Anoda cristata* (L.) Schlecht.
Rare, adventive; waste places, roadsides. RO: cornfield, Vause Station, Bartley s.n., 20 Sep 1953 (OS 46923).

957 *Hibiscus laevis* All.   Halberd-leaved Rose-mallow
Hibiscus militaris Cav.
Rare; swamps, river banks. PC: swamp, Wayne Twp., Cusick 17059, 14 Jul 1977 (OS 122341); swamp, Wayne Twp., Roberts 2237b, 26 Sep 1971 (OS 112533). RO: Green Twp. Bartley & Pontius s.n., 2 Sep 1935 (OS 47014).

958 *Hibiscus moscheutos* L. (including *H. palustris* L.)   Swamp Rose-mallow
Frequent; buttonbush swamps, lake margins, marshes. AB, CU, FA, GE, HL, LK, LI, LR, MH, PE, PC, RO, ST, SU, TR.

959 **Hibiscus trionum* L.   Flower-of-an-hour
Frequent, naturalized; cultivated fields, waste places. AB, CU, FA, HL, KN, LK, LI, LR, MH, MD, PC, PO, ST, WA.

960 **Malva moschata* L.   Musk-mallow
Frequent, naturalized; roadsides, fields. AS, AB, CL, CU, GE, KN, LK, LR, MH, MD, PO, SU, TR.

961 *Malva neglecta* Wallr.   Cheeses, Common Mallow
Common, naturalized; roadsides, fields, abandoned gardens, waste places. All counties except AB, MO, PC.

962 *Malva rotundifolia* L.
Rare, adventive; roadsides, waste places. LR: sheep lot, Russia Twp., Cooperrider 7302, 25 Jul 1960 (KE 4224). MH: roadside, Jackson Twp., Chuey 922, 2 Sep 1967 (BHO 26310).

963 *Napaea dioica* L.   Glade-mallow
Rare, native to western Ohio, adventive elsewhere (Cooperrider, 1985a); moist roadsides, railroads, weedy floodplains. FA: railroad embankment, Berne Twp., Cusick 14895, 21 Jun 1976 (KE 39429); Hocking Twp., Goslin s.n., 2 Jul 1961 (OS 77387). PC: railroad, Circleville Twp., Cusick & Spooner 23835, 3 Aug 1984 (OS 158208); Perry Twp., Bartley & Pontius s.n., 20 Sep 1932 (OS 15893). RO: creek bank, Chillicothe, Rypma s.n., 7 Jul 1983 (BHO 35285).

964 *Sida spinosa* L.   Prickly Mallow
Infrequent, naturalized; fields, railroads, waste places. AB, CU, FA, GE, LK, LR, PC, PO, RO, SU.

MARTYNIACEAE (Martynia Family)

965 *Proboscidea louisianica* (Miller) Thell   Unicorn-plant
Rare, adventive; waste places. RI: Shelby, Bowman s.n., 18 Sep 1903 (OS 32376).

MELASTOMATACEAE (Melastome Family)

966 *Rhexia virginica* L.   Virginia Meadow-beauty
Rare; peaty woodland margins, wet fields. PO: dry sandy field, Hiram Twp., Andreas 4192, 23 Aug 1978 (KE 41921). RO: wet ground, Liberty Twp., Bartley s.n., 25 Aug 1946 (OS 19427). ST: moist peat woods, Jackson Twp., Amann 2006, 27 Sep 1960 (KE 3935). SU: lake margin, Green Twp., Andreas 4512, 6 Aug 1979 (KE 42233).

MENISPERMACEAE (Moonseed Family)

967 *Menispermum canadense* L.   Moonseed
Common; weedy floodplains, thickets, shaded roadside banks, woodland borders. All counties except PC. Braun (1961) indicates records for LK, MO.

MENYANTHACEAE (Bogbean Family)

968 # *Menyanthes trifoliata* L.   Buckbean, Bogbean
Frequent; bog margins, sphagnous thickets. AS, AB, GE, LK, LI, LR, PO, RI, ST, SU, WA.

MORACEAE (Mulberry Family)

969 *Maclura pomifera* (Raf.) Schneider   Osage-orange
Common, naturalized; fence rows, woodland borders, roadsides. ALL COUNTIES.

970 *Morus alba* L.   White Mulberry
Common, naturalized; roadsides, stream margins, fence rows. All counties except MH.

971 *Morus rubra* L.   Red Mulberry
Common; floodplains, fields, moist woods. ALL COUNTIES.

MYRICACEAE (Wax-myrtle Family)

972 *Comptonia peregrina* (L.) Coulter   Sweet-fern
Infrequent; dry sandy banks, pastures, dry clearings. AS, AB, HL, KN, LK, LI, PO.

973 # *Myrica pensylvanica* Loisel.   Bayberry
Rare; shrubby fens. GE: clayey slump, Chardon Twp., Bissell 1982:273, 7 Aug 1982 (CLM 18247) [This record is probably an escape from cultivation.]. PO: weedy fen, Streetsboro Twp., Andreas 5811, 3 Sep 1981 (KE 44058); fen, Streetsboro Twp., Andreas 5110, 3 Jun 1980 (KE 42974); *Larix* community, Streetsboro Twp., Tandy 1340, 2 Jun 1973 (KE 40560); dry sphagnous bog, Streetsboro Twp., Cooperrider 8536, 8 Jul 1961 (KE 3579); open peat bog, Streetsboro Twp., Webb s.n., 23 Jul 1919 (KE 12218). SU: swamp near Hudson, Beardslee s.n., Aug 1928 (OC 94891) [This specimen was probably collected in Portage County.].

NYCTAGINACEAE (Four-o'clock Family)

974 *Mirabilis jalapa* L.   Marvel-of-Peru
Rare, adventive; roadsides, waste places. AB: near Conneaut, Hicks s.n., 5 Sep 1931 (OS 17969). TR: Bloomfield Twp., Shanks s.n., 7 Oct 1934 (OS 17965).

975 *Mirabilis nyctaginea* (Michx.) MacM.   Four-o'clock
Frequent, adventive; railroads, roadsides, waste places. AB, CL, CU, GE, HL, KN, LK, LI, LR, PO, RI, RO, ST, SU, TR.

NYMPHAEACEAE (Water-lily Family)

976 # *Brasenia schreberi* Gmel.   Water-shield
Frequent; ponds, quiet streams, calcareous waters, kettle lakes. AB, GE, LI, LR, MD, PO, RO, ST, SU, TR, WA.

977 *Cabomba caroliniana* Gray   Fanwort
Rare, adventive; ditches, ponds. GE: Cuyahoga River, Troy Twp., Bissell 1982:220, 22 Jul 1982 (CLM 18673). LK: buttonbush bog, Kirtland Twp., Aldrich s.n., 3 Sep 1937 (OS 2592). SU: canal, Akron, Roberts 2243, 30 Oct 1971 (OS 112804 & KE 39508). TR: Mosquito Creek, Howland Twp., Rood 577, 2 Jul 1933 (KE 12074).

978 *Nelumbo lutea* (Willd.) Persoon   Yellow Nelumbo, Pond-nuts
Infrequent, often introduced; ponds, lakes. HL, LK, LI, PE, PC, RO, ST.

979 *Nuphar advena* (Aiton) Aiton f.   Spatter-dock, Pond Lily
Common; ponds, marshes, sphagnous mats, lakes. All counties except PC, RO.

980 *Nymphaea odorata* Aiton (including *N. tuberosa* Paine)   Fragrant Water-lily
Common; ponds, lakes. AS, AB, CL, CU, FA, GE, HL, LK, LI, LR, MH, MD, PE, PO, ST, SU.

NYSSACEAE (Tupelo Family)

981 *Nyssa sylvatica* Marsh.   Black Gum, Sour Gum, Tupelo
Common; dry fields, dry woods, fields, pastures, sphagnous mats, boggy woods. All counties except PC.

OLEACEAE (Olive Family)

982a *Fraxinus americana* L. var. *americana*   White Ash
Common; moist and dry woods, thickets, marsh margins. ALL COUNTIES.

982b *Fraxinus americana* L. var. *biltmoreana* (Beadle) J. Wright
Infrequent; thickets, moist and dry woods. CU, LI, PE, PC, RO, SU.

983 *Fraxinus nigra* Marsh.   Black Ash
Common; wooded floodplains, marshes, lake margins. All counties except FA, LI, LR.

984a *Fraxinus pennyslvanica* Marsh. var. *pennyslvanica*   Red Ash
Common; wooded floodplains, stream banks. All counties except CL, LI, MO.

984b *Fraxinus pennsylvanica* Marsh. var. *subintegerrima* (Vahl) Fern.   Green Ash
Common; wooded floodplains, stream banks. All counties except CL.

985 *Fraxinus quadrangulata* Michx.   Blue Ash
Rare; dry and moist woods. LI: hillside, Hopewell Twp., Youger FR-146, 24 Aug 1970 (OS 105054); Atherton, Kellerman s.n., 25 Jul 1897 (OS 29814). RO: forest, Paint Twp., Youger 102, 13 Oct 1972 (OS 108413); park, Chillicothe, Kellerman s.n., 10 Jul 1906 (OS 29820).

986 *Ligustrum obtusifolium* Sieb. & Zucc.   Privet
Infrequent, naturalized; roadsides, thickets, dry and moist woods. CU, GE, LR, MD, ST, SU, WA.

987 *Ligustrum ovalifolium* Hassk.   California Privet
Rare, adventive; thickets, roadsides. GE: open ground, Newbury Twp., Hawver 1079, 22 Jul 1960 (KE 8761).

988 *Ligustrum vulgare* L.   Common Privet
Common, naturalized; roadside thickets, waste places, dry and moist woods. AS, AB, CL, CU, FA, GE, HL, KN, LK, LI, LR, MD, PE, PO, ST, SU, TR, WA.

989 *Syringa vulgaris* L.   Common Lilac
Infrequent, adventive; roadsides, abandoned homesites. LK, MD, PE, RO, TR.

ONAGRACEAE (Evening-primrose Family)

990 *Circaea alpina* L.   Small Enchanter's Nightshade
Frequent; moist sandstone cliffs, rocky moist slopes, hemlock ravines. AS, AB, CL, CU, GE, HL, KN, LK, LI, LR, MH, MD, PO, SU, TR.

991 *Circaea* x *intermedia* Ehrh. (*C. alpina* x *C. lutetiana*)
Rare; moist rocky wooded slopes. GE: open wooded slope, Bainbridge Twp., Hawver 1123, 25 Jul 1960 (KE 2097). TR: steep moist slope, Mesopotamia Twp., Cooperrider 7924, 11 Aug 1960 (KE 1968).

992 *Circaea lutetiana* L. var. *canadensis* L.   Enchanter's Nightshade
Circaea quadrisulcata (Maxim.) Franch. & Sav.
Common; moist, mesic and dry woods, wooded floodplains, moist and dry thickets. ALL COUNTIES.

993 *Epilobium angustifolium* L. var. *platyphyllum* (Daniels) Fern.   Fireweed
Frequent; railroads, wet thickets, woodland borders, moist roadsides. AS, AB, CL, CU, GE, LK, LR, MD, PO, ST, SU, TR, WA.

994 *Epilobium ciliatum* Raf.
Epilobium glandulosum Lehm.
Infrequent; marshes, stream margins, weedy floodplains. AB, CU, FA, LI, LR, MD, SU, TR.

995 *Epilobium coloratum* Sprengel
Common; stream and pond margins, wet fields, moist waste places. All counties except MO.

996 **Epilobium hirsutum* L.   Great Hairy Willow-herb
Infrequent, naturalized; wet meadows, roadside ditches, thickets, pastures. AB, CL, CU, GE, LK, MH, PO, ST, SU, TR.

997 *Epilobium leptophyllum* Raf.   Narrow-leaved Willow-herb
Infrequent; fens, weedy thickets, wet meadows. AB, CL, GE, HL, KN, LR, PO, RO, ST, SU.

998 # *Epilobium strictum* Muhl.   Simple Willow-herb
Frequent; fens, sedge meadows, sphagnous thickets. CL, CU, GE, LK, LI, LR, MH, PO, ST, SU, WA.

999 *Epilobium* x *wisconsinense* Ugent (*E. ciliatum* x *E. coloratum*)
Rare; thickets, fields. GE: thicket, Lake Kelso, Burton Twp., Hawver 1089, 22 Jul 1960 (KE 8260). LR: weedy field, Cooperrider 7291, 25 Jul 1960 (KE 4230).

1000 *Gaura biennis* L.   Gaura
Common; moist fields, roadside ditches, weedy floodplains, railroads. AS, AB, CL, CU, FA, GE, HL, KN, LK, LI, LR, PC, PO, RI, RO, ST, SU, WA.

1001 **Gaura longiflora* Spach
Gaura biennis L. var. *pitcheri* T. & G.
Rare, adventive; waste places. LR: weed in cultivated field, Russia Twp., Cooperrider 8047, 15 Aug 1960 (KE 4241).

1002 **Gaura parviflora* Douglas   Small-flowered Gaura
Rare, adventive; dry fields. ST: Stark County, Case s.n., 1 Aug 1900 (OS 28125).

1003 *Ludwigia alternifolia* L.   Seedbox
Common; moist fields, pond and stream margins, swamps. AB, CL, CU, FA, GE, HL, KN, LK, LI, LR, PE, PO, RO, ST, SU, TR, WA.

1004 *Ludwigia palustris* (L.) Ell.   Water-purslane
Common; wet ditches, pond and stream margins, shallow waters. All counties except PE.

1005 *Ludwigia polycarpa* Short & Peter
Rare; pond margins, ditches, wet fields. LI: Licking Reservoir, Jones s.n., 7 Jul 1891 (OC 80088). PC: swamp, Jackson Twp., Bartley s.n., 5 Aug 1959 (OS 73627).

1006 *Oenothera biennis* L.   Common Evening Primrose
Common; dry fields, roadside banks, railroads, waste places. All counties except MO, PC.

1007a *Oenothera fruticosa* L. var. *fruticosa*
Rare; calcareous meadows, fields. RO: Liberty Twp., Bartley & Pontius s.n., 29 May 1932 (OS 28070).

1007b **Oenothera fruticosa** L. var. **ambigua** Nutt. (including **O. tetragona** Roth)
Frequent; fields, woodland borders, pastures. AS, AB, CL, CU, GE, FA, HL, MH, PO, RO, ST, SU, TR, WA.

1008 **Oenothera laciniata** Hill
Infrequent; waste places, sandy fields. CU, FA, LK, LR, PC, PO, RO, TR.

1009 **Oenothera perennis** L.
Common; dry and moist fields, woodland borders. AS, AB, CU, FA, GE, KN, LK, LR, MH, PE, PO, RI, RO, ST, SU, TR, WA.

1010 **Oenothera pilosella** Raf.
Infrequent; open woods, meadows, roadsides. AS, AB, HL, LI, LR, PE, RI, RO, SU, TR.

OROBANCHACEAE (Broom-rape Family)

1011 **Conopholis americana** (L.) Wallr. Squaw-root, Cancer-root
Common; rich woods, parasitic on oaks (Valley & Cooperrider, 1966). All counties except MO.

1012 **Epifagus virginiana** (L.) Bart. Beech-drops
Common; rich woods, parasitic or saprophytic on beech roots. ALL COUNTIES.

1013 **Orobanche uniflora** L. One-flowered Cancer-root
Frequent; moist thickets, rich moist woods, floodplains. AB, CL, CU, FA, GE, KN, LK, LR, PO, RO, ST, SU.

OXALIDACEAE (Wood-sorrel Family)

1014 ***Oxalis corniculata** L. Creeping Wood-sorrel
Infrequent, naturalized; fields, gardens, sidewalks, waste places. AS, FA, HL, LK, LR, PE, PC, PO, ST.

1015 **Oxalis dillenii** Jacq.
Oxalis stricta of authors, not L.
Common; waste places, fields, roadsides. AB, CU, FA, HL, KN, LK, LR, MD, MO, PE, PC, PO, RO, SU, TR, WA.

1016 **Oxalis grandis** Smith Large Wood-sorrel
Frequent; woods, shaded slopes, roadside banks. AB, CL, CU, FA, GE, HL, LI, MH, PE, PO, ST, SU.

1017 **Oxalis montana** Raf. White Wood-sorrel
Oxalis acetosella L.
Rare; boggy woods, hemlock woods. AB: hemlock woods, Richmond Twp., Cusick 19411, 29 Jun 1979 (KE 41771). LK: under hemlock (possibly planted), Kirtland Twp., Yates 1978:15, 27 May 1978 (CLM 12666).

1018 **Oxalis stricta** L. Common Wood-sorrel
Oxalis europaea Jord.
Common; roadsides, fields, waste places. ALL COUNTIES.

1019 **Oxalis violacea** L. Violet Wood-sorrel
Common; rich woods, shaded roadside banks. All counties except AB, GE.

PAPAVERACEAE (Poppy Family)

1020 ***Chelidonium majus** L. Celandine
Common, naturalized; disturbed woods, roadside banks, weedy floodplains. All counties except MO, PC.

1021 ***Papaver argemone** L.
Rare, adventive; waste places. RO: Camp Sherman, Bartley & Pontius s.n., 31 May 1931 (OS 13034).

1022 ***Papaver dubium** L.
Rare, adventive; cultivated fields, roadsides, waste places. SU: fields, Akron, Stair s.n., 30 Jun 1896 (CLM 7847).

1023 ***Papaver rhoeas** L. Corn-poppy
Rare, adventive; waste places, roadsides. FA: roadside, Berne Twp., Goslin s.n., 7 Jul 1935 (OS 13025). PO: Portage County, Bartley & Pontius s.n., 5 Jun 1938 (OS 13022).

1024 *Papaver somniferum* L.   Common Poppy
Rare, adventive; abandoned homesites, roadsides, fields, waste places. KN: landfill, Clinton Twp., Pusey 470, 3 Jul 1975 (KE 38542). TR: adventive, Braceville Twp., Rood 2826, 5 Jul 1954 (KE 12087).

1025 *Sanguinaria canadensis* L.   Bloodroot
Common; rich mesic and moist woods, alluvial floodplains. ALL COUNTIES.

1026 *Stylophorum diphyllum* (Michx.) Nutt.   Wood-poppy
Infrequent; rocky woods, wooded floodplains, moist cliffs. FA, HL, LK, MO, PE, RO.

PASSIFLORACEAE (Passion-flower Family)

1027 *Passiflora lutea* L.   Yellow Passion-flower
Rare; thickets, woodland borders. RO: Colerain Twp., Bartley & Pontius s.n., 22 Jun 1930 (OS 17078); Mt. Logan, Kellerman s.n., 12 Aug 1898 (OS 17072).

PHRYMACEAE (Lopseed Family)

1028 *Phyrma leptostachya* L.   Lopseed
Common; rich and dry woods, thickets, wooded roadside banks. ALL COUNTIES.

PHYTOLACCACEAE (Pokeweed Family)

1029 *Phytolacca americana* L.   Pokeweed, Poke
Common; fields, wooded floodplains, moist woods, moist thickets. ALL COUNTIES.

PLANTAGINACEAE (Plantain Family)

1030 *Plantago aristata* Michx.   Bracted Plantain
Common, naturalized; sandy fields, abandoned quarries, waste places. AB, CU, FA, HL, KN, LK, LI, LR, MD, PE, PC, PO, RO, ST, SU, WA.

1031 *Plantago cordata* Lam.   Heart-leaf Plantain
Rare; stream banks in wet woods. LR: Oberlin, Dick s.n., 11 May 1895 (OC 80980 & OS 34773); running brook, Oberlin, Ricksecker s.n., 21 May 1892 (OC s.n.); south woods, Oberlin, Cowles s.n., 21 May 1892 (OC 75777 & KE 30391). MH: slow-moving woodland stream, Boardman Twp., Andreas & Jones 5852, 11 May 1982 (KE 44059).

1032 *Plantago lanceolata* L.   English Plantain
Common, naturalized; roadsides, fields, waste places, lawns. All counties except MO, PC.

1033 *Plantago major* L.   Common Plantain
Common, naturalized; roadsides, fields, waste places. AS, AB, CL, CU, GE, KN, LK, LR, MH, MO, PE, PC, PO, RI, ST, SU, TR. WA.

1034 *Plantago psyllium* L. (including *P. indica* L.)
Rare, adventive; roadsides, railroads, waste places. TR: railroad ballast, Braceville Twp., Rood 2179, 30 Jul 1945 (KE 11955).

1035 *Plantago rugelii* Dcne.   Rugel's Plantain
Common; roadsides, fields, waste places. All counties except MO.

1036 *Plantago virginica* L.   Pale-seeded Plantain, Hoary Plantain
Frequent; dry fields, sandy soils, waste places. CL, CU, FA, HL, KN, LI, PE, PC, RO, ST, SU, TR.

PLATANACEAE (Plane-tree Family)

1037 *Platanus occidentalis* L.   Sycamore
Common; rich floodplains, marshes, wet ravines. ALL COUNITES.

POLEMONIACEAE (Polemonium Family)

1038 *Phlox divaricata* L.   Common Woodland Phlox
Common; rich mesic and dry woods, woodland terraces, pastures. ALL COUNTIES.

1039 *Phlox maculata* L.   Wild Sweet William
Frequent; fens, ditches, sedge meadows, pastures, roadside ditches. AB, CL, CU, FA, HL, KN, LI, MD, PO, RI, ST, SU, WA.

1040 *Phlox paniculata* L.   Fall Phlox
Common; roadsides, stream banks, fields, thickets. All counties except MO, PC.

1041 **Phlox pilosa** L.
Frequent; dry roadside banks, open woods, fens, eroding slopes, prairie remnants. AS, AB, CU, FA, GE, HL, PE, PO, RO, SU, WA.

1042 **Phlox subulata** L. Mountain Phlox, Moss Phlox
Frequent; dry open woods, eroding slopes, sandy banks, roadsides. AS, CU, FA, GE, HL, KN, LK, LI, LR, MD, PO, RO, SU, WA.

1043 ***Polemonium caeruleum** L. Greek Valerian
Rare, adventive; roadsides, waste places. AB: Windsor Twp., Hicks s.n., 18 Jun 1931 (OS 29048).

1044 **Polemonium reptans** L. Jacob's Ladder
Common; rich woods, wooded floodplains, fens, pastures. ALL COUNTIES.

POLYGALACEAE (Milkwort Family)

1045 **Polygala paucifolia** Willd. Gay-wings, Flowering Wintergreen
Rare; dry woods, pine woods. AB: white pine woods, Farnham, Hicks s.n., 10 Aug 1930 (OS 15226). ST: Canton, Case s.n., 1900 (OS 15225); bog, Canton, Stair s.n., 16 May 1899 (CLM s.n.).

1046 **Polygala polygama** Walter Racemed Milkwort
Rare; sandy fields, dry sandy woods. GE: railroad embankment, Bainbridge Twp., Hawver 597, 21 Jun 1960 (KE 8123). LK: Werner s.n., 8 Jul 1923 (OS 15232).

1047 **Polygala sanguinea** L. Red Milkwort
Common; dry fields, sandy soils, open woods, eroding slopes. AB, CL, CU, FA, GE, KN, LK, LI, LR, MH, MD, PE, PO, RO, ST, SU, TR, WA.

1048 **Polygala senega** L. Seneca Snakeroot
Infrequent; shaded eroding slopes, fens, fields. AB, CU, LK, LR, PC, RO, ST, SU.

1049 **Polygala verticillata** L. var. **verticillata**
Common; dry woods, slopes, moist seepage banks, fields. AS, AB, CL, CU, FA, GE, HL, KN, LK, LI, LR, MH, PE, PC, PO, RO, ST, SU, TR, WA.

POLYGONACEAE (Smartweed Family)

1050 ***Fagopyrum esculentum** Moench Buckwheat
Common, adventive; roadsides, fields, waste places. AB, CL, CU, FA, GE, HL, KN, LK, LI, LR, MD, PO, RO, ST, TR, WA.

1051 **Polygonum amphibium** L. (including **P. amphibium** var. **emersum** Michx., **P. amphibium** var. **stipulaceum** Coleman, and **P. coccineum** Willd.)
Water Smartweed
Common; marshes, lake margins, ponds. All counties except MO.

1052 **Polygonum arifolium** L. Halberd-leaved Tearthumb
Common; shrubby floodplains, bog margins, thickets. AS, AB, CL, CU, FA, GE, HL, LK, LR, MH, MD, PO, RI, RO, ST, SU, TR, WA.

1053 ***Polygonum aviculare** L. Knotweed
Common, naturalized; roadsides, fields, waste places, sidewalks, gardens. All counties except AS, MO.

1054 ***Polygonum caespitosum** Blume
Frequent, naturalized; gardens, fields, shaded roadsides, woodland margins. CL, CU, FA, GE, HL, LR, MH, PE, PO, ST, SU, WA.

1055 # **Polygonum cilinode** Michx. Mountain Bindweed
Rare; boggy thickets, rich moist woods. AB: river bank, Farnham, Hicks s.n., 16 Aug 1932 (OS 19032). LK: slope, Kirtland, Twp., Bradt 1549, 9 July 1985 (KE 49306). PO: boggy woods, Shalersville Twp., Andreas 3796, 16 Jun 1979 (KE 422348). SU: rich woods, Copley Twp., Andreas 4027, 9 Jul 1979 (KE 42265).

1056 ***Polygonum convolvulus** L. Black Bindweed
Common, naturalized; fields, fence rows, river banks, roadsides, railroads. AS, AB, CL, CU, FA, GE, HL, KN, LK, LI, LR, MH, MD, PE, PO, RO, ST, SU, TR, WA.

1057 **Polygonum cristatum** Engelm. & Gray Crested Bindweed
Frequent; thickets, fields, railroads, fence rows. AS, FA, GE, HL, KN, MH, PE, PO, ST, SU, TR.

1058 **Polygonum cuspidatum* Sieb. & Zucc.** Japanese Knotweed
Common, naturalized; railroads, roadsides, river banks, moist thickets. AS, AB, CL, CU, FA, GE, HL, KN, LK, LR, MH, MD, PE, PO, RO, ST, SU, WA.

1059 *Polygonum erectum* L.
Frequent; fields, waste places, weedy fields. AB, CL, CU, KN, LK, LI, LR, MH, PC, PO, RI, ST, SU, WA.

1060 *Polygonum hydropiper* L. Common Smartweed, Water-pepper
Common; roadsides, pastures, pond margins, waste places. All counties except MO, RO.

1061 *Polygonum hydropiperoides* Michx. Mild Water-pepper
Common; stream and pond margins, wet meadows, sedge meadows, thickets. All counties except MO, RO.

1062 *Polygonum lapathifolium* L. Dock-leaved Smartweed, Nodding Smartweed
Common; stream and pond margins, swamps, moist waste places. All counties except RI, TR.

1063 **Polygonum orientale* L.** Prince's-feather
Frequent, adventive; waste places, gardens, roadsides. AB, CL, CU, HL, KN, LI, MH, MD, PC, RI, RO, ST, SU, WA.

1064 *Polygonum pensylvanicum* L. Pinkweed
Common; fields, moist meadows, waste places. All counties except MO.

1065 **Polygonum persicaria* L.** Lady's Thumb, Heart's-ease
Common, naturalized; roadsides, pond and swamp margins, stream banks, fields, railroads. ALL COUNTIES.

1066a *Polygonum punctatum* Ell. var. **punctatum** Water Smartweed
Common; shaded stream banks, bogs, roadside ditches, marshes, wet fields. All counties except MO.

1066b **Polygonum punctatum* Ell. var. *robustius* Small**
Polygonum robustius (Small) Fern.
Rare, adventive; streams, ponds. PO: river margin, Webb 612, 24 Aug 1920 (KE 10942).

1067 *Polygonum sagittatum* L. Arrow-leaved Tearthumb
Common; marshes, boggy thickets, weedy floodplains. All counties except PC.

1068 *Polygonum scandens* L. Climbing False Buckwheat
Common; moist thickets, woodland borders, railroads. All counties except MO.

1069 *Polygonum tenue* Michx.
Rare; dry soils, roadsides, fields, railroads. FA: Greenfield Twp., Goslin s.n., 10 Sep 1941 (OS 19546); Kettle Hills, Thomas s.n., 25 Sep 1932 (OS 19548) RO: Colerain Twp., Bartley & Pontius s.n., 2 Sep 1931 (OS 19537).

1070 *Polygonum virginianum* L. Jumpseed
Tovara virginiana (L.) Raf.
Common; rich mesic and dry woods, wooded floodplains, roadside thickets, woodland borders. ALL COUNTIES.

1071 **Rumex acetosella* L.** Sheep Sorrel
Common, naturalized; roadsides, railroads, pastures, fields, gardens, waste places. ALL COUNTIES.

1072 *Rumex altissimus* Wood Pale Dock
Frequent; weedy floodplains, moist soils, river margins. CU, FA, LI, LR, PE, PC, RO, ST, SU, TR, WA.

1073 **Rumex conglomeratus* Murray**
Rare, adventive; waste places. ST: railroad embankment, Louisville, Stair s.n., Jul 1899 (CLM 7554).

1074 **Rumex crispus* L.** Yellow Dock
Common, naturalized; fields, waste places, pastures, roadsides, railroads. ALL COUNTIES.

1075 **Rumex maritimus* L.**
Rare, adventive; waste places, fields, roadsides. PO: muck land, Ravenna Twp., Herrick s.n., 10 Jul 1958 (KE 2735 & OS 60976). SU: muck farm, Tallmadge, Herrick s.n., 8 Aug 1958 (OS 61013); garden, Twinsburg, Herrick s.n., 13 Oct 1958 (OS 61011).

1076 *Rumex obtusifolius* L.   Blunt-leaved Dock
Common, naturalized; marshes, stream and pond margins, fields, moist waste places. ALL
COUNTIES.

1077 *Rumex orbiculatus* Gray   Water Dock
Common; ponds margins, marshes, swamps, wet meadows. AS, AB, CL, CU, FA, GE, HL,
KN, LK, LI, LR, PC, PO, RO, ST, SU, WA.

1078 *Rumex verticillatus* L.   Whorled Swamp Dock
Common; pond and stream margins, weedy floodplains, swamps, wet thickets. AB, CU, FA,
GE, HL, LK, LI, LR, MH, MO, PE, PC, PO, RI, RO, ST, SU, TR, WA.

PORTULACACEAE (Purslane Family)

1079 # *Claytonia caroliniana* Michx.   Carolina Spring Beauty
Rare; wet hemlock woods, wooded floodplains, stream terraces. AB: floodplain, Morgan
Twp., Bissell 1982:13, 4 May 1982 (CLM 17310); woods, Monroe Twp., Bissell s.n., 15 May
1972 (OS 109791); moist woods, Pierpont, Brockett 579, 14 Apr 1968 (KE 19171). GE: woods,
Clarion Twp., Bissell 1982:21, 11 May 1982 (CLM 17154); Montville, Leonard s.n., 25 Apr
1888 (OC 78724). LK: creek bottom, Madison Twp., Fowler s.n., 13 Apr 1963 (OS 71658);
scout camp, Madison Twp., Tyler s.n., 9 May 1937 (OS 17780).

1080 *Claytonia virginica* L.   Spring Beauty
Common; dry, mesic and moist woods, floodplains, pastures, clearings, thickets. ALL
COUNTIES.

1081 *Portulaca oleracea* L.   Common Pigweed
Frequent, adventive; gardens, railroads, sidewalks, waste places. AS, AB, CU, GE, HL, KN,
LR, PO, RI, ST, TR, WA.

PRIMULACEAE (Primrose Family)

1082 *Anagallis arvensis* L.   Scarlet Pimpernel
Frequent, naturalized; lawns, roadsides, fields, waste places. AS, CL, CU, FA, GE, HL, LK,
LR, MH, MO, PO, RO, ST, SU, WA.

1083 *Centunculus minimus* L.   Chaffweed
Rare, adventive; sandy soils. PC: Pickaway Twp., Bartley & Pontius s.n., 9 Jul 1935 (OS
18728).

1084 *Dodecatheon meadia* L.   Shooting Star
Rare; open woods, prairie remnants. FA: quarry, Berne Twp. Goslin s.n., 18 May 1902 (OS
117601). PO: Rootstown, no collector, 29 May 1949 (KE 1351). ST: moist prairies, Kuehnle
828, 5 Apr 1937 (KE 1352).

1085 *Lysimachia ciliata* L.   Fringed Loosestrife
Common; moist roadside ditches, woodland borders, stream and pond margins, marshes.
ALL COUNTIES.

1086 *Lysimachia lanceolata* Walter
Rare; thickets, fens, moist open woods. FA: Hocking Twp., Goslin s.n., 24 Jun 1933 (OS
18538). PE: mesic woods, Reading Twp., Cline 804, 11 Jul 1975 (KE 43220). PC: Pickaway
Twp., Bartley & Pontius s.n., 4 Jul 1929 (OS 18523). RO: Chillicothe, Bower s.n., 1892 (OS
18521).

1087 *Lysimachia nummularia* L.   Moneywort
Common, naturalized; moist meadows, marshes, roadsides, lawns. ALL COUNTIES.

1088 *Lysimachia x producta* (Gray) Fern. (*L. quadrifolia* x *L. terrestris)*
Rare; damp thickets, roadsides. GE: marsh, Herrick s.n., 21 Jun 1955 (OS 54708). LK:
weedy area, Mentor, Tandy 1611, 11 Jul 1973 (KE 34659). PO: grassy lane, Franklin Twp.,
Cooperrider & Brockett 10875, 3 Jul 1969 (KE 26547); disturbed slope, Mantua Twp.,
Brockett 248, 3 Jul 1967 (KE 16843). ST: woodland edge, Lake Twp., Amann 416, 25 Jun
1960 (KE 3256 & KE 3258).

1089 *Lysimachia quadriflora* Sims
Infrequent; fens, sedge meadows. AB, FA, HL, PC, RO, ST, SU.

1090 *Lysimachia quadrifolia* L.   Whorled Loosestrife
Common; dry woods, roadsides, thickets. All counties except PC.

1091 **Lysimachia terrestris** (L.) BSP. Swamp-candles
Common; wet meadows, boggy thickets, marshes. AS, AB, CL, CU, GE, HL, LK, LR, MH, MD, PO, RO, ST, SU, TR, WA.

1092 **Lysimachia thyrsiflora** L. Tufted Loosestrife
Frequent; boggy thickets, swamps, lake and stream margins. AS, AB, CU, GE, LK, LI, LR, MH, MD, PE, PO, RI, ST, SU, TR, WA.

1093 ***Lysimachia vulgaris** L. Garden Loosestrife
Rare, naturalized; marshes, thickets, roadsides. PO: wet marshy area, Suffield Twp., Cooperrider 11215, 27 Jul 1973 (KE 34077). ST: swamp, Lake Twp., Brockett 831, 18 Aug 1968 (KE 20395); Lake Twp., Amann 844, 26 Jul 1960 (KE 3257).

1094 **Samolus parviflorus** Raf. Water Pimpernel
Common; weedy floodplains, stream margins, mudflats, wet pastures. AS, AB, CU, GE, LK, LI, LR, MH, MD, PE, PC, PO, RI, RO, SU, TR, WA.

1095 # **Trientalis borealis** Raf. Star-flower
Infrequent; hemlock slopes, sphagnous thickets, shaded rock ledges. AB, CU, GE, LK, LR, PO, SU, TR.

RANUNCULACEAE (Buttercup Family)

1096 # ***Aconitum noveboracense*** Gray Northern Monkshood
Rare; moist sandstone cliffs, cool shaded ravines. PO: sandstone outcrop, Nelson Twp., Andreas 2457, 2 Aug 1978 (KE 39906); ravine, Garrettsville, Rood & Webb 710, 26 Jul 1908 (KE 12061 & OS 11920). SU: sandstone overhang, Akron, Andreas & Cusick 2452, 27 Jul 1978 (KE 42007); moist ravine, Stoutamire 7809, Akron 14 Sep 1968 (University of Akron Herbarium s.n.).

1097 **Actaea pachypoda** Ell. White Baneberry, Doll's-eyes
Common; rich moist and mesic woods, wooded floodplains, boggy thickets. ALL COUNTIES.

1098 **Actaea rubra** (Aiton) Willd. Red Baneberry
Rare; rich moist woods. AB: Wayne Twp., Hicks s.n., 3 Jul 1932 (OS 12344).

1099 **Anemone canadensis** L. Canada Anemone
Common; wooded floodplains, damp thickets, sedge meadows. All counties except GE, PE, PO.

1100 **Anemone quinquefolia** L. Wood Anemone
Common; rich moist and mesic woods, damp thickets, floodplains. All counties except PC.

1101 **Anemone virginiana** L. Thimbleweed
Common; dry woods, eroding slopes, shaded roadside banks, prairie remnants. ALL COUNTIES.

1102 **Anemonella thalictroides** (L.) Spach Rue-anemone
Common; rich dry, mesic and moist woods, thickets, clearings, roadsides. ALL COUNTIES.

1103 **Aquilegia canadensis** L. Wild Columbine
Common; rich dry and mesic woods, thickets, dry eroding slopes, sandstone ledges. All counties except KN, MH, MO.

1104 ***Aquilegia vulgaris** L. Garden Columbine
Infrequent, adventive; fields, roadsides. AB, CU, GE, PO, SU, TR.

1105 **Caltha palustris** L. Marsh Marigold, Cowslip
Common; marshes, boggy thickets, swamps, fens. All counties except PE.

1106 **Cimicifuga racemosa** (L.) Nutt. Black Snakeroot, Black Cohosh
Common; rich woods, wooded floodplains, shaded roadside banks, weedy railroads. ALL COUNTIES.

1107 ***Clematis dioscoreifolia** Levl. & Vaniot
Rare, naturalized; roadsides, fence rows. RO: railroad yard, Chillicothe, Carr 2065, 20 Aug 1979 (OS 150134). SU: weedy slope, Akron, Cusick 12806, 9 Aug 1972 (OS 150134).

1108 **Clematis virginiana** L. Virgin's Bower
Common; thickets, woodland borders, roadsides. All counties except RO.

1109 ***Consolida ambigua** (L.) Ball & Heywood Rocket Larkspur
Delphinium ajacis L.
Infrequent, adventive; roadsides, waste places, abandoned homesites. AB, HL, KN, LK, LI, MH, RO, WA.

1110 # *Coptis trifolia* (L.) Salisb. Goldthread
Coptis groenlandica (Oeder) Fern.
Infrequent; sphagnous mats, boggy woods, sphagnous thickets. AB, CU, GE, LK, LR, PO, ST, SU, TR.

1111 *Delphinium exaltatum* Aiton Tall Larkspur
Rare; rocky slopes, thickets. ST: Canton, Case s.n., 28 Jul 1900 (OS 11964).

1112 *Delphinium tricorne* Michx. Dwarf Larkspur
Infrequent; rich dry and mesic woods, eroding slopes. CL, FA, LI, MH, MD, MO, PE, PC, RO, WA.

1113 *Hepatica acutiloba* DC. Sharp-lobed Hepatica, Liverleaf
Common; rich mesic woods, shaded hemlock ravines. ALL COUNTIES.

1114 *Hepatica americana* (DC.) Ker Round-lobed Hepatica
Common; rich dry and mesic woods, hemlock slopes. ALL COUNTIES.

1115 *Hydrastis canadensis* L. Golden-seal
Common; rich woods, wooded slopes, stream terraces. AS, AB, CU, FA, HL, LK, LI, LR, MD, MO, PE, PO, RI, RO, ST, SU, TR, WA.

1116 *Isopyrum biternatum* (Raf.) T. &. G. False Rue-anemone
Infrequent; alluvial stream terraces, open moist woods. AS, CU, KN, LK, LR, MD, MO, PC, RI, RO.

1117 *Ranunculus abortivus* L. Kidney-leaf Buttercup
Common; rich woods, wooded floodplains, clearings, shaded slopes. ALL COUNTIES.

1118 *Ranunculus acris* L. Tall Buttercup
Common, naturalized; fields, roadsides. AS, AB, CL, CU, FA, GE, HL, KN, LK, LR, MH, MD, PE, PO, RI, ST, SU, TR, WA.

1119 *Ranunculus allegheniensis* Britton Allegheny Buttercup
Infrequent; rich woods, rocky wooded slopes. CL, HL, MH, PO, RO, TR.

1120 *Ranunculus ambigens* S. Watson Water-plantain Spearwort
Frequent; pond margins, weedy floodplains, oxbows. CU, FA, GE, LK, LI, LR, PO, RI, RO, ST, SU, WA.

1121 *Ranunculus bulbosus* L. Bulbous Buttercup
Infrequent, naturalized; stream banks, wooded roadsides. AB, CL, PO, RI, ST.

1122 *Ranunculus flabellaris* Raf. Yellow Water-buttercup
Infrequent; mudflats, stream banks, ponds. AB, HL, KN, LR, PE, RI, TR, WA.

1123 *Ranunculus hispidus* Michx. Swamp Buttercup
Common; dry and moist woods, wooded stream terraces, weedy floodplains. AS, CL, CU, FA, HL, KN, LI, LR, MH, MD, MO, PE, PC, PO, RO, ST, SU, TR, WA.

1124 *Ranunculus longirostris* Godr. White Water-buttercup
Frequent; shallow waters, ponds, lakes. AB, CU, FA, GE, LK, LI, LR, MO, PC, PO, RO, ST, SU.

1125 *Ranunculus micranthus* Nutt. Small-flowered Buttercup
Infrequent; slopes, dry woods. CL, FA, HL, LI, LR, MO, PE, PC, RO.

1126 *Ranunculus pensylvanicus* L. f. Bristly Crowfoot
Common; pond and lake margins, wet woods, boggy thickets. AS, AB, CL, CU, FA, GE, HL, KN, LK, LI, LR, MH, MD, PC, PO, ST, SU, WA.

1127 *Ranunculus recurvatus* Poiret
Common; rich woods, stream terraces, woodland stream margins, thickets. ALL COUNTIES.

1128 *Ranunculus repens* L. Creeping Buttercup
Frequent, naturalized; roadsides, pastures, waste places. AS, AB, CL, CU, FA, GE, HL, KN, LK, LI, LR, PO, ST, SU, WA.

1129 *Ranunculus sceleratus* L. Cursed Crowfoot
Common; wet fields, woodland streams, marshes, stream terraces. AS, AB, CU, FA, GE, HL, KN, LK, LR, MH, MD, MO, PE, PO, RO, ST, SU, WA.

1130 *Ranunculus septentrionalis* Poiret Swamp Buttercup
Common; wooded floodplains, wet ditches, marshes, stream and pond margins, thickets. ALL COUNTIES.

1131 **Thalictrum dasycarpum** Fisch. & Lall. Purple Meadow-rue
Common; boggy thickets, sedge meadows, fens, wet fields. All counties except PE, PC, RO.

1132 **Thalictrum dioicum** L. Early Meadow-rue
Common; rich woods, shaded slopes, wooded floodplains. ALL COUNTIES.

1133 **Thalictrum pubescens** Pursh Tall Meadow-rue
Thalictrum polygamum Muhl.
Common; roadsides ditches, stream banks, thickets. All counties except MO, SU.

1134 **Thalictrum revolutum** DC. Skunk Meadow-rue, Wax-leaved Meadow-rue
Infrequent; dry woods, thickets, prairie remnants. FA, HL, KN, MH, PE, PO, RI, RO, TR.

1135 # **Trollius laxus** Salisb. Spreading Globe-flower
Rare; wooded stream terraces, thickets, meadows. AB: fen seep, Wayne Twp., Bissell 1984:49,
6 Jun 1984 (KE 47315). MH: swamp woods, Boardman Twp., Cooperrider 12743, 29 Apr
1978 (KE 39836). ST: Canton Bog, Case s.n., 1900 (OS 11851). WA: Wooster cemetery,
Fullerton s.n., 22 Jun 1938 (Miami University Herbarium 42172).

RESEDACEAE (Mignonette Family)

1136 ***Reseda luteola** L. Dyer's Rocket
Rare, adventive; waste places, roadsides. LI: Newark, Kellerman & Kellerman s.n., 5 Jul
1895 (OS 14619).

RHAMNACEAE (Buckthorn Family)

1137 **Ceanothus americanus** L. New Jersey-tea
Common; dry roadside banks, prairie remnants, dry open woods. All counties except RI.

1138 # **Rhamnus alnifolia** L'Her. Alder-leaved Buckthorn
Infrequent; fens, sedge meadows, boggy thickets. AB, CL, CU, GE, LK, PO, ST, SU.

1139 ***Rhamnus cathartica** L. European Buckthorn
Infrequent, naturalized; disturbed fields, woodland borders, fens. KN, LK, LR, PO, SU.

1140 ***Rhamnus frangula** L. European Alder Buckthorn
Common, naturalized; boggy thickets, sphagnous thickets, bogs, fens, wet woods, marshes,
waste places. AS, AB, CL, CU, GE, KN, LK, LI, LR, MH, MD, PO, RI, ST, SU, TR, WA.

1141 **Rhamnus lanceolata** Pursh Buckthorn
Infrequent; thickets, woodland borders. FA, KN, LI, PC, RO.

ROSACEAE (Rose Family)

1142 **Agrimonia gryposepala** Wallr. Tall Hairy Agrimony
Common; rich mesic and moist woods, thickets, roadside banks. All counties except PE,
PC.

1143 **Agrimonia parviflora** Aiton Small-flowered Agrimony
Common; swamps, wet meadows, damp thickets. All counties except PC.

1144 **Agrimonia pubescens** Wallr. Hairy Agrimony
Common; rich dry and mesic woods, dry slopes, moist thickets. All counties except MH,
MO, PC.

1145 **Agrimonia rostellata** Wallr.
Infrequent; open dry and mesic woods. AS, CU, HL, KN, LR, PC, ST, TR, WA.

1146 # **Agrimonia striata** Michx. Hairy Agrimony
Infrequent; woodland borders, stream margins, thickets. AB, CU, GE, LK, PO, TR.

1147 **Amelanchier arborea** (Michx. f.) Fern. Downy Serviceberry
Common; dry slopes, mesic and dry woods, thickets, fens, sphagnous thickets. All counties
except PC.

1148 # **Amelanchier laevis** Wieg. Allegheny Serviceberry
Infrequent; sphagnous thickets, fens, marshes. AB, GE, KN, LK, LR, MD, PO, RI, ST, TR.
Braun (1961) indicates a record from CU.

1149 **Amelanchier spicata** (Lam.) K. Koch
Amelanchier humilis Wieg.
Infrequent; roadsides, fields, rocky woods. AS, AB, CL, PO, ST, SU, TR.

1150 *Aronia prunifolia* (Marsh.) Rehder   Black Chokeberry
 Pyrus melanocarpa (Michx.) Willd.
 Common; boggy thickets, fens, bog margins, dry woods, rock ledges. All counties except PC.

1151 *Aronia prunifolia* (Marsh.) Rehder x *A. arbutifolia* (L.) Ell.   Purple Chokeberry
 Pyrus floribunda Lindley
 Frequent; sphagnous thickets, fens, bog margins, dry clearings. AB, CU, GE, LK, LI, LR,
 PO, RI, ST, SU, TR, WA.

1152 *Aruncus dioicus* (Walter) Fern.   Goat's-beard
 Infrequent; rich woods, ravines, wooded roadside banks. CL, FA, LI, MH, PE, RO.

1153 **Chaenomeles lagenaria* (Loisel.) Koidz.   Japanese Quince
 Rare, adventive; roadsides, abandoned homesites. FA: roadside ditch, Violet Twp., Goslin
 44-76-10, 30 Apr 1971 (BHO 31300). MD: roadside, Montville Twp., Mutz 1104, 3 Jul 1965
 (KE 14615). PE: fence row, Thorn Twp. Cline 1426, 7 May 1976 (KE 43660). WA: West
 Salem, Wilkinson s.n., 30 Apr 1887 (KE 19297).

1154 **Cotoneaster pyracantha* (L.) Spach   Fire-thorn
 Rare, adventive; roadsides, abandoned homesites, waste places. PE: steep roadbank, Hopewell
 Twp., Cline 1260, 15 Sep 1975 (KE 43659).

1155 *Crataegus calpodendron* (Ehrh.) Medicus
 Infrequent; open woods, woodland borders. LK, LR, PE, PO, ST, TR, WA. Braun (1961)
 indicates records for FA, LI.

1156 *Crataegus coccinea* L.   Cockspur Thorn
 Common; open woods, fence rows, pastures. AS, AB, CU, FA, GE, HL, KN, LK, LI, LR,
 MD, PO, RI, ST, SU, TR, WA.

1157 *Crataegus crus-galli* L.   Cockspur Thorn
 Common; fields, thickets, open woods. All counties except GE, HL, MO.

1158 *Crataegus flabellata* (Spach) Kirchner
 Common; fields, clearings. AS, AB, CL, FA, GE, HL, KN, LK, LI, LR, MH, MD, MO, PE,
 PO RI, ST, SU, TR.

1159 *Crataegus intricata* Lange
 Infrequent; thickets, woodland borders. FA, PE, PC, SU, TR. Braun (1961) indicates a
 record for KN.

1160 *Crataegus mollis* Scheele
 Frequent; moist thickets, fields. AS, FA, LK, LR, MD, MO, PC, RI, RO, ST, SU, WA.

1161 **Crataegus monogyna* Jacq.   English Hawthorn
 Infrequent, adventive; pastures, roadsides, fields. AB, CU, GE, LK, LR, MD.

1162 *Crataegus pruinosa* (Wendl.) K. Koch
 Frequent; dry fields, thickets, open woods. AS, AB, FA, GE, LK, LR, MD, PE, PC, PO,
 RO, ST, SU, TR, WA. Braun (1961) indicates records for MH, RI.

1163 *Crataegus punctata* Jacq.   Dotted Hawthorn
 Common; moist wooded slopes, open woods. All counties except PC.

1164 *Crataegus rotundifolia* Moench
 Rare; thickets, rocky slopes, roadsides. GE: roadside, Russell Twp., Hawver 326, 30 Sep
 1960 (KE 2021). TR: Andrews Creek, Mesopotamia Twp., Rood 1939, 25 May 1941 (KE
 123416).

1165 *Crataegus succulenta* Link
 Infrequent; hillsides, thickets. AS, FA, KN, LK, PO, RI, RO, ST, SU, TR, WA.

1166 *# Dalibarda repens* L.   Robin Run-away, Dewdrop
 Rare; damp boggy woods, hemlock woods. AB: woods, Morgan Twp., Andreas 5441, 7 Aug
 1980 (KE 43063); *Sphagnum* moss flats, Rome Twp., Bissell 1979:56, 21 Jun 1979 (CLM
 12365); near Amboy, Hicks s.n., 18 Aug 1932 (OS s.n.). GE: edge of woodland pools, Montville
 Twp., Bissell 1983:95, 16 Jul 1983 (CLM 19717); near Chardon, Hicks s.n., 3 Jul 1928 (OS
 20522).

1167 **Duchesnea indica* (Andr.) Focke   Indian-strawberry
 Infrequent, adventive; waste places, fields. AS, AB, FA, LK, PC, RO, SU.

1168 *Filipendula rubra* (Hill) Robins.   Queen-of-the-prairie
 Infrequent; moist meadows, fens, prairie remnants. AS, AB, CL, CU, FA, HL, LI, LR, MH,
 RO.

1169 *Filipendula ulmaria* (L.) Maxim.   Queen-of-the-meadow
Rare, naturalized; thickets, roadsides, wet meadows. AB: roadside, Kellogsville, Hicks s.n., 16 Jul 1929 (OS 20367). PO: boggy thicket, Freedom Twp., Andreas 2266, 6 Jul 1978 (KE 49684). RI: Mansfield, Wilkinson s.n., Jul 1888 (OS 19365). RO: field, Colerain Twp., Bartley s.n., 19 Jun 1959 (OS 63683).

1170 *Fragaria chiloensis* (L.) Miller   Garden Strawberry
Rare, adventive; waste places. KN: railroad ballast, Howard Twp., Pusey 1571, 5 Oct 1974 (KE 38634).

1171 **Fragaria vesca** L.   Woodland Strawberry
Common; shaded roadside banks, dry open woods, rocky slopes. AS, AB, CL, CU, FA, GE, HL, KN, LK, LI, LR, MH, MD, MO, PO, RI, SU, TR, WA.

1172 **Fragaria virginiana** Miller   Common Wild Strawberry
Common; roadsides, fields, woodland borders, railroads. All counties except PC.

1173 **Geum aleppicum** Jacq.
Frequent; moist thickets, fields. AS, AB, CU, FA, GE, KN, LK, LI, LR, PC, PO, RO, ST, SU, WA.

1174 **Geum canadense** Jacq.
Common; moist and mesic woods, woodland borders, thickets. All counties except RO.

1175 **Geum laciniatum** Murray
Common; damp thickets, meadows, woodland borders. ALL COUNTIES.

1176 # **Geum rivale** L.   Water Avens, Purple Avens
Infrequent; fens, thickets, wet meadows, sphagnous thickets. AB, CU, GE, LK, PO, SU.

1177 **Geum vernum** (Raf.) T. & G.
Common; rich mesic and moist woods, thickets. AS, AB, CU, FA, GE, KN, LI, LR, MD, MO, PE, PC, PO, RI, RO, SU, WA.

1178 **Geum virginianum** L.
Frequent; dry woods, thickets, rocky slopes. AB, CL, CU, FA, GE, LK, LI, LR, MH, MD, PE, RO, SU, WA.

1179 *Kerria japonica* (L.) DC.   Kerria
Rare, adventive; roadsides, fields. MD: pasture border, Granger Twp., Mutz 1034, 9 Jun 1965 (KE 14632).

1180 **Malus angustifolia** (Aiton) Michx.   Narrow-leaved Crab
Pyrus angustifolia Aiton
Once rare, now presumably extirpated from Ohio; woods, thickets. WA: Killbuck bottom, Duvel s.n., 17 May 1898 (OS 96920).

1181 **Malus coronaria** (L.) Miller   Wild Crab
Pyrus coronaria L.
Common; thickets, fields, pastures, open woods. ALL COUNTIES.

1182 *Malus pumila* Miller   Apple
Pyrus malus L.
Common, naturalized; roadsides, thickets, woodland borders, abandoned homesites, fields. ALL COUNTIES.

1183 **Physocarpus opulifolius** (L.) Maxim.   Ninebark
Common; dry slopes, moist floodplains, fens, thickets. All counties except MO.

1184 **Porteranthus stipulatus** (Willd.) Britton   American Ipecac
Gillenia stipulata (Muhl.) Baill.
Rare; railroads, thickets, rocky woods. FA: Berne Twp., Goslin s.n., Jul 1950 (BHO 22415). PE: mesic woods, Bearfield Twp., Cline 709, 26 Jun 1975 (KE 43654). RO: Mt. Logan, Chillicothe, Kellerman s.n., 12 Aug 1898 (OS 20325). SU: railroad, Hudson, Herrick s.n., 28 Jun 1957 (OS 58666).

1185 **Porteranthus trifoliatus** (L.) Britton   Bowman's Root
Gillenia trifoliata (L.) Moench
Rare; rich woods, roadside banks. CL: N-facing slope, Middleton Twp., Andreas 6100, 23 Jun 1983 (KE 46486); wet dripping cliff, Middleton Twp., Cooperrider 5665, 21 Jun 1960 (KE 5382). TR: roadside thicket, Newton Twp., Rood 1075, 27 Jun 1937 (KE 12356).

1186 *Potentilla argentea* L.   Silvery Cinquefoil
Frequent, adventive; dry fields, waste places. AB, CU, GE, HL, KN, LK, LI, PO, RO, ST, SU, WA.

1187 *Potentilla arguta* Pursh   Tall Cinquefoil
Infrequent; sandy fields, alluvial soils, roadsides. CU, LK, LI, PO, RO.

1188 *Potentilla canadensis* L.
Common; dry open soils, roadside banks, dry and mesic woods, rocky fields. ALL COUNTIES.

1189 *Potentilla fruticosa* L.   Shrubby Cinquefoil
Infrequent; fens, sedge meadows. AB, FA, HL, LK, PO, RO, ST, SU.

1190 *Potentilla inclinata* Vill.
Potentilla canescens Bess.
Rare, adventive; roadsides, waste places. PC: cemetery near Circleville, Bartley s.n., 30 May 1968 (OS 104324).

1191 *Potentilla norvegica* L.   Norwegian Five-fingers
Common; dry fields, thickets, roadsides, waste places. ALL COUNTIES.

1192 # *Potentilla palustris* (L.) Scop.   Marsh Five-fingers
Frequent; kettle-hole bogs, wet meadows, marshes, sphagnous thickets. AS, AB, GE, HL, LI, LR, PO, ST, SU, TR, WA.

1193 *Potentilla recta* L.   Upright Cinquefoil
Common, naturalized; dry fields, roadsides, waste places. ALL COUNTIES.

1194 *Potentilla reptans* L.
Rare, adventive; lawns, roadsides. LR: lawns, Oberlin, Jones 71-7-1-269, 1 Jul 1971 (KE 30134).

1195 *Potentilla simplex* Michx.   Field Cinquefoil
Common; dry fields, open woods, roadsides. All counties except MH, PC, RO.

1196 *Prunus americana* Marsh.   Wild Plum
Common; thickets, woodland borders, stream margins. ALL COUNTIES.

1197 *Prunus avium* L.   Sweet Cherry
Common, adventive; roadsides, pastures, woodland borders. All counties except PC.

1198 *Prunus cerasus* L.   Sour Cherry
Infrequent, adventive; roadsides, fields, thickets. CU, GE, HL, LK, LR, MD, PE, ST, TR.

1199 *Prunus mahaleb* L.   Perfumed Cherry
Rare, adventive; woodland borders, roadsides. KN: disturbed area, Liberty Twp., Pusey 853, 12 Sep 1975 (KE 38620). TR: woods, Vernon Twp., Burns 276, 27 Apr 1977 (KE 46311).

1200 *Prunus nigra* Aiton   Canada Plum
Rare; thickets, moist meadows, swamps. Braun (1961) indicates a record for CU.

1201 # *Prunus pensylvanica* L. f.   Pin Cherry, Fire Cherry
Frequent; dry sandy fields, abandoned quarries, burned fields. AS, AB, CL, CU, GE, LK, LR, MD, PO, ST, SU, TR.

1202 *Prunus persica* (L.) Batsch   Peach
Frequent, adventive; roadsides, fields, thickets, pastures. AS, AB, CU, HL, KN, LK, LR, PE, PO, RO, ST, WA.

1203 *Prunus serotina* Ehrh.   Wild Cherry
Common; dry and mesic woods, thickets, fence rows, fields. All counties except PC, RO.

1204 *Prunus virginiana* L.   Choke Cherry
Common; thickets, fence rows, roadsides, woodland borders. All counties except PE.

1205 *Pyrus communis* L.   Pear
Common, adventive; roadsides, pastures, old fields. AS, AB, Cl, CU, GE, KN, LK, LR, MH, MD, MO, PE, PO, RI, ST, SU, TR, WA.

1206 *Rosa blanda* Aiton   Smooth Rose
Infrequent; dry rocky slopes, fields. CU, HL, KN, LK, LR, PO, RI.

1207 *Rosa canina* L.   Dog-rose
Common, naturalized; roadsides, railroads, fields. AS, AB, CL, CU, GE, KN, LK, LI, LR, MH, MD, MO, PE, PC, PO, ST, SU, TR, WA.

1208 **Rosa carolina** L.
Common; dry fields, eroding slopes, woodland borders. ALL COUNTIES.

1209 ***Rosa cinnamomea** L.   Cinnamon Rose
Rare, adventive; roadsides, fence rows. AB: escape, Windsor Twp., Hicks s.n., 11 Jun 1932 (OS 20626).

1210 ***Rosa eglanteria** L.   Sweet-brier
Common, naturalized; thickets, fields, roadsides. All counties except CU, PE.

1211 ***Rosa micrantha** Smith   Small-flowered Rose
Rare, adventive; roadsides, thickets, fields, railroads. AS: Savanah, Hanwood s.n., 14 Jun 1953 (OS 48648). PO: along railroad, Garrettsville, Webb 1645, 16 Aug 1905 (KE 12372). ST: canal towpath, Lawrence Twp., Andreas 3817, 18 Jun 1979 (KE 44330). WA: roadside, Sugar Creek Twp., Andreas 923, 27 May 1977 (KE 49685).

1212 ***Rosa multiflora** Murray   Multiflora Rose
Common, naturalized; pastures, roadsides, waste places, thickets, railroads. All counties except RO.

1213 **Rosa palustris** Marsh.   Swamp Rose
Common; wet fields, swamps, marshes, thickets, fens. All counties except PC.

1214 ***Rosa rugosa** Thunb.
Rare, adventive; roadsides, thickets, waste places. SU: Stow, Hobbs s.n., 7 Jun 1937 (KE 1439).

1215 **Rosa setigera** Michx.   Climbing Rose, Prairie Rose
Common; wet thickets, clearings, roadsides. ALL COUNTIES.

1216 ***Rosa wichuriana** Crepin   Memorial Rose
Rare, adventive; woodland borders, roadsides, abandoned homesites. AB: abandoned dwelling, Windsor Twp., Hicks s.n., 18 Jun 1931 (OS 20579). HL: open hillside, Hardy Twp., Wilson 315, 4 Jul 1970 (KE 29836). LR: upland, Henrietta Twp., Peabody 218, 25 Jul 1971 (KE 32237).

1217 **Rubus allegheniensis** Porter   Blackberry
Common; roadsides, thickets, disturbed woods, waste places. All counties except MH, PC.

1218 **Rubus flagellaris** Willd.   Dewberry
Common; roadsides, fields, open woods, waste places. All counties except CL, MO, PC.

1219 **Rubus hispidus** L. var. **obovalis** (Michx.) Fern.   Small Dewberry
Frequent; boggy thickets, wet woods, fields, woodland borders. AS, AB, CU, GE, LK, LI, LR, MD, PO, ST, SU, TR, WA.

1220 ***Rubus laciniatus** Willd.   Cut-leaf Blackberry
Rare, adventive; roadsides, waste places. PO: roadside, Nelson Twp., Webb 2513, 15 Jun 1903 (KE 12343).

1221 **Rubus occidentalis** L.   Black Raspberry
Common; dry and moist woodland borders, thickets, floodplains, railroads, waste places. ALL COUNTIES.

1222 **Rubus odoratus** L.   Flowering Raspberry
Common; steep shaded slopes, ravines. AS, AB, CL, CU, FA, GE, HL, KN, LK, LR, MH, PE, PO, RO, SU, TR, WA.

1223 **Rubus pensylvanicus** Poiret   Blackberry
Common; thickets, swamps, woodland borders. AS, AB, CL, CU, GE, HL, KN, LK, LR, MH, MO, PE, PO, RI, SU, TR, WA.

1224 **Rubus pubescens** Raf.   Dwarf Red Bramble
Infrequent; sphagnous thickets, fens. AB, CL, GE, LK, LR, PO, ST, SU, TR.

1225 **Rubus setosus** Bigelow
Once rare, now presumably extirpated from Ohio; thickets, open woods. TR: open pasture field, Braceville Twp., Rood 2467, 7 Jun 1949 (KE 12354 & KE 13540).

1226 **Rubus strigosus** Michx.   Wild Red Raspberry
Rubus idaeus L. var. *strigosus* (Michx.) Maxim.
Frequent; fens, sedge meadows, thickets. AB, CL, CU, GE, LI, LR, MD, PO, ST, SU, TR, WA.

1227 *Sanguisorba canadensis* L. Canada Burnet
Frequent; fens, wet fields, thickets. AB, CU, GE, HL, KN, LK, PC, PO, RO, ST, SU.

1228 **Sorbaria sorbifolia* (L.) A. Br. False Spiraea
Infrequent, adventive; moist ravines, dry slopes. LK, PO, ST, SU, WA.

1229 **Sorbus aucuparia* L. Mountain-ash
Pyrus aucuparia (L.) Gaertner
Infrequent, adventive; thickets, pastures, roadsides. AB, CU, GE, LK, LR, MD, PO, SU.

1230 # *Sorbus decora* (Sarg.) Schneider Mountain-ash
Pyrus decora (Sarg.) Hyland
Rare; swampy woods, rocky slopes. AB: woods, Harpersfield Twp., Cusick & Bissell 18822,
13 Sep 1978 (OS s.n.). GE: ravine, Chardon, Bissell 1982-182, 8 Jul 1982 (CLM 17273);
near Burton, Hicks s.n., 8 Aug 1928 (OS 21067). TR: east Johnson Twp., Shanks s.n., 30
Sep 1934 (OS 21065).

1231 *Spiraea alba* Duroi var. *alba* Meadow-sweet
Common; moist fields, thickets, ditches, marshes. All counties except PC, RO.

1232 *Spiraea tomentosa* L. Hardhack, Steeple-bush
Common; dry fields, pond and bog margins, roadside banks. AS, AB, CL, CU, FA, GE, HL,
LK, LI, LR, MH, MD, PE, PO, RO, ST, SU, TR, WA.

1233 **Spiraea* x *vanhouttei* (Briot) Carr. Bridal-wreath
Rare, adventive; roadsides, waste places. MD: roadside, Sharon Twp., Mutz 874, 10 Jun
1965 (KE 14681).

1234 *Waldsteinia fragarioides* (Michx.) Tratt. Barren Strawberry
Frequent; wooded floodplains, rich moist woods, ledges, stream margins. AB, CU, GE, HL,
KN, LK, LI, MD, PO, RI, SU, TR, WA.

RUBIACEAE (Madder Family)

1235 *Cephalanthus occidentalis* L. Buttonbush
Common; swamps, kettle-hole depressions, pond margins, roadside ditches, bogs. ALL
COUNTIES.

1236 *Diodia teres* Walter Buttonweed
Infrequent; thickets, railroads, waste places. AB, CU, FA, LK, PE, PO, RO, ST, SU.

1237 *Galium aparine* L. Cleavers, Goosegrass
Common; rich mesic and moist woods, wooded floodplains, thickets, marshes. ALL COUNTIES.

1238 *Galium asprellum* Michx. Rough Bedstraw
Common; wet woods, floodplains, thickets, fens, woodland borders. AS, AB, CL, CU, FA,
GE, HL, KN, LK, LI, LR, MH, MD, PO, RI, ST, SU, TR, WA.

1239 *Galium boreale* L. Northern Bedstraw
Infrequent; fens, thickets, sedge meadows. AB, CU, GE, PO, RO, ST, SU, TR.

1240 *Galium circaezans* Michx. Wild Licorice
Common; rich dry, mesic and moist woods. ALL COUNTIES.

1241 *Galium concinnum* T. & G. Elegant Bedstraw
Common; dry mesic and moist woods, roadsides, woodland borders. All counties except PC.

1242 # *Galium labradoricum* (Wieg.) Wieg. Bog Bedstraw
Rare; sphagnous thickets, fens. CL: sedge meadow, Butler Twp., Andreas 6063, 31 May 1983
(KE 47234). PO: bog, Streetsboro Twp., Cooperrider 5571, 18 Jun 1960 (KE 1480); bog,
Mantua Twp., Rood 927, 21 May 1933 (KE 16198 & OS 36595).

1243 *Galium lanceolatum* Torrey Lance-leaved Bedstraw
Common; dry woods, rocky slopes. AS, AB, CU, FA, GE, HL, KN, LK, LI, LR, PE, PO,
RI, RO, ST, SU, TR, WA.

1244 **Galium mollugo* L.
Common, naturalized; roadsides, railroads, disturbed woods. AS, AB, CL, CU, GE, HL, KN,
LK, MH, MO, PC, PO, RO, ST, SU, WA.

1245 *Galium obtusum* Bigelow
Common; wet ditches, sedge meadows, fens, wooded floodplains. AS, AB, CL, CU, GE, HL,
KN, LK, LR, MD, MO, PO, RI, ST, SU, TR, WA.

1246 *Galium odoratum* (L.) Scop. Sweet Woodruff
Asperula odorata L.
Rare, naturalized; dry woods, roadsides, rocky woods. LK: secondary woods, North Perry, Bissell 1984:27, 24 May 1984 (CLM 21271). PO: dry rocky woods, Nelson Twp., Andreas & Cooperrider 842, 20 May 1977 (KE 49686). SU: Summit County, Herrick s.n., 1 May 1955 (OS s.n.).

1247 # *Galium palustre* L. Marsh Bedstraw
Rare; calcareous thickets, wet meadows. AB: pasture seep, Richmond Twp., Andreas 5318, 2 Jul 1980 (KE 43133); marshy ground, Pierpont, Cooperrider 6231, 1 Jun 1960 (KE 6321).

1248 *Galium pedemontanum* (Bellardi) All.
Rare, naturalized; thickets, pastures. RO: sandy pasture, Bartley s.n., 4 Jun 1961 (OS s.n.).

1249 *Galium pilosum* Aiton Hairy Bedstraw
Common; dry woods, roadside banks, thickets. AS, AB, CL, CU, FA, GE, HL, KN, LK, LI, LR, MH, PO, RO, ST, WA.

1250 *Galium tinctorium* L.
Common; sedge meadows, fens, marshes, pond margins, moist fields. ALL COUNTIES.

1251 # *Galium trifidum* L. Three-cleft Bedstraw
Infrequent; sphagnous thickets, swamps, fen margins. AB, GE, PO, ST, SU.

1252 *Galium triflorum* Michx. Sweet-scented Bedstraw
Common; rich dry and mesic woods, rocky slopes. ALL COUNTIES.

1253 *Galium verum* L. Yellow Bedstraw
Infrequent, naturalized; roadsides, fields. AB, HL, LK, LR, PC.

1254 *Houstonia caerulea* L. Bluets
Common; dry slopes, open woods, pastures, fields, thickets. ALL COUNTIES.

1255 *Houstonia canadensis* Willd.
Infrequent; dry open woods, rocky hillsides, eroding slopes. AB, CU, GE, LK, LI, PC, RO, SU, WA.

1256 *Houstonia longifolia* Gaertner Long-leaved Bluets
Infrequent; rocky slopes, dry open woods. AS, AB, CL, FA, HL, KN, LI, PE, RO, ST.

1257 *Mitchella repens* L. Partridge-berry
Common; dry and moist woods, boggy thickets. All counties except MO, PC, RO.

1258 *Sherardia arvensis* L. Field-madder
Rare, adventive; fields, disturbed soils, waste places. CU: Berea, Watson s.n., 1894 (OS 36722. LR: disturbed soil, Oberlin, Jones s.n., 27 May 1905 (KE 14477). PO: weed under spruce, Kent, Cooperrider 12466, 18 Jun 1977 (KE 39154).

RUTACEAE (Rue Family)

1259 *Ptelea trifoliata* L. Hop-tree, Wafer-ash
Frequent; moist woods, wooded floodplains. AB, CU, KN, LK, LI, LR, PO, RI, RO, ST, WA.

1260 *Zanthoxylum americanum* Miller Prickly-ash
Common; wooded floodplains, moist thickets. ALL COUNTIES.

SALICACEAE (Willow Family)

1261 *Populus alba* L. Silver Poplar
Common, naturalized; roadsides, fields, abandoned homesites, railroads. ALL COUNTIES.

1262 *Populus balsamifera* L. Balsam Poplar
Rare; river banks, floodplains. AB: thicket along river, Harpersfield Twp., Bissell 1984:58, 10 Jun 1984 (CLM 21479); Ashtabula, Kellerman s.n., 24 Jun 1899 (OS 46306). GE: Thompson Ledges, Tyler s.n., 11 Jun 1901 (OS 46303).

1263 *Populus canescens* (Aiton) J. E. Smith Gray Poplar
Infrequent, adventive; woodland borders, roadsides. CU, LK, LR, PE, PO, RI.

1264 *Populus deltoides* Marsh. Cottonwood
Common; wooded floodplains, swamps. ALL COUNTIES.

1265 *Populus* x *gileadensis* Rouleau (*P. balsamifera* x *P. deltoides*)   Balm-of-Gilead
Infrequent, adventive; roadsides, waste places. AS, AB, HL, LK, LR, MD, PO, TR.

1266 *Populus grandidentata* Michx.   Big-tooth Aspen
Common; clearings, woodland blow-downs, fens, waste places, fields. ALL COUNTIES.

1267 *Populus heterophylla* L.   Swamp Cottonwood
Infrequent; inundated swamps, bottomlands. AS, KN, LK, LR, MD, RI, ST.

1268 *Populus nigra* L.   Black Poplar, Lombardy Poplar
Infrequent, adventive; sandy fields, beaches, railroads, abandoned homesites. AS, KN, LK,
LR, ST, TR.

1269 *Populus tremuloides* Michx.   Quaking Aspen
Common; fens, pond margins, abandoned fields, woodland borders. All counties except RO.

1270 *Salix alba* L.   White Willow
Common, naturalized; roadsides, pond and stream margins, wet thickets. AS, AB, CL, CU,
FA, HL, KN, LK, LI, LR, MH, MO, RO, ST, SU, TR, WA.

1271 *Salix amygdaloides* Andersson   Peach-leaf Willow
Frequent; bog and lake margins, swamps. AB, CU, GE, LK, LR, PO, RI, ST, SU, TR, WA.

1272 *Salix babylonica* L.   Weeping Willow
Frequent, adventive; ponds, lake margins, streams, abandoned homesites. AB, FA, HL, KN,
LK, LR, MD, PO, ST, TR, WA.

1273 *Salix bebbiana* Sarg.   Long-beaked Willow, Bebb's Willow
Infrequent; fens, muck soils. AB, LK, LR, ST, SU.

1274 # *Salix candida* Fluegge   Sage-leaf Willow, Hoary Willow
Infrequent; fens, thickets. CL, LR, PO, ST, SU.

1275 *Salix discolor* Muhl.   Pussy Willow
Common; swamps, marshes, moist thickets, fens. All counties except RI.

1276 *Salix eriocephala* Michx.   Heart-leaved Willow
Salix rigida Muhl.
Common; pond and bog margins, wet meadows, fen thickets. All counties except MO, PC,
RO.

1277 *Salix exigua* Nutt.   Sandbar Willow
Salix interior Rowlee
Common; mudflats, stream margins, thickets. All counties except MO, RI.

1278 *Salix fragilis* L.   Crack Willow, Brittle Willow
Common, naturalized; thickets, pond margins, stream banks. AB, CL, CU, FA, GE, HL,
LK, LI, LR, MD, PO, RI, ST, SU, TR, WA.

1279 *Salix* x *glatfelteri* Schneider (*S. amygdaloides* x *S. nigra*)
Rare; lake margins, sedge meadows, fens. AB: Ashtabula County, Griggs s.n., 7 Sep 1905
(OS 26999). PO: lake margin, Mantua Twp., Cooperrider 8541, 22 Jul 1961 (KE 6461); bog
margin, Streetsboro Twp., Cooperrider 8538, 8 Jul 1961 (KE 6334).

1280 *Salix humilis* Marsh. (including *S. tristis* Aiton)   Upland Willow, Prairie Willow
Frequent; dry eroding slopes, fens, dry thickets. AS, CL, FA, GE, KN, LK, LI, LR, MD,
PO, RI, RO, SU, TR, WA.

1281 *Salix lucida* Muhl.   Shining Willow
Frequent; fens, thickets, sedge meadows. AS, AB, CL, CU, GE, KN, LK, LR, MO, PO, RI,
ST, SU, TR, WA.

1282 *Salix nigra* Marsh.   Black Willow
Common; ponds, lake and stream margins, moist thickets. All counties except CL, MH, PC.

1283 # *Salix pedicellaris* Pursh   Bog Willow
Infrequent; fens, sphagnous thickets. AS, AB, LI, PE, PO, ST, SU, WA.

1284 *Salix petiolaris* J. E. Smith   Slender Willow
Rare; moist meadows, thickets. AB: Morgan Swamp, Hicks s.n., 20 Jun 1931 (OS 81581).
LR: Camden Lake, Camden Twp., Mossman s.n., 12 Sep 1970 (OS 107115). Braun (1961)
indicates a record for SU.

1285 *Salix purpurea* L.   Purple Osier, Basket Willow
Infrequent, adventive; wet grounds, thickets. AB, CL, FA, GE, LK, LR, RO, ST.

1286 **Salix sericea** Marsh.   Silky Willow
Common; sphagnous thickets, fens. AS, AB, CL, CU, FA, GE, HL, KN, LK, LI, LR, MD,
PE, PO, ST, SU, TR, WA.

1287 **Salix x rubens** Schrank (**S. alba** x **S. fragilis**)
Rare; thickets. AB: moist seep, Morgan Twp., Bissell 1982:71, 28 May 1982 (CLM 18172).

1288 **# Salix serissima** (Bailey) Fern.   Autumn Willow
Infrequent; fens, thickets. AB, CL, GE, PO, ST, SU.

1289 **Salix x subsericea** (Andersson) Schneider (**S. petiolaris** x **S. sericea**)
Rare; fens, thickets. AB: Ashtabula, Kellerman s.n., 24 Jun 1899 (OS 27308). LR: Lorain
County, Grover s.n., 14 Mar 1903 (OS 27305). ST: marshy ground, Jackson Twp., Andreas
5055, 23 May 1980 (KE 43052). SU: fen, Green Twp., Cusick 10126, 4 Jul 1969 (KE 28323).

SANTALACEAE (Sandalwood Family)

1290 **Comandra umbellata** (L.) Nutt.   Bastard Toadflax
Common; dry oak woods, eroding slopes, sphagnous thickets. All counties except MO, RI.

SARRACENIACEAE (Pitcher-plant Family)

1291 **# Sarracenia purpurea** L.   Pitcher-plant
Frequent; bogs, sphagnous thickets, fens. AS, AB, GE, LK, LI, LR, PO, RI, ST, SU, WA.

SAURURACEAE (Lizard's-tail Family)

1292 **Saururus cernuus** L.   Lizard's Tail
Common; wooded floodplains, moist thickets, stream margins. AB, CU, FA, GE, HL, LK,
LI, LR, MD, PC, PO, RI, RO, ST, SU, TR.

SAXIFRAGACEAE (Saxifrage Family)

1293 **Chrysosplenium americanum** Hooker   Golden Saxifrage
Common; wooded shallow streams, seepage banks, moist thickets. All counties except PE,
PC, RO.

1294 **Heuchera americana** L.   Alumroot
Common; rich mesic and moist woods, wooded floodplains, fens, roadside banks. ALL
COUNTIES.

1295 **Hydrangea arborescens** L.   Hydrangea
Frequent, native in southern counties, adventive elsewhere; ravines, rich moist woods. AS,
CL, FA, GE, HL, KN, LI, MH, MO, PE, RI, RO, SU, WA.

1296 **Mitella diphylla** L.   Bishop's-cap
Common; dry, mesic, and moist woods, seepage areas, floodplains. ALL COUNTIES.

1297 **Parnassia glauca** Raf.   Grass-of-Parnassus
Frequent; fens, calcareous seepage slopes. CL, CU, FA, GE, LK, LR, PC, PO, RO, ST, SU.

1298 **Penthorum sedoides** L.   Ditch Stonecrop
Common; roadside ditches, marshes, wet meadows. ALL COUNTIES.

1299 ***Philadelphus coronarius** L.   Mock-orange
Infrequent, adventive; roadside thickets. CL, CU, MD, PE, ST, WA.

1300 **Ribes americanum** Miller   Wild Black Currant
Common; boggy thickets, swamps, moist shaded slopes, fens, cat-tail marshes, thickets. All
counties except PC.

1301 **Ribes cynosbati** L.   Wild Gooseberry
Common; rich dry and mesic woods, ravines. All counties except PC.

1302 **# Ribes glandulosum** Grauer   Skunk Currant
Once rare, now presumably extirpated from Ohio; wet woods. AB: boggy swamp, eastern
Monroe Twp., Hicks s.n., 1933 (OS s.n.).

1303 ***Ribes grossularia** L.   European Gooseberry
Rare, adventive; thickets, roadsides. LR: fence row, Oberlin, Cowles s.n., 2 Jul 1892 (OC
73370).

1304 **# Ribes hirtellum** Michx. Smooth Gooseberry
Infrequent; fens, sphagnous thickets, calcareous marshes. AB, GE, HL, LK, PO, ST, SU, TR, WA. Braun (1961) indicates a record for LR.

1305 ***Ribes odoratum** Wendland f. Buffalo Currant
Rare, naturalized; roadsides, thickets. LR: woods, Lorain County, Hardy s.n., May 1902 (OC 73344). WA: roadside bank, Milton Twp., Andreas 2645, 15 Aug 1978 (KE 49687).

1306 ***Ribes sativum** Syme Garden Currant, Red Currant
Infrequent, adventive; roadsides, thickets. AB, CU, LK, LR, MH, SU, TR.

1307 **# Ribes triste** Pallas Swamp Red Currant
Rare; moist woods, boggy thickets. AB: till plain, Denmark Twp., Bissell 1983:26, 19 May 1983 (CLM 19009). GE: swamp forest, Montville Twp., Bissell 1984:7, 9 May 1984 (KE 47317).

1308 **Saxifraga pensylvanica** L. Swamp Saxifrage
Common; boggy thickets, wooded and grassy floodplains, wet meadows, fens. AB, CL, CU, FA, GE, HL, LK, LR, MH, PO, RI, RO, ST, SU, TR, WA.

1309 **Saxifraga virginiensis** Michx. Early Saxifrage
Common; dry wooded slopes, rocky woods, dry roadside banks. All counties except LR, MO, PC.

1310 **Tiarella cordifolia** L. False Miterwort, Foamflower
Common; rich mesic and moist woods, ravines. AB, CL, CU, FA, GE LK, LR, MH, MD, MO, PE, PO, RO, ST, SU, TR.

SCROPHULARIACEAE (Figwort Family)

1311a **Agalinis purpurea** (L.) Pennell var. **parviflora** (Benth.) Biovin
Northern Purple Foxglove
Gerardia paupercula (Gray) Britton var. *borealis* (Pennell) Deam
Rare; fens, meadows. ST: marl fen, Jackson Twp., Andreas 5510, 10 Sep 1980 (KE 42867); bogs, Kuehnle 984, 14 Aug 1937 (KE 33529); prairie meadow, Massillon, Stair s.n., Aug 1898 (CLM s.n.).

1311b **Agalinis purpurea** (L.) Pennell var. **purpurea** Purple Foxglove
Gerardia purpurea L. var. *purpurea*
Infrequent; meadows, fens. FA, LK, LI, PC, RO, ST, WA.

1312 **Agalinis tenuifolia** (Vahl) Raf.
Gerardia tenuifolia Vahl
Common; dry fields, oak woods, woodland borders, shaded roadside banks. AS, AB, CL, CU, FA, GE, HL, KN, LK, LI, LR, MH, PE, PC, PO, RO, ST, SU, WA.

1313 ***Antirrhinum majus** L. Snapdragon
Rare, adventive; roadsides, gardens, waste places. MD: rubbish pile, Hinckley Twp., Jones 71-7-14-398, 14 Jul 1971 (OC 214,590). Cooperrider (1985a) indicates records for MH.

1314 **Aureolaria flava** (L.) Farw. Yellow False Foxglove
Gerardia flava L.
Frequent; dry sandy woods. AB, CU, FA, GE, HL, KN, LK, LR, MH, PO, RO, SU.

1315 **Aureolaria virginica** (L.) Pennell Downy False Foxglove
Gerardia virginica (L.) BSP.
Common; dry oak woods, sandy ridges. AS, AB, CL, CU, FA, GE, HL, KN, LK, LI, MH, MD, PE, PO, ST, SU, WA.

1316 **Castilleja coccinea** (L.) Sprengel Scarlet Painted-cup
Infrequent; dry eroding slopes, shaded roadside banks, dry open woods, sphagnous thickets, fens. AS, CL, CU, FA, GE, KN, PC, PO, SU.

1317 ***Chaenorrhinum minus** (L.) Lange Dwarf Snapdragon
Common, naturalized; railroads, parking lots. ALL COUNTIES.

1318 **Chelone glabra** L. Turtlehead
Common; stream margins, shaded roadside ditches, moist thickets, fens. ALL COUNTIES.

1319 **Collinsia verna** Nutt. Blue-eyed Mary
Frequent; rich woods, shaded ravines, thickets. CL, CU, LI, LR, MH, MD, PE, PO, RI, RO, ST, SU, TR.

1320 *Cymbalaria muralis* Gaertner, Meyer & Schreber   Kenilworth Ivy
Infrequent, naturalized; waste places, sidewalks, lawns. CU, FA, GE, HL, PO, SU.

1321 **Dasistoma macrophylla** (Nutt.) Raf.   Mullein Foxglove
Seymeria macrophylla Nutt.
Infrequent; rich woods, stream banks. AS, LI, LR, PE, PC, RO.

1322 ***Digitalis grandiflora*** Miller   Yellow Foxglove
Rare, adventive; roadsides, fields. PO: roadside ditch, Ravenna Twp., Andreas 2264, 6 Jul 1978 (KE 49765); grassy area, Kent, Cooperrider 12,362, 29 Jun 1974 (KE 35053 & KE 35054).

1323 ***Digitalis lanata*** Ehrh.   Wooly Foxglove
Rare, adventive; waste places, roadsides. PC: Florence Cemetery, Bartley s.n., 16 Jun 1952 (OS 114762). TR: creek bank, Braceville Twp., Rood 2656, 2 Jul 1952 (KE 11422).

1324 **Gratiola neglecta** Torrey   Hedge-hyssop
Common; stream and pond margins, moist roadside ditches, woodland pools. All counties except MO, RI.

1325 ***Kickxia elatine*** (L.) Dumort.   Canker-root
Rare, adventive; roadsides, lawns. PO: parking lot, Kent, Cooperrider 12513, 6 Jul 1982 (KE 47796).

1326 **Leucospora multifida** (Michx.) Nutt.
Conobea multifida (Michx.) Benth.
Infrequent; mudflats, rocky hillsides. AB, FA, PC, RO.

1327 ***Linaria vulgaris*** Hill   Butter-and-eggs
Common, naturalized; roadsides, waste places, railroads. ALL COUNTIES.

1328 **Lindernia dubia** (L.) Pennell   False Pimpernel
Common; pond margins, mudflats, wet fields, ditches. ALL COUNTIES.

1329 **Melampyrum lineare** Desr.   Cow-wheat
Infrequent; rocky dry woods. AB, CU, GE, LK, LI, LR, PO, SU.

1330 **Mimulus alatus** Aiton   Winged Monkey-flower
Common; stream banks, fields, pastures, thickets. All counties except MH, RO.

1331 **Mimulus ringens** L.   Monkey-flower
Common; marshes, pond and stream margins, pastures, fields. ALL COUNTIES.

1332 **Pedicularis canadensis** L.   Wood-betony, Common Lousewort
Common; rich dry and mesic woods, wooded floodplains, shaded roadside banks. All counties except RO.

1333 **Pedicularis lanceolata** Michx.   Fall Lousewort
Common; fens, sedge meadows, marshes. All counties except AS, PE, RI.

1334 **Penstemon calycosus** Small
Infrequent; meadows, fields. CL, FA, PO, RO, ST.

1335 **Penstemon digitalis** Nutt.   Beard-tongue
Common; grassy fields, woodland borders, roadside banks. ALL COUNTIES.

1336 **Penstemon hirsutus** (L.) Willd.
Common; fields, woodland borders, eroding slopes. AS, AB, CL, CU, FA, GE, KN, LK, LI, LR, MH, PC, PO, RO, ST, SU, WA.

1337 **Penstemon pallidus** Small   Downy White Beard-tongue
Rare; moist woods, roadsides. LR: Oberlin, Dick s.n., 15 Jun 1895 (OS 94968 & OC 75666); edge of woods, Oberlin, Ricksecker s.n., 12 Jun 1895 (OS 31378). PO: Portage County, Hopkins s.n., 22 Jun 1913 (OS 42417).

1338 **Scrophularia lanceolata** Pursh   Lance-leaved Figwort
Frequent; railroads, open woods, thickets. AS, AB, CL, CU, GE, HL, LK, MH, MD, PO, ST, SU, WA.

1339 **Scrophularia marilandica** L.   Maryland Figwort
Common; roadsides, rich woods, stream banks. ALL COUNTIES.

1340 ***Verbascum blattaria*** L.   Moth Mullein
Common, naturalized; roadsides, railroads, fields. ALL COUNTIES.

1341 *Verbascum thapsus* L. Common Mullein
Common, naturalized; railroads, roadsides, fields, waste places. ALL COUNTIES.

1342 *Veronica agrestis* L. Field Speedwell
Rare, adventive; fields. SU: field, Twinsburg Twp., Herrick s.n., 31 May 1955 (OS 55516).

1343 *Veronica americana* (Raf.) Schwein. American Brooklime
Common; stream banks, wet fields, pastures, ditches, marshes, fens. AS, AB, CL, CU, GE, HL, KN, LK, LI, LR, MD, PO, RI, RO, ST, SU, WA.

1344 *Veronica anagallis-aquatica* L. Water Speedwell
Infrequent; moist fields, woodland streams, ditches. AS, HL, KN, LR, PC, PO, SU, WA.

1345 *Veronica arvensis* L. Corn Speedwell
Common, naturalized; roadsides, fields, railroads, waste places. All counties except MO.

1346 *Veronica catenata* Pennell
Veronica comosa Richter
Rare; ditches, marshes, fields. RO: canal, Union Twp., Roberts 4468, 25 Aug 1974 (KE 37116).

1347 *Veronica chamaedrys* L.
Rare, adventive; fields, open woods, waste places. LR: Oberlin, Tracy s.n., 1 Jun 1904 (OC 75720). Cooperrider (1985) indicates a record for MH.

1348 *Veronica filiformis* Smith Blue-eyed Speedwell
Frequent, naturalized; lawns, waste places. AS, CU, GE, LK, LI, LR, MH, PO, RI, ST, SU, WA.

1349 *Veronica latifolia* L.
Rare, adventive; waste places. MD: Medina, Thomas s.n., no date (OS 31929).

1350 *Veronica officinalis* L. Common Speedwell
Common, naturalized; fields, open woods, thickets, roadsides. ALL COUNTIES.

1351 *Veronica peregrina* L. Purslane Speedwell
Common; stream banks, roadside ditches, pastures. AS, AB, CL, CU, FA, GE, HL, KN, LK, LI, LR, MD, PE, PC, PO, RO, ST, SU, TR, WA.

1352 *Veronica persica* Poiret Bird's-eye Speedwell
Infrequent, naturalized; railroads, roadsides, waste places. AS, AB, CU, GE, LK, LR, PO, RO, ST, SU, TR.

1353 *Veronica polita* Fries. Wayside Speedwell
Infrequent, naturalized; lawns, waste places. FA, HL, LR, PC, PO, RO.

1354 *Veronica scutellata* L. Marsh Speedwell
Common; marshes, sedge meadows, sphagnous thickets. AB, CL, CU, GE, KN, LK, LI, LR, MH, PE, PC, PO, ST, SU, TR, WA.

1355 *Veronica serpyllifolia* L. Thyme-leaved Speedwell
Common, naturalized; fields, roadsides, lawns, clearings. ALL COUNTIES.

1356 *Veronicastrum virginicum* (L.) Farw. Culver's-root
Common; shaded roadside banks, thickets, railroads, prairie remnants. AB, CL, CU, FA, GE, HL, KN, LK, LI, MH, MD, PO, RI, RO, ST, SU, TR, WA.

SIMAROUBACEAE (Quassia Family)

1357 *Ailanthus altissima* (Miller) Swingle Tree-of-heaven
Common, naturalized; waste places, roadsides, city lots. ALL COUNTIES.

SOLANACEAE (Nightshade Family)

1358 *Datura stramonium* L. Jimson-weed
Frequent, naturalized; abandoned cultivated fields, roadsides, waste places. AS, CU, FA, GE, HL, KN, LK, LI, LR, PO, RI, RO, ST, TR.

1359 *Lycium halimifolium* Miller Common Matrimony-vine
Lycium chinense Miller
Infrequent, adventive; waste places. AB, CU, FA, LK, LR, SU.

1360 *Lycopersicon esculentum* Miller   Tomato
Infrequent, adventive; abandoned gardens, waste places, roadsides. AB, CU, FA, GE, HL, KN, WA.

1361 *Nicandra physalodes* (L.) Gaertner
Rare, adventive; waste places. TR: yard, Braceville, Rood 2848, 9 Sep 1896 (KE 11397). WA: Applecreek, Bogue 2514, 26 Jul 1900 (OS 89717).

1362 *Petunia* x *hybrida* Vilm.   Garden Petunia
Infrequent, adventive; gardens, waste places, roadsides. AB, CL, CU, HL, LR, SU, TR.

1363 *Physalis alkekengii* L.   Chinese Lantern
Rare, adventive; abandoned homesites, waste places. LK: weedy ground, Painesville, Cooperrider 12293, 1 Oct 1973 (KE 34134). PO: abandoned homesite, Paris Twp., Andreas 2844, 12 Sep 1978 (KE 46735).

1364 *Physalis heterophylla* Nees   Ground Cherry
Common; roadsides, stream banks, railroads. All counties except AB.

1365 *Physalis longifolia* Nutt.
Physalis subglabrata Mackenz. & Bush
Common; roadsides, waste places, railroads. AS, CL, CU, FA, GE, HL, KN, LK, LI, LR, MO, PE, PC, PO, RO, ST, SU, TR, WA.

1366 *Physalis pubescens* L.   Hairy Ground Cherry
Infrequent, adventive; waste places. LK, LR, PC, RO, TR.

1367 *Physalis pumila* Nutt. (including *P. lanceolata* Michx.)
Rare, adventive; waste places, railroads. ST: railroad ballast, Lawrence Twp., Amann 390, 21 Jun 1960 (KE 9449).

1368 *Physalis virginiana* Miller
Rare; dry woods. PC: Pickaway County, Pontius s.n., 6 Jun 1931 (BHO 13282).

1369 *Solanum carolinense* L.   Horse-nettle
Common; pastures, roadsides, waste places. All counties except MO, PC.

1370 *Solanum dulcamara* L.   Deadly Nightshade
Common, naturalized; fields, marshes, disturbed woods, waste places. ALL COUNTIES.

1371 *Solanum nigrum* L. (including *S. americanum* Miller)
Common; barnyards, disturbed woodlots, roadsides, waste places. AS, AB, CL, CU, FA, GE, HL, KN, LK, LI, LR, PO, RO, ST, SU, WA.

1372 *Solanum rostratum* Dunal   Buffalo-bur
Frequent, naturalized; waste places, thickets. AB, CU, HL, KN, LK, LR, PE, PC, PO, SU, TR, WA.

1373 *Solanum sarachoides* Sendtner
Rare, adventive; waste places. TR: garden weed, Braceville Twp., Rood 2533, 5 Sep 1951 (KE 11400).

1374 *Solanum tuberosum* L.   Potato
Rare, adventive; roadsides, waste places. AB: roadside, Ashtabula, Kellerman s.n., 24 Jun 1899 (OS 1139).

STAPHYLEACEAE (Bladdernut Family)

1375 *Staphylea trifolia* L.   Bladdernut
Common; wooded floodplains, moist thickets. ALL COUNTIES.

THYMELAEACEAE (Mezereum Family)

1376 *Dirca palustris* L.   Leatherwood, Moosewood
Common; wooded floodplains, thickets, moist woods. AS, AB, CL, CU, GE, HL, LK, LI, LR, MH, MD, MO, PE, PO, RI, RO, SU, WA.

TILIACEAE (Linden Family)

1377 *Tilia americana* L.   Basswood, Linden
Common; rich woods, stream margins, thickets. ALL COUNTIES.

ULMACEAE (Elm Family)

1378 *Celtis occidentalis* L.   Hackberry
Common; thickets, wooded floodplains, marshes. AS, FA, HL, KN, LI, LR, MD, MO, PE, PC, PO, RI, RO, SU, TR, WA.

1379 *Ulmus americana* L.   American Elm, White Elm
Common; stream banks, woodland borders, thickets, roadsides. ALL COUNTIES.

1380 *Ulmus rubra* Muhl.   Red Elm, Slippery Elm
Common; roadsides, wet woods, thickets. ALL COUNTIES.

1381 *Ulmus thomasii* Sarg.   Cork Elm, Rock Elm
Rare; wet woods. CU: Chagrin Falls, Phillips s.n., 1893 (OS 25068). LR: Oberlin, Ricksecker s.n., 2 Jun 1894 (OS 25072). RO: Bowser tract, Crowl s.n., 14 Aug 1937 (OS 25122).

UMBELLIFERAE / APIACEAE (Parsley Family)

1382 **Aegopodium podagraria* L.   Goutweed
Frequent, naturalized; roadsides, waste places, woodlots. AB, CU, GE, HL, LK, PC, PO, RO, ST, SU, WA.

1383 *Angelica atropurpurea* L.   Purple Angelica
Common; sedge meadows, marshes, moist thickets, fens. AS, AB, CL, CU, FA, GE, HL, KN, LK, LI, LR, MH, MD, PE, PO, RI, ST, SU, TR, WA.

1384 *Angelica venenosa* (Greenway) Fern.
Frequent; swamps, thickets, moist openings. CL, FA, HL, KN, LK, LI, LR, PE, PO, RI, RO, ST, SU.

1385 **Carum carvi* L.   Caraway
Infrequent, adventive; abandoned homesites, fields. AS, CL, CU, LK, LI, LR, TR.

1386a *Chaerophyllum procumbens* (L.) Crantz var. *procumbens*   Chervil
Common; rich mesic and moist woods, wooded floodplains. AS, CL, CU, FA, GE, HL, KN, LK, LI, LR, MD, MO, PE, PC, PO, RO, ST, SU, WA.

1386b *Chaerophyllum procumbens* (L.) Crantz var. *shortii* T. & G.
Rare; mesic woods, floodplains, roadside banks. FA: riverbank, Clear Creek Twp., Cusick 18181, 30 May 1978 (OS 163604). PC: roadside, Circleville Twp., Bartley s.n., 1 May 1964 (OS 74345). RO: roadside, Delano, Bartley s.n., 26 May 1951 (BHO 19520).

1387 *Cicuta bulbifera* L.   Bulblet Water-hemlock
Common; wet fields, swamps, moist thickets, roadside ditches. All counties except FA, MO, RI.

1388 *Cicuta maculata* L.   Spotted Water-hemlock
Common; marshes, thickets, wet fields, shrub swamps, roadside ditches. All counties except PC, RO.

1389 # *Conioselinum chinense* (L.) BSP.   Hemlock-parsley
Once rare, now presumably extirpated from Ohio; thickets, open wooded slopes. SU: cliff at falls, Cuyahoga Falls, Krebs s.n., 1891 (OS 842090).

1390 **Conium maculatum* L.   Poison Hemlock
Frequent, naturalized; ponds, thickets, roadside ditches, floodplains. AS, AB, CU, FA, HL, KN, LK, LI, LR, PC, RI, RO, ST, WA.

1391 *Cryptotaenia canadensis* (L.) DC.   Wild Chervil
Common; rich mesic and moist woods, moist thickets, wooded roadside banks. All counties except PC.

1392 **Daucus carota* L.   Queen Anne's Lace, Wild Carrot
Common, naturalized; fields, roadsides, waste places. ALL COUNTIES.

1393 *Erigenia bulbosa* (Michx.) Nutt.   Harbinger-of-spring
Common; wooded floodplains, moist alluvial woods. All counties except KN, RI.

1394 *Eryngium yuccifolium* Michx.   Rattlesnake-master
Rare; prairie remnants, dry open woods. FA: Kettle Hills, Berne Twp., Goslin s.n., Aug 1933 (OS 35147).

1395 *Heracleum lanatum* Michx. Cow-parsnip
Heracleum maximum Bartr.
Common; wet woods, thickets, fields. AS, AB, CL, CU, FA, GE, HL, KN, LK, LI, LR, MO, PC, PO, RI, ST, SU, TR, WA.

1396 # *Hydrocotyle americana* L. American-pennywort
Frequent; boggy sphagnous thickets, moist fields, fens. AS, CL, CU, GE, HL, KN, LK, LI, PO, RO, ST, SU, WA.

1397 **Hydrocotyle ranunculoides* L. f.
Rare, naturalized; shallow water, pond margins. CL: shallow water, West Twp., Carr 5905, 25 Sep 1983 (KE 46531).

1398 # *Hydrocotyle umbellata* L. Navelwort
Rare; shallow calcareous waters, lake margins. GE: thicket, Lake Punderson, Cusick 3513, 21 Oct 1966 (KE 20653); Lake Punderson, Hicks & Chapman s.n., 15 Jun 1933 (OS 35131 & OS 70729). PO: lake margin, Franklin Twp., Andreas & Emmitt 5730, 8 Jul 1981 (KE 44037); Twin Lakes, Webb 647, 1 Aug 1920 (KE 12801); Twin Lakes, Claassen s.n., 1895 (OS 35129). ST: Congress Lake, Sterki s.n., 22 Sep 1900 (OS 35128); Canton, Case s.n., 1900 (OS 25127).

1399 *Osmorhiza claytonii* (Michx.) C. B. Clarke Sweet Jervil
Common; rich mesic and moist woods, thickets. ALL COUNTIES.

1400 *Osmorhiza longistylis* (Torrey) DC. Anise-root
Common; rich mesic and moist woods, thickets, eroding slopes. All counties except RO.

1401 *Oxypolis rigidior* (L.) Raf. Cowbane, Water-dropwort
Frequent; marshes, fens, sedge meadows. CL, CU, FA, GE, HL, LK, LI, LR, PO, RO, ST, SU, WA.

1402 **Pastinaca sativa* L. Parsnip
Common, naturalized; roadsides, old fields, thickets, waste places. AS, CL, CU, FA, GE, HL, KN, LI, LR, MH, MD, MO, PE, PO, RI, RO, ST, SU, TR, WA.

1403 *Perideridia americana* (Nutt.) Reichenb. Perideridia
Once rare, now presumably extirpated from Ohio; rich woods, prairies. PC: wet woods, Salt Creek Twp., Bartley s.n., 11 May 1950 (OS 48018).

1404 *Sanicula canadensis* L. Black Snakeroot
Common; rocky woods, dry wooded slopes, roadside banks. All counties except MO.

1405 *Sanicula gregaria* Bickn. Sanicle
Common; rich mesic and moist woods, wooded floodplains. All counties except RO.

1406 *Sanicula marilandica* L.
Frequent; wooded slopes, rich mesic and dry woods. AB, CL, GE, HL, LK, LI, LR, MD, PC, PO, RI, ST, SU, TR, WA.

1407 *Sanicula trifoliata* Bickn.
Common; rich mesic and moist woods. All counties except ST.

1408 *Sium suave* Walter Water Parsnip
Common; wet thickets, marshes, meadows, fens, roadside ditches. All counties except MO, PE.

1409 *Taenidia integerrima* (L.) Drude Yellow Pimpernel
Common; thickets, dry woods, slopes. AS, AB, CL, CU, GE, HL, KN, LK, LI, LR, MH, MD, PE, PC, PO, RO, ST, SU, WA.

1410 *Thaspium barbinode* (Michx.) Nutt. Meadow-parsnip
Frequent; moist woods, seeps. AS, AB, CU, FA, LI, LR, PE, PC, PO, RO, ST, SU, WA.

1411 *Thaspium trifoliatum* (L.) Gray Three-leaved Meadow-parsnip
Common; rich mesic and moist woods. ALL COUNTIES.

1412 **Torilis japonica* (Houtt.) DC.
Frequent, naturalized; roadsides, waste places. AS, CL, CU, HL, LK, MH, MD, PO, SU, WA.

1413 *Zizia aurea* (L.) W. D. J. Koch Golden Alexanders
Frequent; thickets, moist woods. AB, CL, KN, LK, LI, LR, MH, MO, PC, RI, SU, TR.

URTICACEAE (Nettle Family)

1414 *Boehmeria cylindrica* (L.) Swartz   Bog Hemp, False Nettle
Common; moist roadside ditches, floodplains, stream margins, moist woods. ALL COUNTIES.

1415 *Laportea canadensis* (L.) Wedd.   Wood Nettle
Common; moist woods, wooded floodplains, stream margins. ALL COUNTIES.

1416 *Parietaria pensylvanica* Willd.   Pellitory
Common; dry woods, gravelly and sandy soils. AS, CL, CU, FA, GE, HL, KN, LK, LI, LR, MD, PE, PO, RI, RO, ST, SU, WA.

1417 *Pilea fontana* (Lunell) Rydb.   Coolwort
Infrequent; seeps, marshes, fens, moist thickets. AB, CL, GE, KN, LK, LI, PO, RO, ST.

1418 *Pilea pumila* (L.) Gray   Clearweed
Common; shaded roadsides, wooded floodplains, hemlock ravines, moist rock ledges, moist woods, thickets. ALL COUNTIES.

1419 *Urtica dioica* L. (including varieties *U. dioica* var. *dioica* and var. *procera* (Willd.) Willd.)

       Stinging Nettle
Common; roadsides, woodland margins, thickets. All counties except MO, PC.

VALERIANACEAE (Valerian Family)

1420 **Valeriana officinalis* L.   Garden Heliotrope
Infrequent, naturalized; roadsides, thickets. AB, CU, GE, LK, PO, SU.

1421 *Valeriana pauciflora* Michx.
Rare; wooded floodplains, stream terraces. FA: Rush Creek Gorge, Rush Creek Twp., Shaw & Kitz sn., 1 Jun 1967 (BHO 29504).

1422 *# Valeriana uliginosa* (T. & G.) Rydb.   Swamp Valerian
Rare, presumably extirpated from Ohio; fens, calcareous thickets. ST: south swamp, Canton, Steele s.n., 25 Jun 1912 (OS 157049 & KE 46349); boggy ground, Canton, Stair s.n., 16 May 1899 (CLM 11666); low ground, Canton, Stair s.n., 23 May 1896 (CLM 11667).

1423 *Valerianella chenopodifolia* (Pursh) DC.   Lamb's Lettuce
Infrequent; fields, floodplains. CL, FA, HL, KN, LR, PE, RO, ST, SU.

1424 ** Valerianella locusta* (L.) Laterrade (including *V. olitoria* (L.) Poll.)   Corn-salad
Infrequent, naturalized; waste places, roadsides. CU, FA, LK, LR, RO, WA.

1425 ** Valerianella radiata* (L.) Dufr.
Rare, adventive; meadows, fields, roadsides. HL: railroad ballast, Prairie Twp., Wilson 1733, 6 Jun 1971 (KE 29979).

1426 *Valerianella umbilicata* (Sulliv.) Wood (including *V. intermedia* Dyal)   Corn-salad
Common; stream banks, fields, roadside ditches, meadows, woodland borders. All counties except MH, TR.

VERBENACEAE (Vervain Family)

1427 *Lippia lanceolata* Michx.   Fog-fruit
Common; damp roadsides, moist fields, marshes. AS, AB, CU, GA, GE, KN, LI, LR, MH, PE, PC, PO, RO, ST, SU, WA.

1428 *Verbena bracteata* Lag. & Rodr.
Infrequent; railroads, sandy fields, prairie remnants, waste places. AB, CU, HL, LK, PC, RI, RO.

1429 *Verbena canadensis* (L.) Britton   Rose Vervain
Rare; sandy fields, roadsides, railroads. CL: gravel, Fairfield Twp., Mureat s.n., 17 May 1979 (Youngstown State University Herbarium 16143). HL: railroad embankment, Hardy Twp., Wilson 1860, 24 Jun 1971 (KE 29978). RO: Frankford, Drushel s.n., 1900 (OS 33112).

1430 *Verbena x engelmannii* Moldenke (*V. hastata* x *V. urticifolia*)
Rare; moist fields, floodplains. CL: marsh, Butler Twp., Cusick s.n., 27 Aug 1961 (OS 66717). PO: floodplain, Kent, Dumke 766-III, 13 Aug 1967 (KE 20333).

1431 *Verbena hastata* L.   Blue Vervain
Common; thickets, marshes, stream banks, fens. ALL COUNTIES.

113

1432 **Verbena simplex** Lehm.
Infrequent; dry sandy fields, roadsides. CU, LK, LR, MH, PC, TR.

1433 **Verbena stricta** Vent   Hoary Vervain
Infrequent; dry roadside banks, fields, prairie remnants. AB, CU, HL, LK, LI, PC, RO, ST, TR.

1434 **Verbena urticifolia** L.   White Vervain
Common; thickets, moist fields, woodland borders, marshes. All counties except PC.

VIOLACEAE (Violet Family)

1435 **Hybanthus concolor** (Forster) Sprengel   Green Violet
Frequent; rich mesic and moist woods, slopes, ravines. AB, FA, HL, LK, LI, LR, MO, PE, PC, RO, SU, WA.

1436 **Viola affinis** LeConte
Common; bogs, woods, pastures, fields. All counties except AB, LK, LI.

1437 ***Viola arvensis** Murray   European Field Pansy
Infrequent, naturalized; fields, cemeteries, roadsides. CU, GE, LK, PO, SU.

1438 **Viola blanda** Willd.   Sweet White Violet
Common; hemlock slopes, moist cliffs, sphagnous thickets, moist woods. AS, AB, CL, CU, FA, GE, HL, LK, LI, LR, MH, PO, RI, RO, ST, SU, TR, WA.

1439 **Viola x bissellii** House (**V. cucullata** x **V. sororia**)
Rare; moist woods, ravines. CL: ravine slope, Knox Twp., Sluss 27, 20 May 1968 (KE 19129). GE: open woods, Chester Twp., Hawver 1224, 5 Aug 1960). ST: moist field, Kuehnle 857, 9 Apr 1938 (KE 33709).

1440 **Viola x brauniae** Cooperrider ex Grover (**V. rostrata** x **V. striata**)
Frequent; stream terraces, moist woods, meadows. AS, AB, CL, GE, HL, KN, LR, PO, RI, RO, ST, SU, TR, WA.

1441 **Viola candensis** L.   Canada Violet
Common; rich woods, rocky slopes, wooded floodplains. All counties except PC, ST, WA.

1442 **Viola conspersa** Reichenb.   American Dog Violet
Frequent; moist woods, thickets, stream margins. AB, CU, GE, LK, LR, MH, MD, PO, RI, ST, SU, TR, WA.

1443 **Viola cucullata** Aiton   Blue Marsh Violet
Common; wet woods, fens, sedge meadows, moist fields, wooded seeps. AS, AB, CL, CU, FA, GE, LK, LI, LR, MH, MD, PE, PO, RI, RO, ST, SU, TR, WA.

1444 **Viola fimbriatula** Smith   Ovate-leaved Violet
Frequent; woodland borders, sandy fields. AS, AB, CL, CU, HL, KN, LK, LI, PO, RI, ST, SU, WA.

1445 # **Viola hastata** Michx.   Halberd-leaved Violet
Frequent; rich moist woods, hemlock slopes. AB, CL, CU, GE, LK, LR, MH, MD, PO, SU, SU, TR.

1446 **Viola hirsutula** Brainerd   Southern Wood Violet
Infrequent; dry woods, thickets. AS, FA, LK, LI, PE, RO.

1447 **Viola incognita** Brainerd
Common; woodland thickets, roadsides, hemlock woods. AB, CU, FA, GE, HL, KN, LK, LI, LR, MH, MD, PO, ST, SU, TR, WA.

1448 **Viola lanceolata** L.   Lance-leaved Violet
Infrequent; open sandy fields, pastures, prairie remnants, stream and pond margins. CU, FA, LK, LR, RO.

1449 ***Viola odorata** L.   English Violet
Infrequent, naturalized; lawns, roadsides, fields. AB, LK, LR, MH, PO, RI, RO, ST, SU, WA.

1450 **Viola pallens** (Banks) Brainerd   Northern White Violet
Frequent; sphagnous hummocks, thickets, wet woods. AB, CU, FA, GE, LK, LR, PO, ST, SU, TR, WA.

1451 *Viola palmata* L. Palmate-leaved Violet, Wood Violet
Frequent; rich mesic and moist woods. CU, FA, GE, KN, LI, MH, PE, PC, PO, SU, TR,
WA.

1452 *Viola primulifolia* L. Primrose-leaved Violet
Rare; bogs, meadows. PO: bog, Brady Lake, Hopkins s.n., 25 May 1915 (OS 16819).

1453 *Viola pubescens* Aiton (including *V. pensylvanica* Michx.) Common Yellow Violet
Common; rich dry and mesic woods, woodland borders, meadows. ALL COUNTIES.

1454 *Viola rafinesquii* Greene Wild Pansy
Infrequent; dry fields, roadsides. PO: field, Windham Twp., Rood & Brobst 2931, 8 May
1931 (OS 106602). RO: Springfield Twp., Bartley & Pontius s.n., 23 Mar 1930 (OS 16721);
Salt Creek Twp., Crowl s.n., 16 Apr 1938 (OS 16720).

1455 *Viola rostrata* Pursh Long-spurred Violet
Common; rich dry woods, mesic and moist woods, woodland borders, moist open floodplains.
All counties except PC.

1456 *#Viola rotundifolia* Michx. Round-leaved Yellow Violet
Infrequent; hemlock slopes, moist wooded ridges, boggy woods. AB, CU, GE, LK, MD, PO,
SU, TR.

1457 *Viola sagittata* Aiton Arrow-leaved Violet
Common; dry sandy banks, fields, open woods. AS, AB, CU, FA, GE, HL, KN, LK, LR,
PE, PO, RO, ST, SU, TR, WA.

1458 *Viola sororia* Willd. (including *V. papilionacea* Pursh) Common Blue Violet
Common; rich mesic and moist woods, meadows, wooded floodplains. ALL COUNTIES.

1459 *Viola striata* Aiton Striped Violet
Common; boggy thickets, wooded floodplains, fields, roadsides. ALL COUNTIES.

1460 **Viola tricolor* L. Johnny Jump-up
Infrequent, adventive; fields, lawns, gardens. CL, CU, FA, KN, LK, MH, PO, SU.

1461 *Viola triloba* Schwein. Three-lobed Violet
Common; rich mesic and moist woods, moist thickets. ALL COUNTIES.

1462 *Viola walteri* House Walter's Violet
Rare; rocky ledges, calcareous woods. RO: Springfield Twp., Bartley & Pontius s.n., 23 Mar
1930 (OS 16677).

VISCACEAE (Mistletoe Family)

1463 *Phoradendron serotinum* (Raf.) Johnston American Mistletoe
Phoradendron flavescens (Pursh) Nutt.
Rare; parasitic on branches of deciduous trees. RO: Liberty Twp., Bartley & Pontius s.n.,
Feb 1931 (OS 28448); near Chillicothe, Werner s.n., 8 Jun 1895 (OS 28477).

VITACEAE (Grape Family)

1464 *Parthenocissus inserta* (Kerner) Fritsch
Frequent; woods, thickets, roadside banks. AS, AB, CL, CU, HL, KN, LK, LR, MH, MD,
MO, RI, ST, SU, TR.

1465 *Parthenocissus quinquefolia* (L.) Planch. Virginia Creeper
Common; railroads, thickets, roadsides, woodland borders. ALL COUNTIES.

1466 *Vitis aestivalis* Michx. Summer Grape
Common; rich dry and mesic woods, thickets, woodland borders, railroads. ALL COUNTIES.

1467 *Vitis labrusca* L. Fox Grape
Infrequent; roadside thickets, dry and mesic woods. AB, LK, LR, MH, ST, TR.

1468 *Vitis riparia* Michx. Summer Grape
Common; river and stream margins, wet woodland borders. All counties except RO.

1469 *Vitis vulpina* L. Frost Grape
Frequent; moist woods, railroads, thickets, marshes. AS, AB, CL, FA, GE, KN, LI, LR,
MD, PE, PC, RI, RO, ST, WA.

ZYGOPHYLLACEAE (Caltrop Family)

1470 *Tribulus terrestris* L.   Caltrop
Rare, adventive; dry waste places, railroads. LI: cindery soil, Newark, Carr 860, 6 Aug 1979 (KE 45667).

Division Magnoliophyta

CLASS LILIOPSIDA (MONOCOTYLEDONS)

ALISMATACEAE (Water Plantain Family)

1471 *Alisma plantago-aquatica* L.   Mud Plantain
Alisma subcordatum Raf. (in Braun, 1967)
Common; shallow water, muddy shores, sphagnous meadows, marshes, roadside ditches. ALL COUNTIES.

1472 *Echinodorus rostratus* (Nutt.) Engelm.   Bur-head
Rare; muddy shores. PC: mudflat, Pickaway Twp., Carr 2389, 17 Sep 1979 (KE 45715); swampy place, Pickaway Twp., Bartley s.n., 21 Aug 1948 (OS 106686). RO: pond, Vauce Station, Bartley s.n., 25 Jun 1937 (BHO 1064).

1473 *Sagittaria brevirostra* Mack. & Bush   Short-beaked Arrowhead
Rare; meadows, pond and lake shores. RO: calcareous meadow, Green Twp., Stuckey 9721, 23 Oct 1977 (OS 125578). Braun (1967) indicates a record for FA.

1474 *Sagittaria cuneata* Sheldon   Wapato
Rare; marshes, muddy shores, lake margins. LI: Buckeye Lake, Abbott s.n., Aug 1938 (OS 45166); WA: Browns Lake, Duvel 209, 7 Sep 1899 (OS 92462). Braun (1967) indicates a record for PC.

1475 *Sagittaria graminea* Michx.   Grass-leaf Arrowhead
Braun (1967) indicates a record for CU.

1476 *Sagittaria latifolia* Willd.   Duck Potato
Common; shallow water in ponds, lakes, marshes, sphagnous meadows, fens, roadside ditches. ALL COUNTIES.

1477 *Sagittaria rigida* Michx.
Frequent; calcareous ditches, shallow water, fens, muddy lake and pond shores. AB, CL, LK, LI, LR, MD, PE, PC, PO, RO, SU.

ARACEAE (Arum Family)

1478 *Acorus calamus* L.   Sweetflag
Acorus americanus Raf. (in Braun, 1967)
Common; marshes, roadside ditches, pastures. ALL COUNTIES.

1479 *Arisaema dracontium* (L.) Schott   Green Dragon
Common; wooded floodplains, stream terraces, marshes. ALL COUNTIES.

1480 # *Arisaema stewardsonii* Britton   Swamp Jack-in-the-pulpit
Rare; boggy woods, thickets, fen margins. AB: moist woods, Kingsville Twp., Brockett 326, 15 Jul 1961 (KE 17949). CL: fen thicket, Butler Twp., Andreas 6060, 31 May 1983 (KE 46484). GE: wooded area, Troy Twp., Cooperrider & Brockett 10408, 22 Jun 1967 (KE 16540). PO: in *Sphagnum* under willow, Franklin Twp., Andreas 7298, 27 May 1988 (KE 50264); under alders, Mantua Twp., Cusick 10813, 20 May 1970 (KE 27588); deep shade, Streetsboro Twp., Cusick 10773, 18 May 1970 (OS 116829).

1481 *Arisaema triphyllum* (L.) Schott (including *A. atrorubens* (Aiton) Blume)
                                                                    Jack-in-the-pulpit
Common; mesic and moist woods, wooded floodplains. ALL COUNTIES.

1482 # *Calla palustris* L.   Wild Calla
Infrequent; bog margins, shallow water in kettle-hole depressions, wet thickets, buttonbush swamps, beaver ponds. AB, CU, GE, PO, ST, SU, TR.

1483 *Peltandra virginica* (L.) Schott & Endl.   Arrow-arum
Common; river margins, marshes, shallow ponds. AB, CU, FA, GE, HL, LK, LI, LR, MH, MD, PE, PO, ST, SU, TR, WA.

1484 *Symplocarpus foetidus* (L.) Nutt.   Skunkcabbage
Common; marshes, wet wooded seeps, floodplains, pastures, fens, thickets. All counties except PC.

COMMELINACEAE (Spiderwort Family)

1485 ***Commelina communis* L.   Dayflower
Common, naturalized; waste places, roadside ditches, lawns, gardens. All counties except RI.

1486 ** Commelina diffusa* Burm. f.
Rare, naturalized; streams, river banks. RO: Chillicothe, Cusick 14500, 15 Sep 1975 (KE 38905).

1487 ***Tradescantia bracteata* Small   Bracted Spiderwort
Rare, adventive; roadsides, waste places. PO: railroad cinders, Franklin Twp., Cusick 9903, 14 Jun 1969 (KE 27603 & OS 102706).

1488 *Tradescantia ohiensis* Raf.   Ohio Spiderwort, Glaucous Spiderwort
Common; wet prairies, eroding hillsides, wet thickets, fens. AS, AB, CL, CU, FA, GE, HL, LI, LR, MH, MD, PO, RI, RO, ST, SU, TR, WA.

1489 *Tradescantia subaspera* Ker   Zigzag Spiderwort
Rare; open woods, hillsides. PC: Pickaway Twp., Pontius s.n., 15 Aug 1931 (BHO 7957).

1490 *Tradescantia virginiana* L.   Virginian Spiderwort
Infrequent; floodplains, thickets, roadsides. CU, FA, LI, LR, RO, ST, SU.

CYPERACEAE (Sedge Family)

1491 *Bulbostylis capillaris* (L.) Clarke
Infrequent; dry sandy fields, cinders, roadsides, railroads. FA, HL, KN, LK, MH, PE, PC, ST, SU.

1492 *Carex aggregata* Mackenz.   Crowded Sedge
Infrequent; mesic and moist woods, gravelly thickets, fields. AS, AB, CU, LI, LR, PC, PO, SU, WA.

1493 # *Carex alata* T. & G.   Broad-winged Sedge
Infrequent; fens, wet open woods, marshy fields. AS, AB, CU, LK, PO, ST, SU.

1494 *Carex albolutescens* Schwein.   Pale Straw Sedge
Rare; bogs, wet woods, thickets, marshy fields. SU: sphagnous mat, Norton Twp., Andreas & Stoutamire 3897, 20 Jun 1979 (KE 42274).

1495 *Carex albursina* Sheldon
Frequent; woods, shaded ravines. AS, CU, FA, LH, KN, LK, LI, LR, MH, MO, PE, RI, RO, SU, WA.

1496 *Carex amphibola* Steud. var. *turgida* Fern.
Common; rich moist woods, thickets, wooded floodplains, stream banks. AS, AB, CL, CU, FA, GE, HL, KN, LK, LI, LR, MH, MD, MO, PC, PO, SU, WA.

1497a *Carex annectens* (Bickn.) Bickn. var. *annectens*
Frequent; wet fields, pond margins, marshes, roadside ditches. AS, AB, CL, CU, GE, KN, LR, MH, MD, PC, PO, RO, ST, SU, TR.

1497b *Carex annectens* (Bickn.) Bickn. var. *xanthocarpa* (Bickn.) Wieg.
Infrequent; wet fields, thickets, marshes. AS, LK, LI, PC, ST, SU, TR.

1498 # *Carex arctata* Boott   Drooping Wood Sedge
Rare; rocky woods, thickets. PO: rocky wooded slope, Nelson Twp., Cooperrider 6021, 27 Jun 1960 (KE 4496). ST: bogs, Canton, Stair s.n., 16 May 1899 (CLM 13305). SU: rock ledges, Boston Twp., Andreas 6910, 8 May 1986 (KE 49767); Cuyahoga Falls, Werner s.n., 4 Jun 1892 (OS 4879 & OS 4880).

1499 # *Carex argyrantha* Tuckerman   Silvery Sedge
Rare; dry woods, rock exposures, dry clearings. GE: Thompson Ledges, Tyler s.n., 15 Jul 1901 (OS 4302). SU: ledges, Boston Twp., Andreas 6957, 23 May 1986 (KE 50261); sandstone ledges, Boston Twp., Cusick 11727, 16 Jun 1971 (OS 118684).

1500 *Carex artitecta* Mackenz.
Common; dry woods, woodland borders, rocky slopes. AS, AB, CU, FA, GE, HL, KN, LK, LI, LR, MH, MO, PC, PO, ST, SU, TR, WA.

117

1501 **Carex atherodes** Sprengel   Wheat Sedge
Rare; sphagnous meadows, calcareous meadows. FA: Onion Island, Buckeye Lake, Thomas
s.n., 17 May 1932 (OS 5041). PC: Pickaway Twp., Bartley s.n., 18 Jun 1936 (BHO 17738).

1502 **Carex atlantica** Bailey   Prickly Bog Sedge
Carex incomperta Bickn. (in Braun, 1967)
Frequent; calcareous sedge meadows, fens, sphagnous thickets. AB, CU, GE, HL, LK, LI,
LR, PE, PO, RO, ST, SU, WA.

1503 **Carex aurea** Nutt.   Golden-fruited Sedge
Rare; eroding calcareous slopes, sedge meadows, fens. AB: clay slumps, Harpersfield Twp.,
Bissell 1981:55, 12 Jun 1981 (CLM 15615 & KE 46358). GE: fen seep, Burton Twp., Andreas
& Knoop 6366, 15 Apr 1985 (KE 48567). LK: steep slumps, Kirtland Hills, Bissell 1983:56,
9 Jun 1983 (CLM 19686). SU: eroding slope, Boston Twp., Andreas 7054, 5 Jun 1986 (KE
50269); clay slope, Boston Twp., Cusick 10792, 24 May 1967 (KE 25425 & OS 102700);
Akron, Claypole s.n., 13 May 1888 (University of Akron Herbarium s.n.).

1504 **# Carex bebbii** (Bailey) Fern.   Bebb's Sedge
Rare; fens, sedge meadows, thickets. PO: hummock in fen, Streetsboro Twp., Andreas 5594,
11 Jun 1981 (KE 46501); shrubby bog, Streetsboro Twp., Cooperrider 5586, 18 Jun 1960
(KE 4644); peat bog, Mantua Twp., Rood 1408, 14 Jul 1918 (KE 10488). ST: Myer's Lake,
Plain Twp., Brown s.n., 5 Jul 1936 (OS 47622).

1505 **Carex blanda** Dewey
Common; rich mesic and moist woods, wet thickets, wooded floodplains. All counties except
CL, KN.

1506 **Carex bromoides** Schkuhr   Brome Sedge
Frequent; open moist woods, fen margins, marshes. AB, CL, CU, FA, GE, KN, LK, LR,
MO, PO, RO, ST, SU, TR, WA. Braun (1967) indicates a record for MD.

1507 **# Carex brunnescens** (Persoon) Poiret   Brownish Sedge
Infrequent; boggy woods, dry wooded hummocks. AB, GE, LK, PO, TR.

1508 **Carex buxbaumii** Wahlenb.   Brown Bog Sedge
Infrequent; sphagnous meadows, marly fens, wet calcareous soils. FA, HL, LR, MD, MO,
PO, ST, SU.

1509 **# Carex canescens** L.   Glaucous Sedge
Frequent; wooded swamps, bogs, beaver ponds, disturbed sphagnous areas, open boggy
woods. AB, CU, FA, GE, HL, LK, LR, MD, PO, RI, ST, SU, TR, WA.

1510 **Carex careyana** Dewey   Carey's Sedge
Infrequent; moist woods, wooded floodplains, moist meadows. FA, HL, LK, LI, LR, MD,
MO, RI.

1511 **Carex caroliniana** Schwein.   Carolina Sedge
Infrequent; swamps, woods, floodplains, moist meadows. AB, CU, GE, LI, PE, PO, SU, TR,
WA.

1512 **# Carex cephalantha** (Bailey) Bickn.   Little Prickly Sedge
Infrequent; sedge meadows, sphagnous meadows, thickets. AB, HL, LK, LR, PO. Braun
(1967) shows a record from ST.

1513 **# Carex cephaloidea** Dewey   Thin-leaf Sedge
Infrequent; rich mesic woods, hemlock ravines, thickets. CU, LR, MD, SU, TR, WA.

1514 **Carex cephalophora** Muhl.
Common; dry and mesic woods, steep wooded slopes. AS, AB, CL, CU, FA, GE, HL, KN,
LK, LI, LR, MH, PC, PO, SU, TR, WA. Braun (1967) indicates a record for RO.

1515 **Carex communis** Bailey
Common; mesic woods, roadside banks, thickets. AS, AB, CL, CU, FA, GE, HL, KN, LI,
LI, LR, MH, MD, PE, PC, PO, RI, SU, TR.

1516 **Carex comosa** Boott
Common; sedge meadows, thickets, bog margins, pond and lake margins, swamps. AS, AB,
CL, CU, GE, HL, KN, LK, LI, LR, MD, PE, PC, PO, RO, ST, SU, TR, WA.

1517 **Carex conjuncta** Boott
Infrequent; fens, sedge meadows, open woods. AS, AB, FA, HL, KN, MO, PC, ST, SU, TR.

1518 *Carex conoidea* Willd. Field Sedge
Rare; marshes, fens, sedge meadows. CU: Strongsville, Selby s.n., 16 May 1889 (OS 93064). ST: bogs, Canton, Stair s.n., 16 May 1897 (CLM 5516). SU: marsh, Boston Twp., Cusick 1736, 30 May 1969 (KE 25398).

1519 *Carex convoluta* Mackenz.
Common; sandstone ledges, rocky roadside banks, dry sandy woods. All counties except RI.

1520 *Carex crawei* Dewey Crawe's Sedge
Rare; calcareous soils in fens, sedge meadows. PO: marsh, Streetsboro Twp., Tandy 507, 15 Jun 1972 (KE 39624); marly area, Mantua Twp., Cusick 9908, 16 Jun 1969 (KE 25389). ST: marly seep, Jackson Twp., Andreas 5050, 23 May 1980 (KE 43130). SU: marly area, Green Twp., Cusick 10132, 4 Jul 1969 (KE 25388). RO: sedge meadow, Green Twp., Andreas and Knoop 7000, 4 Jun 1987 (KE 49764).

1521 *Carex crinita* Lam.
Common; swamp borders, wet fields, sedge meadows, fens, thickets, marshes. All counties except MO, PC, RO.

1522 *Carex cristatella* Britton
Common; swamps, marshes, moist woods. AS, AB, CL, CU, FA, KN, LK, LI, LR, MD, PE, PO, RI, RO, ST, SU, WA.

1523 *Carex crus-corvi* Kunze Raven-foot Sedge
Rare; swampy thickets, woods. RO: near Chillicothe, Hall & Bartley 2415, Jul 1951 (BHO 2176); Scioto Twp., Bartley s.n., 14 Jun 1936 (OS 47684).

1524 *Carex cryptolepis* Mackenz. Little Yellow Sedge
Infrequent; open marl in fens, sedge meadows, calcareous thickets. AB, LK, PO, ST, SU, TR.

1525 *Carex davisii* Schwein. & Torrey Davis's Sedge
Infrequent; weedy wooded floodplains, moist calcareous woods, sedge meadows. CL, LK, LI, LR, PC, RI, RO, WA.

1526 # *Carex debilis* Michx. var. *rudgei* Bailey Weak Sedge
Infrequent; rich woods, boggy floodplains, rocky slopes, ravines. AS, AB, CU, GE, LK, LR, MD, PO, SU, TR.

1527 *Carex decomposita* Muhl. Cypress-knee Sedge
Rare; wooded swamps. LI: Cranberry Island, Detmers s.n., 7 Jul 1910 (OS 4025).

1528 # *Carex deweyana* Schwein. Dewey's Sedge
Once rare, now presumably extirpated from Ohio; dry open woods. PO: dry soil, top of rocks, Nelson Twp., Webb 438, 25 May 1923 (KE 10440 & OS 4095).

1529 *Carex diandra* Schrank Lesser Panicled Sedge
Infrequent; calcareous thickets, fens, sedge meadows. LK, LI, LR, PO, SU.

1530 *Carex digitalis* Willd. Finger Sedge
Common; dry woods. AS, AB, CL, CU, FA, GE, HL, KN, LK, LR, MD, PE, PO, ST, SU, TR, WA.

1531 *Carex eburnea* Boott Bristle-leaf Sedge
Rare; eroding calcareous slopes, sand and gravel banks. GE: seepage slope, Chardon Twp., Bissell 1979:89, 23 Aug 1979 (CLM 12453). SU: eroding slope, Boston Twp., Andreas 7053, 5 June 1986 (KE 50270); exposed clay slope, Boston Twp., Cusick 9722, 10 May 1969 (KE 25609); Cuyahoga Falls, Werner s.n., 4 Jun 1892 (CLM 5538).

1532 # *Carex emmonsii* Dewey Emmon's Sedge
Infrequent; dry woods, woodland borders, open floodplains. AB, CU, GE, LR, SU.

1533 *Carex emoryi* Dewey Emory's Sedge
Frequent; wooded floodplains, swamps, stream margins. AB, CL, CU, FA, GE, LK, LR, MD, PO, SU, TR.

1534 *Carex festucacea* Willd. Fescue Sedge
Frequent; marshes, weedy thickets, moist roadsides. AS, AB, CU, GE, HL, LK, MH, MD, PC, PO, RO, ST, TR.

1535 *Carex flava* L. Yellow Sedge
Rare; open marl in fens, sedge meadows. AB: open fen, Wayne Twp., Bissell 1984: 48, 6 Jun 1984 (CLM 20915); Plymouth marshes, Hicks s.n., 11 Jun 1930 (OS 5051). ST: marl seep, Jackson Twp., Andreas 1837, 27 May 1978 (KE 42404); Lake O' Springs, Brown s.n.,

5 Jun 1937 (OS 5047); Canton, Stair s.n., 16 May 1897 (CLM 5536). PO: *Potentilla* meadow, Streetsboro Twp., Cusick 10053, 16 Jun 1969 (KE 25947). RO: marly seep, Union Twp., Andreas 4652, 3 May 1982 (KE 46521).

1536 # *Carex folliculata* L.   Long Sedge
Infrequent; boggy woods, thickets, tamarack swamps. AB, CU, GE, PO, SU, TR.

1537 *Carex frankii* Kunth   Frank's Sedge
Frequent; fen meadows, open woods, thickets. AS, CL, CU, FA, HL, KN, LK, LI, LR, MO, PE, PC, RI, RO, WA.

1538 *Carex glaucodea* Tuckerman   Blue-green Sedge
Infrequent; woods, open meadows, moist fields. CU, LK, ST, TR, WA.

1539 *Carex gracilescens* Steud.
Common; moist woods, wooded floodplains. AS, AB, CL, CU, FA, GE, HL, KN, LK, LI, LR, MH, MD, MO, PO, RI, SU, TR, WA.

1540 *Carex gracillima* Schwein.
Common; moist woods, wooded floodplains, swamps. AS, AB, CL, CU, FA, GE, HL, KN, LK, LI, LR, MD, MO, PO, ST, SU, TR.

1541 *Carex granularis* Willd.
Common; marshes, ditches, moist fields, moist thickets. All counties except AB, GE, MH.

1542 *Carex gravida* Bailey   Heavy Sedge
Rare; dry roadside banks, open woods. PC: Pickaway Twp., Drake s.n., 10 Jun 1935 (OS 3899). RO: Scioto Twp., Bartley s.n., 27 May 1945 (BHO 17770).

1543 *Carex grayi* Carey
Common; moist woods, roadside ditches, swamps, temporary woodland pools, shaded pastures. All counties except CL, PC, RO.

1544 # *Carex haydenii* Dewey   Hayden's Sedge
Once rare, now presumably extirpated from Ohio; meadows, thickets. TR: roadside ditch, Southington Twp., Rood s.n., 30 May 1913 (OS 4303).

1545 *Carex hirsutella* Mackenz.
Common; dry, mesic and moist open woods, roadsides, fields, meadows. All counties except MO.

1546 *Carex hirtifolia* Mackenz.
Common; floodplains, rich mesic woods, thickets, wet meadows. CL, CU, FA, GE, HL, KN, LK, LI, LR, MD, MO, PE, PO, RO, ST, SU, WA.

1547 *Carex hitchcockiana* Dewey   Hitchcock's Sedge
Infrequent; rich moist woods, wooded floodplains. AB, FA, LK, LR, MD, PO.

1548 # *Carex howei* Mackenz.   Howe's Sedge
Frequent; sphagnous bogs, boggy woods, thickets. AB, GE, LR, PO, LK, LI, PE, PO, ST, SU, WA.

1549 *Carex hyalinolepis* Steudel
Infrequent; calcareous thickets, fens, marshes. AS, LR, MO, PC, ST, WA. Braun (1967) indicates a record for RO.

1550 *Carex hystricina* Muhl.   Bottle-brush Sedge
Common; sedge meadows, seeps, marshes, fens, ditches, stream margins, sphagnous thickets. All counties except MH, PE, RI.

1551 *Carex interior* Bailey   Inland Sedge
Frequent; calcareous fens, sedge meadows, sphagnous thickets. AB, CL, CU, GE, HL, LK, LR, PO, RO, ST, SU, TR.

1552 *Carex intumescens* Rudge
Frequent; sphagnous thickets, boggy woods, buttonbush swamps. AS, AB, CL, CU, FA, GE, HL, LK, LI, LR, MD, PO, ST, SU, TR.

1553 *Carex jamesii* Schwein.   James' Sedge
Infrequent; rich woods, shaded wooded slopes. AS, CL, FA, LK, LR, MD, PO, SU, TR.

1554 *Carex lacustris* Willd.
Frequent; neutral to calcareous thickets, fen margins, marshes, open swamps. AS, AB, CL, CU, FA, GE, HL, KN, LR, PO, RO, ST, SU, TR.

1555 *Carex laevivaginata* (Kukenth) Mackenz.
Common; fens, sedge medows, fields, wet open woods. AB, CL, CU, FA, GE, HL, KN, LK, LI, LR, MH, MD, PO, ST, SU, TR.

1556 *Carex lanuginosa* Michx.
Frequent; sedge meadows, marly thickets, marshes. AS, AB, FA, HL, LK, LR, MO, PC, PO, RO, ST, SU, TR, WA.

1557 *Carex lasiocarpa* Ehrh.   Slender Sedge
Infrequent; sedge meadows, peaty meadows, marsh margins. CL, LK, LI, LR, PO, RI, ST, SU, WA. Braun (1967) indicates a record for MD.

1558 *Carex laxiculmis* Schwein.
Common; rich moist woods, wooded floodplains. AS, AB, CL, CU, FA, GE, HL, LK, LI, LR, MD, PO, ST, SU, TR, WA.

1559 *Carex laxiflora* Lam.   Loosely-flowered Sedge
Common; dry, mesic and moist woods, wooded floodplains, rocky slopes, thickets. All counties except MH, RO.

1560 *Carex leavenworthii* Dewey   Leavenworth's Sedge
Rare; open woods, meadows. FA: Berne Twp., Goslin s.n., 16 Jun 1935 (OS 3894). PC: Pickaway Twp., Drake s.n., 2 Jun 1935 (OS 3892). Braun (1967) indicates a record for AB.

1561 *Carex leptalea* Wahlenb.   Delicate Sedge
Infrequent; calcareous seeps, sedge meadows, fens, thickets, sphagnous thickets. AB, CL, CU, GE, HL, LK, PO, RO, ST, SU.

1562 # *Carex leptonervia* Fern.   Nerveless Sedge
Infrequent; wet alluvial woods, thickets. AB, GE, LK, SU, TR.

1563 # *Carex limosa* L.   Mud Sedge
Infrequent; sphagnous thickets, marly peat. AS, AB, FA, LI, WA.

1564 # *Carex louisianica* Bailey   Louisiana Sedge
Once rare, now presumably extirpated from Ohio; wet woods, swamps. AB: wet woods, Richmond Twp., Cooperrider 5937, 24 Jun 1960 (KE 1950).

1565 *Carex lupuliformis* Sartwell   False Hop Sedge
Infrequent; swamps, wet meadows, marsh margins. CU, FA, KN, LK, LI, LR, MD, PO, TR, WA.

1566 *Carex lupulina* Willd.   Hop Sedge
Common; roadside ditches, wet fields, moist open woods, thickets, swamps. ALL COUNTIES.

1567 *Carex lurida* Wahlenb.
Common; wet fields, ditches, sedge meadows, pond margins, thickets. All counties except PC.

1568 *Carex mesochorea* Mackenz.   Midland Sedge
Once rare, now presumably extirpated from Ohio; dry woods, thickets. LI: ditch, Adams Mills, Cusick s.n., 6 Jun 1963 (OS 72165). RO: Green Twp., Bartley & Pontius s.n., 18 May 1932 (OS 3878).

1569 *Carex molesta* Mackenz.
Infrequent; sedge meadows, sphagnous thickets, bog margins. CL, LK, LR, MH, PC, PO, RO, ST, SU.

1570 *Carex muhlenbergii* Willd.   Muhlenberg's Sedge
Infrequent; dry woods, wooded openings, sandy fields. CU, FA, LK, LR, PO, RO, SU, WA.

1571 *Carex muskingumensis* Schwein.   Muskingum Sedge
Rare; stream banks, swamps. AB: Grand River, Hicks s.n., 26 Jun 1930 (OS 4239). KN: swamp, Fredericktown, Hicks 28 Jun 1929 (OS 54234). PC: Washington Twp., Bartley s.n., 10 Jul 1941 (BHO 17794); Washington Twp., Drake s.n., 15 Jul 1935 (OS 4232).

1572 *Carex normalis* Mackenz.
Common; moist woods, moist thickets, fields. AS, CL, CU, FA, GE, HL, KN, LK, LR, MH, PE, PC, PO, RO, SU, TR, WA.

1573 *Carex oligocarpa* Willd.
Frequent; calcareous mesic woods, thickets. CU, CL, GE, HL, KN, LK, LI, LR, PC, RI, RO, SU, WA.

121

1574 # *Carex oligosperma* Michx.   Few-seeded Sedge
Rare; boggy depressions, sphagnous thickets, swamps. PO: sphagnous bog, Brimfield Twp.,
Andreas 5396, 29 Jul 1980 (KE 43176); bog, Cooperrider 10804, 27 Aug 1968 (KE 20251);
boggy depression, Ravenna Twp., Andreas & Stoutamire 5167, 12 Jun 1980 (KE 43028);
sphagnous depression, Franklin Twp., Andreas 5348, 15 Jul 1980 (KE 43177); shrubby bog,
Rootstown Twp., Cooperrider 6402, 11 Jul 1970 (KE 4500). ST: sphagnous mat, Sugar Creek
Twp., Andreas 5789, 5 Aug 1981 (KE 44102).

1575 # *Carex ormostachya* Wieg.   Stiff Broad-leaved Sedge
Rare; rocky woods, clearings. CL: lightly wooded slope, Middleton Twp., Coooperrider 5690,
21 Jun 1960 (KE 4499).

1576 # *Carex pallescens* L.   Pale Sedge
Infrequent; swamps, roadside ditches, meadows, woodland borders, thickets. AB, CU, GE,
LK, LR, PO, SU, TR.

1577 # *Carex pedunculata* Muhl.   Long-stalked Sedge
Infrequent; hemlock woods, ravine rims, steep slopes, wet woods. AB, CU, GE, LK, PO,
SU.

1578 *Carex pensylvanica* Lam.
Common; dry and mesic woods, dry slopes, woodland margins, fields. All counties except
PC.

1579 *Carex plantaginea* Lam.   Plantain-leaf Sedge
Frequent; mesic and moist woods, wooded floodplains, stream terraces, rocky slopes, steep
ravines. AS, AB, CU, FA, GE, HL, LK, LR, MH, MD, PO, SU, TR, WA.

1580 *Carex platyphylla* Carey
Common; moist and mesic woods, wooded floodplains, rocky slopes. AS, AB, CL, CU, FA,
HL, KN, LK, LI, LR, PE, PC, PO, ST, SU, WA.

1581 **Carex praegracilis* W. Boott
Infrequent, adventive; dry, gravelly saline soils. AB, CU, LI, MD, MO, SU, WA.

1582 *Carex prairea* Dewey   Prairie Sedge
Infrequent; sedge meadows, thickets, fens. CL, CU, GE, HL, LK, PO, ST, SU.

1583 *Carex prasina* Wahlenb.
Common; rich mesic and moist woods, wooded floodplains. AB, CU, FA, GE, HL, KN, LK,
LI, LR, MO, PE, PO, ST, SU, TR, WA.

1584 # *Carex projecta* Mackenz.   Necklace Sedge
Infrequent; thickets, damp woods, marsh margins. CL, CU, GE, LK, LR, MD.

1585 # *Carex radiata* (Wahlenb.) Dewey   Radiate Sedge
Infequent; mesic and moist woods, meadows, thickets. AB, GE, LK, LR, PO, SU, TR.

1586 *Carex retroflexa* Muhl.   Reflexed Sedge
Rare; dry fields, rocky woods, roadside banks. KN: sandstone slump, Jefferson Twp., Burns
3112, 7 Jun 1983 (KE 46703). RO: open, dry field, Liberty Twp., Bartley & Pontius s.n.,
10 May 1938 (OS 3830). WA: cemetery, Wooster, Detmers s.n., 12 Jun 1925 (OS 93264).

1587 *Carex rosea* Schkuhr
Common; mesic woods, moist slopes, sphagnous thickets. All counties except RI.

1588 # *Carex rostrata* Stokes   Beaked Sedge
Infrequent; sphagnous thickets, swamp margins, stream margins, marshes. CL, CU, GE, LK,
LI, LR, PO, ST, SU, WA.

1589 *Carex sartwellii* Dewey   Sartwell's Sedge
Rare; calcareous sedge meadows, fens. RO: sedge meadow, Green Twp., Andreas 6337, 4
May 1985 (KE 48571); Holderman Prairie, Colerain Twp., Bartley s.n., 31 May 1948 (BHO
17815).

1590 # *Carex scabrata* Schwein.   Rough Sedge
Infrequent; wet woods, wet thickets, wooded seeps. CL, CU, GE, LK, LR, MD, PO, SU,
TR.

1591 *Carex scoparia* Willd.   Broom Sedge
Common; moist open woods, sphagnous thickets, sedge meadows, marshes, floodplains. AS,
AB, CU, GE, HL, KN, LK, LR, MH, MD, PE, PC, PO, RI, ST, SU, TR, WA.

122

1592 **# Carex seorsa** Howe   Weak Bog Sedge
Infrequent; sphagnous bogs, swamps, boggy woods. AB, CU, GE, LR, PO, SU, TR.

1593 **Carex shortiana** Dewey   Short's Sedge
Frequent; secondary and mature woods, wooded floodplains. AS, AB, FA, KN, LK, LI, LR,
MO, PE, PC, RI, RO, SU, TR, WA.

1594 **Carex sparganioides** Muhl.
Common; mesic and moist woods, wooded floodplains, wooded ravines. All counties except
MH, PE.

1595 **Carex sprengelii** Dewey   Sprengel's Sedge
Rare; dry woods, dry ridges in swamps. PO: dry soil, Nelson Twp., Webb 437, 25 May 1923
(KE 10527 & OS 4881). ST: Canton, Wolfe s.n., 18-(OS 77717).

1596 **Carex squarrosa** L.
Common; marshes, floodplains, ditches, open wet fields. AS, CU, HL, KN, LK, LI, LR,
MH, MD, PE, PO, RI, ST, SU, TR, WA.

1597 **Carex sterilis** Willd.   Fen Sedge
Infrequent; calcareous soils in seeps, fens, sedge meadows. CL, PO, RO, ST, SU.

1598 **Carex stipata** Muhl.
Common; marshes, roadside ditches, thickets, wet meadows, fields, swamps. ALL COUNTIES.

1599 **Carex straminea** Willd.   Straw Sedge
Rare; moist and mesic woods. AB: swamp forest, Windsor Twp., Bissell 1985: 113, 1 Jun
1985 (CLM 21784).

1600 **Carex stricta** Lam.   Hummock Sedge
Common; wet open woods, marshes, fens, sedge meadows, moist thickets. All counties except
KN, PE.

1601 **Carex suberecta** (Olney) Britton   Prairie Straw Sedge
Rare; prairie remnants, fens, sedge meadows. FA: Berne Twp., Goslin s.n., 9 Jun 1935 (OS
4294 & 74890). RO: gorge, Paxton Twp., Bartley s.n., 29 Jun 1941 (OS 17825 & 43083).

1602 **Carex swanii** (Fern.) Mackenz.   Swan's Sedge
Common; mesic and dry woods, rocky slopes, dry fields. AS, AB, CL, CU, FA, GE, HL,
KN, LK, LR, MD, MO, PO, RI, RO, SU, TR, WA.

1603 **Carex tenera** Dewey   Slender Straw Sedge
Frequent; wet fields, thickets, sedge meadows. AB, CU, FA, GE, HL, LK, PO, RO, ST, SU,
TR, WA.

1604 **Carex tetanica** Schkuhr   Slender Millet Sedge
Infrequent; fens, sedge meadows, calcareous thickets. CL, HL, LK, PO, RO, ST, SU.

1605 **Carex texensis** (Torrey) Bailey
Rare; rocky and sandy woods, fields. RO: Kingston, Green Twp., Bartley s.n., 6 May 1951
(OS 58864). Braun (1967) states that this species is perhaps adventive to Ohio.

1606 **Carex torta** Boott   Twisted Sedge
Common; wooded stream margins, shaded marsh borders, wet woods. AS, AB, CL, CU, FA,
GE, HL, KN, LK, LR, MD, PE, PO, ST, SU, TR.

1607 **Carex tribuloides** Wahlenb.
Common; wet woods, swamps, thickets, wet meadows. All counties except MO, PC, RI.

1608 **Carex trichocarpa** Schkuhr   Hairy-fruited Sedge
Frequent; sedge meadows, fens, marshes, sphagnous thickets. AS, CU, FA, KN, LK, LI, LR,
PC, RO, ST, SU, WA.

1609 **# Carex trisperma** Dewey   Three-seeded Sedge
Frequent; boggy woods, sphagnous thickets, tall shrub bogs. AS, AB, GE, HL, LK, LR, PO,
ST, SU, TR, WA.

1610 **Carex tuckermannii** Dewey   Tuckerman's Sedge
Infrequent; wet woods, swamps, thickets. AB, CU, KN, LK, LI, LR, MH, PE, PO, TR, WA.

1611 **Carex typhina** Michx.   Cat-tail Sedge
Frequent; meadows, wooded floodplains, thickets. AB, CU, FA, GE, HL, KN, LK, LR, PO,
SU, TR. Braun (1967) indicates a record for MO.

123

1612 **Carex umbellata** Willd. Clustered Sedge
Infrequent; dry open soils, sandstone exposures, steep slopes, dry sandy fields. AB, CL, CU, FA, LK, LI, PO, ST, SU.

1613 **Carex vesicaria** L.
Infrequent; fens, thickets, sedge meadows. AS, AB, CU, GE, LK, PC, PO, SU, TR, WA.

1614 **Carex virescens** Willd.
Frequent; dry woods, open thickets, clearings. AS, AB, CL, CU, FA, GE, HL, KN, LK, LI, LR, MD, PO, TR, SU, WA.

1615 **Carex viridula** Michx. Little Green Sedge
Rare; open calcareous soils in seeps, fens, thickets. PO: marly pasture, Ravenna Twp., Andreas 6221, 24 May 1984 (KE 47756). SU: fen, Green Twp., Andreas & Stoutamire 3546, 31 May 1979 (KE 42270); fen, Green Twp., Cusick 10135, 4 Jul 1969 (KE 25849).

1616 **Carex vulpinoidea** Michx. Fox Sedge
Common; marshes, wet fields, pond margins, roadside ditches. All counties except RO.

1617 **Carex willdenowii** Schkuhr Willdenow's Sedge
Rare; dry ridges, dry woods. AS: oak woods, Hanover Twp., Cusick 10756, 17 May 1970 (KE 25844). CU: dry ravine bluff, Brecksville, Bissell 1982:55, 22 May 1982 (CLM 17186); Strongsville, Fullmer s.n., 2 Jul 1941 (OS 44950); clearing in woods, Parma, Watson s.n., 18 May 1896 (OS 47669). MD: rocky woods, Medina, Segedi s.n., 22 Jun 1974 (CLM 3689).

1618 **Carex woodii** Dewey Wood's Sedge
Infrequent; mesic and moist woods, ridges. AB, CU, GE, LK, LR, MH, MD, RO, SU, TR.

1619 **Cladium mariscoides** (Muhl.) Torrey Twig Rush
Rare; calcareous soils in fens, open marl, sedge meadows. GE: Geauga County, Dachnowski s.n., 1911 (OS 57764). PO: peat bog, Mantua Twp., Rood & Campbell 1349, 28 Jun 1932 (KE 10311 & OS 3735); bog, Mantua Twp., Brown s.n., 20 Jul 1939 (OS 3736). ST: marl seep, Jackson Twp., Andreas 5404, 30 Jul 1980 (KE 43162).

1620 **Cyperus acuminatus** Torrey & Hooker Pale Umbrella-sedge
Rare; wet calcareous soils. PC: Pickaway Twp., Bartley & Pontius s.n., 21 Jul 1935 (OS 49393). RO: North Union Twp., Bartley s.n., 29 Jul 1945 (OS 46092); Green Twp., Bartley s.n., 23 Jul 1933 (OS 3624).

1621 **Cyperus diandrus** Torrey Low Umbrella-sedge
Rare; gravelly soils, stone quarries, sandy pond margins. CL: glacial pool, Butler Twp., Cusick s.n., 12 Sep 1959 (OS 64040). PO: Suffield Bog, Bartley s.n., 9 Sep 1944 (BHO 17846).

1622 **Cyperus engelmannii** Steudel Engelmann's Umbrella-sedge
Frequent; wet sphagnous depressions, lake margins, thickets. AB, CU, GE, LI, LR, MH, PC, PO, ST, SU, WA.

1623 **Cyperus erythrorhizos** Muhl. Red-rooted Umbrella-sedge
Common; mudflats, moist meadows, ditches, thickets. AS, AB, CL, CU, GE, HL, LK, LI, LR, MH, MD, PE, RI, RO, ST, SU, TR, WA.

1624 **Cyperus esculentus** L. Yellow Nut-grass
Common; damp disturbed soils, cultivated fields, ditches, marshes. AS, AB, CL, CU, FA, HL, KN, LK, LI, LR, MH, MO, PC, PO, RI, RO, SU, TR, WA.

1625 **Cyperus ferruginescens** Boeckl. Rusty Umbrella-sedge
Frequent; damp soils, fields, pond margins. AB, GE, LK, LI, LR, MH, MD, PC, PO, RO, TR.

1626 **Cyperus filiculmis** Vahl Thread-like Umbrella-sedge
Frequent; dry sandy soils, roadside banks, fields. AB, CU, FA, GE, HL, LK, LI, MH, PO, RO, SU.

1627 **Cyperus flavescens** L. var. **poaeformis** (Pursh) Fern.
Infrequent; moist sandy soils, fields, pond margins. AS, CL, FA, LK, LI, PC, PO, RI, RO, SU. Braun (1967) indicates a record for GE.

1628 **Cyperus inflexus** Muhl.
Infrequent; wet soils, damp sandy fields, pond margins, waste places. LK, LI, PC, PO, RO, SU.

124

1629 *Cyperus rivularis* Kunth   Stream Umbrella-sedge
Common; roadsides, fields, mudflats, moist steep slopes, pond margins. AS, AB, CL, CU,
FA, GE, HL, KN, LK, LI, LR, MO, PC, PO, RO, ST, SU, TR, WA.

1630 *Cyperus schweinitzii* Torrey   Schweinitz's Umbrella-sedge
Rare; sandy shores, fields, waste places. RO: railroad yard, Chillicothe, Bartley s.n., 11 Sep
1962 (OS 80020).

1631 *Cyperus strigosus* L.   Common Umbrella-sedge
Common; meadows, thickets, lake and stream margins, roadside ditches, seeps. ALL
COUNTIES.

1632 *Cyperus tenuifolius* (Steudel) Dandy   Thin-leaved Umbrella-sedge
Infequent; damp disturbed sandy soils, muddy shores. CU, FA, PC, PO, RO.

1633 *Dulichium arundinaceum* (L.) Britton   Three-way Sedge
Common; wet meadows, swamps, pond margins, sphagnous meadows. AS, AB, CU, GE, HL,
LK, LI, LR, MD, PE, PC, PO, RI, RO, ST, SU, TR, WA.

1634 *Eleocharis acicularis* (L.) R. & S.   Needle Spike-rush
Common; lake margins, shallow sunny pools, lawns, wet waste places, thickets, marshes. All
counties except KN, MO, RI.

1635 *Eleocharis elliptica* Kunth.   Yellow-seeded Spike-rush
Rare; fens, sandy soils. AB: fen seep, Wayne Twp., Bissell 1984:98, 21 Jul 1984 (CLM 20920).
LR: railroad ditch, Oberlin, Cowles s.n., 28 May 1892 (OC 71212). SU: calcareous alluvium,
Boston Twp., Andreas, Knoop & Nekola 7307, 14 Aug 1988 (KE 50265).

1636 *Eleocharis erthyropoda* Steudel   Creeping Spike-rush
Eleocharis calva Torrey (in Braun, 1967)
Common; ditches, swamps, sedge meadows, sphagnous thickets. All counties except MH,
RI.

1637 *Eleocharis intermedia* Schultes   Matted Spike-rush
Infequent; wet calcareous soils, open shallow pools, sphagnous meadows, seeps. AB, CL, FA,
LI, ST, SU.

1638 *Eleocharis obtusa* Schultes   Blunt Spike-rush
Common; roadside ditches, moist waste places, abandoned wet fields, muddy lake margins.
ALL COUNTIES.

1639 *Eleocharis olivacea* (Poiret) Urban   Olivaceous Spike-rush
Rare; wet sands, sphagnous meadows, beaver paths in wetlands. CU: Cleveland, Claassen
s.n., 1894 (OS 3337). GE: swampy sphagnous meadow, Burton Twp., Bissell 79-98, 30 Aug
1979 (CLM s.n.).

1640 *Eleocharis pauciflora* (Lightf.) Link   Few-flowered Spike-rush
Rare; damp calcareous soils, fens. RO: Immell's Bog, Green Twp, Bartley & Pontius s.n.,
31 May 1937 (OS 3244 & BHO 17897).

1641 *Eleocharis quadrangulata* (Michx.) R. & S.   Four-angled Spike-rush
Rare; pools, creek banks, lake margins. AS: Round Lake, Hopkins s.n., 5 Jul 1912 (OS 6429
& OS 42465). PO: Twin Lakes, Webb 474, 23 Jul 1915 (KE 10347); shallow water, East
Twin Lake, Hopkins 148, 16 Aug 1915 (OS 3292 & OS 3293). SU: Cuyahoga Falls, Kellerman
s.n., 3 Jul 1899 (OS 59191).

1642 *Eleocharis rostellata* Torrey   Beaked Spike-rush
Infrequent; fens, seeps, sphagnous meadows, sedge meadows. CL, HL, PO, ST, SU, WA.

1643 *Eleocharis smallii* Britton   Small's Spike-rush
Common; sphagnous meadows, sedge meadows, ditches, marshes, lake and stream margins.
AB, CL, CU, GE, HL, KN, LK, LI, LR, MD, PE, PO, RO, ST, SU, TR, WA.

1644 *Eleocharis tenuis* (Willd.) Schultes   Slender Spike-rush
Rare; wet sands, gravels, meadows. GE: roadbed, Newbury Twp., Cusick 19574, 6 Aug 1979
(OS 131479).

1645 # *Eriophorum virginicum* L.   Tawny Cotton-grass
Infrequent; sphagnous bogs, sphagnous thickets, leatherleaf bogs. AS, AB, GE, LI, LR, PO,
RI, ST, SU, WA.

1646 # *Eriophorum viridi-carinatum* (Engelm.) Fern.   Green Cotton-grass
Infrequent; calcarous fens, sedge meadows, thickets. AB, CU, GE, LK, LI, PO, ST, SU.

1647 *Fimbristylis autumnalis* (L.) R. & S.
Rare; sandy soils. PC: Pickaway Twp., Bartley & Pontius s.n., 26 Aug 1934 (OS 3276). RO: Liberty Twp., Bartley & Pontius s.n, 2 Oct 1932 (OS 3273 & OS 3274).

1648 *Hemicarpha micrantha* (Vahl) Pax
Rare; wet sandy soils, thickets. PC: swampy place, Pickaway Twp., Bartley s.n., 22 Aug 1948 (OS 44071 & BHO 17912).

1649 # *Rhynchospora alba* (L.) Vahl   White Beak-rush
Frequent; sphagnous bogs, sedge meadows, sphagnous thickets. AS, AB, GE, HL, LI, LR, PO, RI, ST, SU, WA.

1650 *Rhynchospora capillacea* Torrey   Slender Beak-rush
Infrequent; marl meadows, fens. CL, GE, PO, RO, ST, SU.

1651 *Rhynchospora capitellata* (Michx.) Vahl   Small-headed Beak-rush
Infrequent; calcareous soils, damp lake shores. AB, FA, LK, PO, RO, TR. Braun (1967) indicates a record for SU.

1652 *Scirpus acutus* Bigelow   Hard-stemmed Bull-rush
Frequent; ditches, sedge meadows, fens, lake margins. AB, CL, FA, GE, HL, KN, LK, PO, RO, ST, SU, WA.

1653 *Scirpus americanus* Persoon   Three-square Bull-rush
Frequent; wet sandy fields, fens, lake margins, seeps. AS, AB, CU, GE, HL, LK, LI, LR, MD, PC, PO, RI, RO, ST, SU.

1654 *Scirpus atrovirens* Willd. (including *S. hattorianus* Mak.)
Common; moist fields, roadside ditches and banks, marshes. ALL COUNTIES.

1655 *Scirpus cyperinus* (L.) Kunth (including *S. pedicellaris* Fern.)   Wool-grass
Common; wet fields, meadows, fens, swamps, abandoned beaver ponds, marshes, ditches, thickets. ALL COUNTIES.

1656 # *Scirpus expansus* Fern.   Woodland Wool-grass
Rare; meadows, wet open woods, fens. HL: fen, Washington Twp., Wilson 2284, 30 Jul 1971 (KE 28928). KN: Sapp's Run, Brown Twp., Pusey 752, 2 Jul 1975 (KE 37933); Knox Lake, Keil et al. 9645, 17 Aug 1973 (OS 111845). RI: Mansfield, Wilkinson s.n., 27 Jun 1892 (OS 3120). Braun (1967) cites a record for WA from "Lakeville" which is located in HL.

1657 *Scirpus fluviatilis* (Torrey) Gray   River Bull-rush
Infrequent; lake and stream margins. AB, CU, FA, LK, LI, LR, MD, PC, RO, WA.

1658 *Scirpus pendulus* Muhl.
Scirpus lineatus Michx. (in Braun, 1967)
Common; calcareous soils in sedge meadows, fens, seeps, thickets. AS, AB, CU, GE, HL, KN, LK, LI, LR, MD, PE, PC, RI, RO, ST, SU, TR, WA.

1659 *Scirpus polyphyllus* Vahl   Many-leaved Bull-rush
Common; swamps, wet woods, marshes, fields. AS, AB, CL, CU, FA, GE, HL, KN, LK, LI, LR, MH, PO, RI, ST, SU, TR, WA.

1660 *Scirpus purshianus* Fern.   Pursh's Bull-rush
Rare; peaty soils, swamps, muddy shores. PC: swampy place, Pickaway Twp., Bartley s.n., 21 Aug 1948 (OS 43906). SU: Summit County, Claassen s.n., 1891 (OS 3229). Braun (1967) indicates a record for RO.

1661 *Scirpus validus* Vahl   Great Bull-rush, Soft-stemmed Bull-rush
Common; marshes, lake margins, fens, wet thickets. ALL COUNTIES.

1662 *Scirpus verecundus* Fern.
Infrequent; dry woods, clearings. AS, FA, HL, LK, RO. Braun (1967) indicates records for KN, LI.

1663 *Scleria verticillata* Willd.   Whorled Nut-grass
Rare; calcareous peats, sedge meadows, thickets, marshes. PO: marly area, Streetsboro Twp., Cusick 10195, 31 Jul 1969 (KE 25957). RO: bog, Colerain Twp., Bartley s.n., 9 Sep 1961 (OS 66788); bog, Frankfort, Bartley s.n., 14 Aug 1932 (OS 3741). ST: meadow, Massillon, Stair s.n., Aug 1898 (CLM 6499).

DIOSCOREACEAE (Yam Family)

1664 *Dioscorea batatas* Dcne.   Chinese Yam
Rare, adventive; thickets, roadsides. PO: roadside, Windham Twp., Andreas 3054, 3 Sep 1978 (KE 48095). RO: fence, Colerain Twp., Bartley s.n., 6 Aug 1961 (OS 67401).

1665 *Dioscorea villosa* L.   Wild Yam
Common; roadside banks, thickets, dry and moist woodland margins. ALL COUNTIES.

ERIOCAULACEAE (Pipewort Family)

1666 # *Eriocaulon septangulare* With.   White Buttons, Duckgrass
Once rare, now presumably extirpated from Ohio; shallow waters, peaty soils. PO: Lake Brady, Shilling s.n., 26 Oct 1935 (OS 10354); sandy shore, Muzzy Lake, Rootstown Twp., Webb 507, 23 Jul 1915 (KE 10736 & OS 10353). SU: near Turkeyfoot Lake, Sterki s.n., 26 Sep 1913 (OS 10352).

GRAMINEAE / POACEAE (Grass Family)

1667 *Aegilops cylindrica* Host   Goat Grass
Rare, adventive; waste places, railroads. RO: waste place, Colerain Twp., Bartley s.n., 7 Jun 1903 (OS 73348); roadside, Kingston, Bartley s.n., 6 Jun 1960 (OS 44104).

1668 *Agropyron repens* (L.) Beauv.   Quack Grass
Common, naturalized; disturbed places, roadsides, railroads. All counties except MO.

1669 *Agropyron smithii* Rydb.   Western Wheat Grass
Infrequent, adventive; railroads, roadsides. CU, FA, GE, LK, PO, ST, SU.

1670 *Agropyron trachycaulum* (Link) Malte   Bearded Wheat Grass
Infrequent; fens, sedge meadows, wet thickets. AS, AB, CU, HL, LR, PO, SU, TR.

1671 *Agrostis gigantea* Roth   Red Top
Agrostis alba L. (in Braun, 1967)
Common, naturalized; roadsides, disturbed fields, damp thickets. ALL COUNTIES.

1672a *Agrostis hyemalis* (Walter) BSP. var. *hyemalis*   Hair-grass
Infrequent; roadsides, sandy fields, thin woods, woodland margins. AS, AB, CU, FA, GE, HL, LK, PC, PO, SU, TR, WA.

1672b *Agrostis hyemalis* (Walter) BSP. var. *tenuis* (Tuckerman) Gleason
Hair-grass, Fly-away Grass
Agrostis scabra Willd. (in Braun, 1967)
Infrequent; dry sandy fields, thin woods, roadsides. CL, CU, HL, KN, LK, LR, PE, PO, SU, WA.

1673 *Agrostis perennans* (Walter) Tuckerman   Upland Bent-grass, Autumn Bent-grass
Common; dry to moist woods, ledges, sphagnous hummocks, fields. ALL COUNTIES.

1674 *Agrostis tenuis* Sibth.   Colonial Bent-grass
Frequent, naturalized; roadsides, fields. AB, CL, CU, GE, HL, LK, MH, PC, PO, SU, TR.

1675 *Alopecurus aequalis* Sobol   Meadow Foxtail
Frequent; ditches, disturbed wet ground, pond margins. AB, CL, KN, LK, LI, LR, MH, PE, PC, PO, RO, ST, SU, TR.

1676 *Alopecurus carolinianus* Walter   Carolina Foxtail
Rare; waste places, fields. PC: Pickaway Twp., Bartley & Pontius s.n., 20 May 1931 (OS 6804).

1677 *Alopecurus pratensis* L.   Meadow Foxtail
Rare, adventive; fields, waste places. LR: North Ridgeville, Evans s.n., 20 Jun 1925 (OC 82209). TR: field, Liberty Twp., Burns 1169, 12 Jun 1979 (KE 44694).

1678 *Andropogon elliottii* Champ.   Elliott's Beard Grass
Rare; old fields, pastures. FA: Hocking Twp., Thomas & Hicks s.n., 9 Dec 1929 (OS 98937).

1679 *Andropogon gerardii* Vitman   Big Bluestem, Turkey-foot
Common; prairies, open dry fields, calcareous seeps, fens, sphagnous thickets, roadsides, railroads. All counties except MH, TR.

1680 *Andropogon scoparius* Michx.   Little Bluestem
Common; roadsides, sandy banks, sphagnous thickets, fens. ALL COUNTIES.

1681 **Andropogon virginicus** L. Broom-sedge
Common; dry sandy fields, waste places, railroads, roadsides, abandoned pastures and fields. ALL COUNTIES.

1682 ***Anthoxanthum odoratum*** L. Sweet Vernal Grass
Common, naturalized; stream margins, moist woods, roadsides, fields, woodland margins. AS, AB, CL, CU, FA, GE, HL, LK, LI, MH, MD, MO, PC, PO, ST, SU, TR, WA.

1683 **Aristida dichotoma** Michx. Poverty-grass
Frequent; dry open ground, railroads, cinders. AS, CU, FA, HL, KN, LI, LR, MH, PE, RO, ST, SU. Braun (1967) indicates records for AB, CL.

1684 **Aristida longispica** Poiret
Infrequent; sandy soil in fields, railroads. CU, FA, KN, LI, LR, PO, RO, TR.

1685 **Aristida oligantha** Michx. Triple-awn Grass
Common; dry waste places, railroads. AB, CL, CU, FA, GE, HL, KN, LK, LI, LR, MD, MO, PE, PC, PO, RO, ST, SU, WA.

1686 **Aristida purpuracens** Poiret Purple Triple-awn Grass
Rare; sandy fields, dry open slopes. FA: Berne Twp., Goslin s.n., 11 Nov 1958 (BHO 21170). PC: Pickaway County, Bartley & Pontius s.n., 31 Aug 1934 (OS 7099). RO: dry open hillside, Jefferson Twp., MacBeth 894, 19 Sep 1954 (OS 43866); Immell's Bog, Green Twp., Bartley s.n., 10 May 1937 (OS 17548).

1687 ***Arrhenatherum elatius*** Presl Tall Oat-grass
Common, naturalized; roadsides, railroads, dry fields, wildlife areas, waste places. All counties except AB, KN.

1688 ***Arundinaria gigantea*** (Walter) Chapm. Bamboo, Cane
Rare, naturalized; roadsides. FA: spreading from cultivation, Revenge, Cook s.n., 14 Nov 1937 (OS 5394); escape, Berne Twp., Cook s.n., 28 Aug 1938 (OS 5393); Violet Twp., Goslin 44-128-18, 30 Apr 1971 (BHO 31133).

1689 ***Aspera spica-venti*** (L.) Beauv.
Agrostis spica-venti L. (in Braun, 1967)
Rare, adventive; roadsides, waste places. WA: Wooster, Thatcher s.n., 6 Jul 1936 (OS 92679). Braun (1967) indicates a record for LR.

1690 ***Avena fatua*** L. Wild Oats
Infrequent, adventive; railroads, waste places. CU, FA, HL, PC, PO, ST, SU.

1691 ***Avena sativa*** L. Common Oats
Frequent, adventive; roadsides, railroads, ditches. AS, CL, CU, HL, KN, LK, LR, MD, PE, SU, TR.

1692 **Bouteloua curtipendula** (Michx.) Torrey Side-oats Grama
Rare; dry ridges, prairies. RO: Green Twp., Bartley & Pontius s.n., 2 Sep 1935 (OS 47019).

1693 **Brachyelytrum erectum** (Roth) Beauv.
Common; moist or dry woods, woodland margins, shaded roadside banks. ALL COUNTIES.

1694 **Bromus ciliatus** L. Fringed Brome Grass
Infrequent; sphagnous thickets, stream banks, bog margins, fens. AS, HL, LK, LR, PO, ST, SU, TR.

1695 ***Bromus commutatus*** Schrader
Common, naturalized; dry roadsides, railroads, waste places, old fields. AS, AB, CL, CU, FA, GE, HL, KN, LK, LI, LR, MD, PE, PO, RO, ST, SU, TR, WA.

1696 ***Bromus inermis*** Leysser Hungarian Brome Grass
Common, naturalized; fields, waste places, roadsides. All counties except LK, RO.

1697 ***Bromus japonicus*** Murray Japanese Brome Grass
Common, naturalized; cultivated fields, railroads, roadsides, waste places. AB, CL, CU, HL, GE, KN, LK, LI, LR, MH, MD, PC, PO, RI, ST, SU, WA.

1698 **Bromus kalmii** Gray Kalm's Brome Grass
Infrequent; fens, thickets, moist sedge meadows, calcareous slopes. CL, PC, PO, ST, SU.

1699 **Bromus latiglumis** (Shear) Hitchc. Woodland Brome Grass
Common; shaded roadside banks, mesic woods, ravines, wooded floodplains. AS, CL, CU, GE, HL, KN, LK, LI, LR, MO, PE, PC, PO, RI, RO, ST, SU, TR, WA.

128

1700 *__Bromus mollis__ L.   Soft Chess
     Rare, adventive; dry waste places, roadsides. LR: Oberlin, Ricksecker s.n., 30 May 1895 (OS
     5410). WA: farm, Wooster, Duvel 1331, 18 Jun 1899 (OS 92648).

1701 *__Bromus nottowayanus__ Fern.
     Rare, adventive; wet roadsides. CU: railroad, Brecksville Twp., Andreas 6611, 5 Aug 1985
     (KE 49768). SU: roadside, Boston Twp., Herrick s.n., 29 Jul 1960 (KE 2015).

1702 __Bromus pubescens__ Willd.
     __Bromus purgans__ L. (in Braun, 1967)
     Common; rich woods, floodplains, rocky slopes. All counties except RI. Braun (1967) indicates
     a record for AB.

1703 *__Bromus secalinus__ L.   Cheat, Chess
     Common, naturalized; fields, disturbed areas. All counties except RO.

1704 *__Bromus sterilis__ L.
     Infrequent, adventive; waste places, disturbed fields, roadsides. CU, FA, LK, LI, PC.

1705 *__Bromus tectorum__ L.   Downy Brome Grass
     Common, naturalized; fields, weedy ground, railroads, roadsides. ALL COUNTIES.

1706 __Calamagrostis canadensis__ (Michx.) Beauv.   Blue-joint
     Common; wet meadows, fens, marshes, ditches, thickets. AB, CL, CU, FA, GE, HL, LK,
     LI, LR, MH, MD, PO, RO, ST, SU, TR, WA.

1707 __Calamagrostis cinnoides__ (Muhl.) Bart.   Cinna-like Bentgrass
     Rare; damp sandy soils. LI: Granville, Jones s.n., [no date] (OS 6626).

1708 __Calamagrostis inexpansa__ Gray   Northern Reed Grass
     Rare; damp woods, fens, thickets. RO: drained bog, Colerain Twp., Bartley s.n., 19 Jul 1961
     (OS 67719 & OS 67717).

1709 __Cenchrus longispinus__ (Hackel) Fern.   Field Sandbur
     __Cenchrus pauciflorus__ Benth. (in Braun, 1967)
     Frequent; roadsides, sandy banks, dry sandy fields, beaches. AB, CU, FA, GE, KN, LK,
     LI, LR, MH, PO, RO, ST, SU.

1710 __Cinna arundinacea__ L.   Wood Reedgrass
     Common; rich woods, ravines, shaded swamps, roadside ditches. ALL COUNTIES.

1711 # __Cinna latifolia__ (Trev.) Griseb.   Broad-leaf Northern Wood Reedgrass
     Rare; shady ravines. AB: east Monroe Twp., Hicks s.n., 4 Aug 1929 (OS 6794). SU: shaded
     seepage bank, Akron, Andreas 2451, 26 Jul 1978 (KE 40753).

1712 *__Cynodon dactylon__ (L.) Persoon   Bermuda Grass
     Infrequent, naturalized; fields, pastures, roadsides, waste places. AB, CU, FA, KN, LK, MD,
     PC, RI, RO.

1713 *__Cynosurus cristatus__ L.   Crested Dog's-tail
     Rare, adventive; lawns, roadsides. LR: yard, Oberlin, Grover s.n., 1 Jul 1925 (CLM 4024).

1714 *__Cynosurus echinatus__ L.
     Braun (1967) indicates a record for RO.

1715 *__Dactylis glomerata__ L.   Orchard Grass
     Common, naturalized; fields, roadsides, waste places. ALL COUNTIES.

1716 # __Danthonia compressa__ Aust.   Flattened Wild Oat Grass
     Infrequent; shaded banks, rocky ledges, dry woodlands. AS, AB, FA, GE, HL, LK, MH,
     PO, SU.

1717 __Danthonia spicata__ R. & S.   Poverty Grass
     Common; dry fields, open woods, roadsides, disturbed fields. ALL COUNTIES.

1718 __Deschampsia caespitosa__ (L.) Beauv.   Tufted Hair Grass
     Rare; damp calcareous soils, fens. ST: marl, Jackson Twp., Andreas 5362, 7 Jul 1980 (KE
     43005). RO: Paxton Twp., Bartley & Crist s.n., 29 Jun 1941 (OS 43850); Immell's Bog,
     Green Twp., Bartley & Pontius s.n., 17 Jul 1937 (OS 6482).

1719 # __Deschampsia flexuosa__ (L.) Beauv.   Wavy Hair Grass
     Rare; shaded roadside banks, dry woods, cliffs. FA: cliff base, Mt. Pleasant, Bartley s.n.,
     10 Jun 1945 (BHO 17551). GE: woodland edge, Bainbridge Twp., Herrick 430, Jul 1959
     (OS 64680). PO: gravelly kame, Streetsboro Twp., Andreas and Knoop 6258, 17 Jun 1984

(KE 47783); boggy woods, Franklin Twp., Andreas 5664, 17 Jun 1981 (KE 44115). SU: dry peaty woods, Cuyahoga Falls, Andreas 5691, 24 Jun 1981 (KE 44116).

1720 *Diarrhena americana* Beauv.
Rare; floodplains, shaded banks, woods. RO: along creek, Paxton Twp., Herrick s.n., 1 Sep 1964 (OS 75302); station, Bartley & Pontius s.n., 3 Jul 1932 (OS 6319); Chillicothe, Kellerman s.n., 12 Aug 1898 (OS 6318). Braun (1967) indicates a record for LI.

1721 *Digitaria filiformis* (L.) Koeler   Slender Crabgrass
Rare; roadsides, prairies, sandy fields. PC: Pickaway Twp., Bartley & Pontius s.n., Aug 1934 (OS 8201). TR: roadside, Phalanx Mills, Rood 3116, 25 Oct 1959 (KE 1873).

1722 **Digitaria sanguinalis* (L.) Scop.   Crab-grass
Common, naturalized; weedy ground, lawns, railroads, roadsides, waste places. ALL COUNTIES.

1723a **Echinochloa crusgalli* (L.) Beauv. var. *crusgalli*   Barnyard Grass
Common, naturalized; damp soil in waste places, roadside ditches, fields. All counties except MO.

1723b **Echinochloa crusgalli* var. *frumentacea* (Link) Wright   Japanese Millet
Rare, adventive; waste places. AB: near Dorset, Hicks s.n., 20 Aug 1932 (OS 72620).

1724 *Echinochloa muricata* (Beauv.) Fern.
Echinochloa pungens (Poiret) Rydb. (in Braun, 1967)
Common; fields, roadsides, waste places. ALL COUNTIES.

1725 *Echinochloa walteri* (Pursh) Heller   Walter's Barnyard Grass
Infrequent; marshes, wet fields. AB, LK, LI, LR, PE.

1726 **Eleusine indica* (L.) Gaertner   Goose Grass
Common, naturalized; railroads, roadsides, waste places. ALL COUNTIES.

1727 *Elymus canadensis* L.   Canada Wild-rye
Frequent; railroads, rocky banks, roadsides. AS, AB, CU, GE, HL, KN, LK, LR, PO, RI, ST, SU, TR, WA.

1728 *Elymus riparius* Wieg.   River Wild-rye
Common; stream banks, weedy floodplains, moist thickets. All counties except FA, MO.

1729 *Elymus villosus* Willd.
Common; wooded slopes, floodplains, rocky moist woods. ALL COUNTIES.

1730 *Elymus virginicus* L.   Virginia Wild-rye
Common; mesic to moist woods, floodplains, marshes, roadsides. ALL COUNTIES.

1731 *Eragrostis capillaris* (L.) Nees   Capillary Love Grass, Lace-grass
Infrequent; dry sandy hillsides, rocky slopes. CL, FA, LK, LI, MO, PC, ST, SU, TR.

1732 **Eragrostis cilianensis* (All.) Mosher   Strong-scented Love Grass
Common, naturalized; railroads, roadsides, waste places. All counties except FA, MH, PE.

1733 **Eragrostis curvula* (Schrader) Nees   Weeping Love Grass
Rare, adventive; waste places. PO: woodlot, Kent State University, Kent, Bricker s.n., 28 Jun 1955 (OS 55155).

1734 *Eragrostis frankii* Steudel   Frank's Love Grass
Frequent; roadsides, cultivated ground, weedy places. CU, FA, LK, LI, LR, PC, PO, RI, RO, ST, SU, WA.

1735 *Eragrostis hypnoides* (Lam.) BSP.
Common; mudflats, pond margins, shores. AB, CL, CU, FA, HL, KN, LK, LI, LR, MH, MO, PE, PO, RI, RO, ST, SU, TR, WA.

1736 *Eragrostis pectinacea* (Michx.) Nees
Common; sandy waste places, roadsides, railroads, cultivated fields. ALL COUNTIES.

1737 **Eragrostis pilosa* (L.) Beauv.   India Love Grass
Rare, adventive; roadsides, waste places, fields. PC: Pickaway Twp., Bartley & Pontius s.n., 21 Jul 1940 (OS 56081). WA: Applecreek, Selby 1326, 12 Aug 1895 (OS 92713).

1738 **Eragrostis poaeoides* R. & S.
Frequent, naturalized; fields, waste places, railroads, cinders. AS, CL, CU, FA, HL, KN, LR, MH, PC, PO, RO, ST, SU, TR, WA.

1739 *Eragrostis spectabilis* (Pursh) Steudel   Purple Love Grass
Common; roadsides, dry sandy fields, waste places. All counties except RI.

1740 *Festuca arundinacea* Schreber   Alta Fescue
Frequent, adventive; roadside banks, bridge embankments, widely planted along highways.
AB, CU, HL, LK, MO, PE, PO, RI, RO, SU, TR.

1741 *Festuca obtusa* Biehler   Nodding Fescue
Common; moist woods, wooded floodplains, ravines. ALL COUNTIES.

1742 *Festuca octoflora* Walter   Six-weeks Fescue
Infequent; sandy disturbed soils. AS, AB, FA, GE, LK, LI, LR, PC, ST.

1743 *Festuca ovina* L.   Sheep Fescue
Infrequent, naturalized; dry fields, rocky slopes. AB, FA, GE, HL, KN, LK, SU, WA.

1744 *Festuca pratensis* Hudson   Meadow Fescue
Festuca elatior L. (in Braun, 1967)
Common, naturalized; roadsides, lawns, fields, waste places. ALL COUNTIES.

1745 *Festuca rubra* L.   Red Fescue
Infrequent; woods, fields. CU, GE, HL, LK, PC, PO, RO, ST, SU, TR.

1746 # *Glyceria acutiflora* Torrey   Sharp-glumed Manna Grass
Once rare, now presumably extirpated from Ohio; pond margins, swamps. HL: dry pond
bed, Washington Twp., Wilson 2270, 30 Jul 1971 (KE 29023). LI: swamp, Jones s.n., 28
May 1891 (OC 77494). MO: swamp woods, Troy Twp., Herrick s.n., 7 Jun 1962 (OS 71526).

1747 # *Glyceria canadensis* (Michx.) Trin.   Rattlesnake Grass
Frequent; bog margins, fens, roadside ditches, sphagnous thickets, abandoned beaver ponds,
swamps. AB, CL, CU, GE, LK, LR, MH, PO, ST, SU, TR, WA.

1748 # *Glyceria grandis* S. Watson   Tall Manna Grass
Infrequent; bog margins, sphagnous thickets, stream banks, wet meadows, fens. AB, CL,
GE, PO, ST, WA.

1749 #*Glyceria melicaria* (Michx.) F. T. Hubbard   Long Manna Grass
Infrequent; sphagnous thickets, moist woods, shaded stream banks, swales, ravine bottoms.
AB, CU, FA, GE, LK, PO, SU, TR.

1750 *Glyceria septentrionalis* Hitchc.   Floating Manna Grass
Common; wet meadows, pastures, bog margins, moist woods. ALL COUNTIES.

1751 *Glyceria striata* (Lam.) Hitchc.   Fowl Manna Grass
Common; swamps, abandoned beaver ponds, wet thickets, wet meadows, moist woods, fen
and bog margins. ALL COUNTIES.

1752 *Hierochloe odorata* (L.) Beauv.   Sweet Grass, Vanilla Grass
Infrequent; calcareous soils in fens, marl meadows. FA, HL, PC, PO, RO, ST, SU, TR.

1753 *Holcus lanatus* L.   Velvet Grass
Common, naturalized; roadsides, fields, railroads, waste places. All counties except PC.

1754 *Hordeum jubatum* L.   Squirrel-tail Barley
Common, naturalized; fields, waste places, roadsides. AS, AB, CL, CU, FA, GE, HL, LK,
LI, LR, MH, MD, MO, PC, PO, RI, ST, SU, TR, WA.

1755 *Hordeum pusillum* Nutt.   Little Barley
Rare, adventive; disturbed open ground, roadsides. PO: disturbed lawn, Kent, Wilson 1842,
20 Jun 1971 (KE 26025).

1756 *Hordeum vulgare* L.   Barley
Infrequent, adventive; roadsides, ditches, waste places. HL, KN, PE, PO, ST, TR.

1757 *Hystrix patula* Moench   Bottle-brush Grass
Common; moist woods, wooded floodplains, shaded roadside banks. ALL COUNTIES.

1758 *Leersia oryzoides* (L.) Swartz   Rice Cutgrass
Common; wet meadows, swales, roadside ditches, abandoned beaver ponds, stream and pond
margins, swamps. ALL COUNTIES.

1759 *Leersia virginica* Willd.   White Grass
Common; moist woods, shaded roadside banks, wooded floodplains, thickets. ALL COUNTIES.

1760 *Leptochloa fasicularis* (Lam.) Gray   Sprangletop, Salt Meadow Grass
Infrequent, naturalized; halophytic roadside berms, dry run-off ditches, industrial tailings.
AS, AB, CU, LK, LR, MD, PC, RO, SU, WA.

1761 **Leptoloma cognatum** (Schultes) Chase   Fall Witch Grass
Infrequent; dry sandy soils, railroad cinders. AB, CU, FA, LK, RO, ST, SU.

1762a ***Lolium perenne** L. var. **perenne***   Perennial Rye-grass
Common, naturalized; roadsides, fields, railroads, waste places. ALL COUNTIES.

1762b ***Lolium perenne** L. var. **aristatum** Willd.*   Italian Rye-grass
Lolium multiflorum Lam. (in Braun, 1967)
Frequent, naturalized; fields, roadsides, borrow pits, waste places. AS, AB, CL, CU, FA,
GE, HL, KN, LK, LI, LR, PC, PO, RO, ST, SU, TR.

1763 **Milium effusum** L.   Millet Grass
Common; rich mesic and moist woods, thickets. AS, AB, CL, CU, GE, HL, LK, LI, LR,
MD, MO, PO, ST, SU, TR, WA.

1764 ***Miscanthus sinensis** Andersson*   Eulalia
Infrequent, adventive; roadsides, thickets, fields. AB, LK, LR, PC, PO, SU.

1765 ***Muhlenbergia asperifolia** (Nees & Meyer) Parodi*   Scratch Grass
Infrequent, adventive; railroads, waste places. CU, GE, LK, ST, SU.

1766 **Muhlenbergia curtisetosa** (Schribner) Bush
Rare; woodland thickets. PC: Pickaway Twp., Bartley & Pontius s.n., 11 Aug 1934 (OS
6984).

1767 **Muhlenbergia frondosa** (Poiret) Fern.
Common; roadsides, moist woods, stream banks, thickets. All counties except PE.

1768 **Muhlenbergia glomerata** (Willd.) Trin.
Infrequent; sphagnous thickets, fens. CL, CU, GE, HL, LI, PO, RO, ST, SU, WA.

1769 **Muhlenbergia mexicana** (L.) Trin.
Common; swamps, roadsides, thickets, fens. All counties except FA, MO, PE.

1770 **Muhlenbergia schreberi** Gmelin   Nimblewill
Common; dry woods, thickets, waste places. ALL COUNTIES.

1771 **Muhlenbergia sobolifera** (Muhl.) Trin.
Infrequent; dry gravelly woods, shaded slopes, ledges. CL, KN, LI, PE, RO, WA.

1772 **Muhlenbergia sylvatica** Torrey
Infrequent; mesic and moist woods, damp thickets, woodland stream margins. AB, CU, GE,
LK, MD, MO, PC, RO, SU, TR.

1773 **Muhlenbergia tenuiflora** (Willd.) BSP.
Infrequent; moist woods, shaded cliffs. FA, HL, LK, MD, PE, PO, SU, TR, WA.

1774 # **Oryzopsis asperifolia** Michx.   Large-leaved Mountain Rice Grass
Rare; shaded, well-drained wooded slopes, dry ridges. AB: dry gravelly kame, Richmond
Twp., Andreas 5134, 6 Jun 1980 (KE 43184). CU: beech woods, Brecksville, Andreas 6755,
16 Apr 1986 (KE 50260); dry hogback, Brecksville, Bissell 1982:6, 29 Apr 1982 (CLM 17322).

1775 **Oryzopsis racemosa** (Smith) Hitchc.   Mountain Rice Grass
Infrequent; rocky woods. GE, HL, LR, MD, PO, SU. Braun (1967) indicates records for
CU, LK.

1776 **Panicum bicknellii** Nash   Bicknell's Panic Grass
Rare; dry thickets. LR: Oak Point, Grover s.n., 6 Jun 1922 (OC 70664).

1777 **Panicum boreale** Nash   Northern Panic Grass
Rare; stream banks, fields, thickets. PO: sandy, mossy open-wooded area along river, Franklin
Twp., Cooperrider 6736, 14 Jul 1960 (KE 6745); bog, Mantua Twp., Rood 1332, 30 Jun
1935 (KE 9722).

1778 **Panicum boscii** Poiret   Bosc's Panic Grass
Frequent; rich mesic woods, dry thickets. CL, CU, FA, GE, HL, KN, LK, LI, PE, RO, SU.

1779 **Panicum capillare** L.   Witch Grass
Common; fields, roadsides, railroads, waste places. ALL COUNTIES.

1780 **Panicum clandestinum** L.
Common; moist fields, thickets, floodplains, meadows, open woods. All counties except PC.

1781 **Panicum columbianum** Scribner   American Panic Grass
Infrequent; dry sandy woods. CU, FA, LI, PO, ST, SU.

1782 *Panicum commutatum* Schultes
Infrequent; dry open woods, rocky woods, dry slopes. CL, CU, FA, HL, LK, LR, MH, PE, RO, TR.

1783 *Panicum depauperatum* Muhl. Starved Panic Grass
Rare; dry open woods, sandy hillsides. FA: Hocking Twp., Goslin s.n., 7 Jun 1936 (OS 68177). PO: dry sterile bank, Windham Twp., Rood & Webb 389, 19 Jun 1910 (KE 9724).

1784 *Panicum dichotomiflorum* Michx. Fall Panic Grass
Common; railroads, roadsides, waste places. ALL COUNTIES.

1785 *Panicum dichotomum* L.
Common; dry fields, railroads, roadsides, dry open woods. All counties except MO, PC, RI.

1786 *Panicum flexile* (Gattinger) Scribner
Rare; calcareous soils in open woods, thickets. PC: Kibler Bog, Bartley s.n., 11 Aug 1960 (OS 71560). RO: swamp, Colerain Twp., Bartley s.n., 24 Aug 1961 (OS 67713); Immell Bog, Green Twp., Bartley s.n., 12 Sep 1957 (OS 59165).

1787 *Panicum gattingeri* Nash Gattinger's Panic Grass
Infrequent; roadsides, waste places, moist open woods, fields. AS, FA, GE, KN, LI, LR, PO, SU, TR, WA.

1788 *Panicum lanuginosum* Ell.
Common; open sandy fields, dry woods, gravel pits, woodland borders. ALL COUNTIES.

1789 *Panicum latifolium* L.
Common; railroads, woodland borders, roadside banks, fields, dry woods. AS, AB, CL, CU, FA, GE, HL, KN, LK, LI , LR, MH, MD, MO, PC, ST, SU, TR, WA.

1790 *Panicum linearifolium* Britton
Common; roadsides, dry woodland borders, railroads, fields. AS, AB, CL, CU, FA, HL, KN, LK, LI, LR, MH, PO, RO, ST, SU, TR, WA.

1791 *Panicum microcarpon* Ell.
Infrequent; wet fields, thickets, moist woods. CU, FA, GE, LI, LR, PE, PO, SU, TR, WA.

1792 **Panicum miliaceum* L. Broom-corn Millet
Infrequent, adventive; fields, roadsides, waste places. AS, HL, KN, LK, PE, PO, RI, SU, WA.

1793 *Panicum oligosanthes* Schultes var. *scribnerianum* (Nash) Fern. Prairie Panic Grass
Rare; dry hillsides, open woods. CU: dry hillside, East Cleveland, Stair s.n., Jun 1897 (OS 7816). RO: Green Twp., Bartley & Pontius s.n., 6 Jun 1934 (OS s.n.).

1794 *Panicum philadelphicum* Trin. Philadelphia Panic Grass
Infrequent; roadside banks, thin woods, dry slopes. AS, HL, LR, PO, ST, SU, TR.

1795 *Panicum polyanthes* Schultes
Rare; dry open woods, thickets. FA: dry hill, Hocking Twp., Walen s.n., 27 Oct 1940 (OS 7888); Hocking Twp., Goslin s.n. 2 Sep 1935 (OS 62345).

1796 *Panicum rigidulum* Nees
Panicum agrostoides Sprengel (in Braun, 1967)
Infrequent; pond margins, marshes, swamps, pond margins. AB, LK, LI, RO, TR.

1797 *Panicum sphaerocarpon* Ell.
Infrequent; dry open soils, thin woods. CU, FA, HL, LK, LR, PO, RO, SU, TR, WA.

1798 *Panicum stipitatum* Nash
Common; roadside ditches, swamps, stream and lake margins, wet fields. AS, AB, CL, CU, FA, GE, KN, LK, LI, LR, MH, PE, PO, ST, SU, TR, WA.

1799 *Panicum virgatum* L. Switch Grass
Common; roadsides, railroads, sandy soils, prairie remnants. All counties except MO, LI.

1800 *Paspalum ciliatifolium* Michx.
Infrequent; dry or moist fields, pastures, thickets. AB, CU, FA, HL, LK, PE, PC, RO.

1801 *Paspalum laeve* Michx.
Rare; damp fields, thickets. RO: Liberty Twp., Bartley & Pontius s.n., 4 Sep 1932 (OS 8341).

1802 *Paspalum pubiflorum* Rupr.
Rare; roadsides, fields. FA: roadside, Hocking Twp., Goslin s.n., 22 Oct 1967 (OS 96261).

PC: Pickaway Twp., Bartley & Pontius s.n., 15 Oct 1935 (OS 8337); Washington Twp., Bartley & Pontius s.n., 10 Oct 1935 (OS 8336).

1803 **Phalaris arundinacea** L.   Reed Canary Grass
Common; wet meadows, stream and pond margins, marshes, abandoned beaver ponds, swamps, ditches. ALL COUNTIES.

1804 ***Phalaris canariensis** L.   Canary Grass
Infrequent, adventive; roadsides, waste places. CU, PE, PC, SU, WA.

1805 ***Phleum pratense** L.   Timothy
Common, naturalized; fields, roadsides, waste places. All counties except PC.

1806 **Phragmites australis** (Cav.) Steudel   Reed Grass
Phragmites communis Trin. (in Braun, 1967)
Common; wet ditches, pond margins, swales, fens, alluvial fans. AB, CU, FA, GE, HL, LK, LI, LR, MH, PE, PC, PO, ST, SU, TR, WA.

1807 **Poa alsodes** Gray
Common; rich mesic woods, thickets, wooded floodplains, stream banks. AS, AB, CL, CU, GE, HL, KN, LK, LR, MD, MO, PO, RI, RO, ST, SU, TR, WA.

1808 ***Poa annua** L.   Annual Bluegrass
Common, naturalized; waste places, fields, roadsides, lawns. ALL COUNTIES.

1809 ***Poa bulbosa** L.
Rare, adventive; lawns, dry fields. PO: campus, Kent, Tandy 1226, 11 May 1973 (KE 39726).

1810 ***Poa compressa** L.   Canada Bluegrass
Common, naturalized; dry fields, roadsides, waste places. ALL COUNTIES.

1811 **Poa cuspidata** Nutt.
Common; rich dry and mesic woods, rocky slopes, ravines. AB, CL, CU, FA, GE, HL, LK, LI, LR, MH, MD, PE, PO, RO, ST, SU, TR.

1812 **Poa languida** Hitchc.   Weak Spear Grass
Infrequent; dry rocky woods, dry oak woods, ledges. CU, LK, PC, PO, ST, SU, TR.

1813 ***Poa nemoralis** L.   Wood Bluegrass
Infrequent, naturalized; open woods, rocky slopes, thickets. FA, GE, LK, PO, SU.

1814 **Poa palustris** L.   Fowl-meadow Grass
Frequent; rich woods, wooded floodplains, meadows, shaded roadside banks. AB, CU, FA, GE, HL, KN, LK, LR, PC, PO, ST, SU, TR, WA.

1815 ***Poa pratensis** L.   Kentucky Bluegrass
Common, naturalized; meadows, fields, lawns, roadsides, railroads. ALL COUNTIES.

1816 **Poa saltuensis** Fern. & Wieg.
Rare; wooded hillsides. GE: slump, Chardon Twp., Bissell 1984:23, 23 May 1984 (CLM 21601).

1817 **Poa sylvestris** Gray
Common; rich woods, shaded slopes, wooded terraces. ALL COUNTIES.

1818 **Poa trivialis** L.   Rough Bluegrass
Common; rich woods, swamps, stream margins. All counties except FA, PE, RI.

1819 ***Puccinellia distans** (Jacq.) Parl   Alkali-grass
Infrequent, naturalized; halophytic areas including roadsides, ditches. AS, AB, CU, LK, MD, PO, SU, TR, WA.

1820 **Puccinellia pallida** (Torrey) Clausen   Pale Manna Grass
Glyceria pallida (Torrey) Trin. (in Braun, 1967)
Frequent; pond margins, swamps, bog margins, seeps. AB, CL, CU, GE, HL, LK, LR, MH, PO, ST, SU, TR.

1821 ***Secale cereale** L.   Rye
Infrequent, adventive; waste places. AB, HL, MD, MO, SU, TR, WA.

1822 ***Setaria faberi** Herrm.   Giant Foxtail Grass
Frequent, naturalized; roadsides, waste places, railroads. AS, CU, HL, KN, LI, LR, MD, PE, PC, PO, SU, TR, WA.

1823 ***Setaria glauca** (L.) Beauv.   Yellow Foxtail Grass
Common, naturalized; roadsides, fields, waste places. ALL COUNTIES.

1824 *Setaria italica* (L.) Beauv. Hungarian Millet
Infrequent, naturalized; waste places, roadsides. AB, CU, FA, HL, KN, LK, LR, WA.

1825 *Setaria verticillata* (L.) Beauv.
Frequent, naturalized; cultivated fields, waste places, roadsides. AB, CU, GE, KN, LK, LI, LR, MO, PC, PO, RI, RO, SU, WA.

1826 *Setaria viridis* (L.) Beauv. Green Foxtail Grass
Common, naturalized; fields, roadsides, railroads, disturbed soils. ALL COUNTIES.

1827 *Sorghastrum nutans* (L.) Nash Indian Grass
Frequent; roadsides, dry calcareous slopes, prairie remnants, fens. AS, AB, CL, CU, FA, HL, KN, LK, PO, RO, ST, SU, WA.

1828 *Sorghum bicolor* (L.) Moench Sorghum
Sorghum vulgare Persoon (in Braun, 1967)
Infrequent, adventive; fields, railroads, roadsides. HL, PE, RI, ST, WA.

1829 *Sorghum halepense* (L.) Persoon Johnson Grass
Frequent, naturalized; old fields, waste places, roadside ditches. AB, CL, CU, FA, KN, LK, LI, LR, PE, PC, RO, WA.

1830 *Spartina pectinata* Link Prairie Cord Slough Grass
Frequent; roadsides, marshes, wet prairie remnants. AB, CL, CU, FA, HL, LK, LI, LR, MD, PC, PO, RI, ST, SU, WA.

1831 *Sphenopholis intermedia* (Rydb.) Rydb.
Common; creek banks, woods, fens, meadows. ALL COUNTIES.

1832 *Sphenopholis nitida* (Biehler) Scribner Shiny Wedge Grass
Frequent; dry woods, dry rocky slopes. AS, AB, CL, CU, FA, GE, HL, KN, LK, LI, PO, RO, ST, SU.

1833 *Sporobolus asper* (Michx.) Kunth Rough Dropseed
Common; sandy fields, roadsides, open woods. All counties except CL.

1834 *Sporobolus neglectus* Nash
Infrequent; railroads, fields, roadsides. CU, LK, LR, MH, PC, RO, SU, TR, WA.

1835 *Sporobolus vaginiflorus* (Torrey) Wood Poverty-grass
Common; meadows, roadsides, dry banks. ALL COUNTIES.

1836 *Tridens flavus* (L.) Hitchc. Tall Redtop, Purpletop
Triodia flavens (L.) Smyth (in Braun, 1967)
Common; roadsides, dry fields, woodland borders. ALL COUNTIES.

1837 *Triplasis purpurea* (Walter) Champ. Purple Sandgrass
Rare, possibly adventive; sandy dry soils. RO: railroad yards, Chillicothe, Bartley s.n., 12 Sep 1965 (OS 105467).

1838 # *Trisetum pensylvanicum* (L.) Beauv. Swamp Oats
Infrequent; fens, sphagnous thickets, alder thickets, swales. AB, CL, GE, LK, PO, RO, SU.

1839 *Triticum aestivum* L. Wheat
Frequent, adventive; old fields, roadsides, railroads. AS, CU, GE, HL, LK, LR, MD, MO, PO, SU, TR, WA.

1840 *Uniola latifolia* Michx. Wild Oats
Rare; dry roadsides, fields. RO: Ross County, Pontius s.n., 19 Jul 1932 (BHO 2102).

1841 *Zea mays* L. Maize, Indian Corn
Rare, adventive; fields, roadsides, refuse areas. HL: dump, Monroe Twp., Wilson 2464, 20 Aug 1971 (KE 29024). KN: roadbank, Harrison Twp., Pusey 999, 17 Sep 1975 (KE 37979).

1842 *Zizania aquatica* L. Annual Wild Rice
Infrequent; quiet clear waters in canals, creek margins. AB, LK, LI, LR, PE, PC, ST, SU.

HYDROCHARITACEAE (Frog's-bit Family)

1843 *Elodea canadensis* Michx. Common Waterweed
Common; streams, lakes, ponds, roadside ditches. AS, AB, CL, CU, GE, HL, KN, LI, LR, MD, MO, PO, RO, ST, SU, TR, WA. Braun (1967) indicates a record for PC.

1844 **# Elodea nuttallii** (Planchon) St. John　　　　　　　　　　Nuttall's Waterweed
Frequent; ponds, lakes, streams, ditches. CL, FA, GE, HL, KN, MH, MD, MO, PE, PO, SU, TR, WA.

1845 **Vallisneria americana** Michx.　　　　　　　　　　　Eelgrass, Tape-grass
Infrequent; lakes, ponds, quiet streams, oxbows. AB, CU, GE, KN, LK, LI, LR, PO, ST, SU. Braun (1967) indicates a record for FA.

IRIDACEAE (Iris Family)

1846 **Iris brevicaulis** Raf.　　　　　　　　　　　　　　Leafy Blue Flag
Rare; stream terraces, floodplains, shores. RO: Green Twp., Bartley & Pontius s.n., 4 Jun 1933 (OS 10594). Braun (1967) indicates a record for PC.

1847 **Iris cristata** Aiton　　　　　　　　　　　　　　Crested Dwarf Iris
Infrequent; rich mesic woods, ravines, floodplains. CU, FA, GE, LR, PO, RO, SU, TR.

1848 ***Iris germanica** L.
Rare, adventive; waste places, roadsides. HL: roadside, Ripley Twp., Wilson 1821, 17 Jun 1971 (KE 29165).

1849 ***Iris pseudacorus** L.　　　　　　　　　　　　　Yellow Water Flag
Common, naturalized; marshes, meadows, pond and lake margins, open floodplains. AS, AB, CL, CU, GE, HL, KN, LK, LR, MH, MD, PO, RI, ST, SU, TR, WA.

1850 **Iris versicolor** L.　　　　　　　　　　　　　Blue Flag, Poison Flag
Frequent; marshes, wet meadows, stream and pond margins. AS, AB, CU, GE, LK, LR, PO, ST, SU, TR, WA.

1851 **Iris virginica** L. var. **shrevei** (Small) E. Andersson　　　Southern Blue Flag
Iris shrevei Small (in Braun, 1967)
Infrequent; marshes, ditches, shallow water. AS, FA, HL, KN, PC, RI, RO, WA.

1852 **Sisyrinchium albidum** Raf.
Infrequent; sandy fields, open woods. AS, CU, KN, PC, PO, RO, WA.

1853 **Sisyrinchium angustifolium** Miller　　　　　　　　　Blue-eyed Grass
Common; wet meadows, pastures, marshes, thickets, open woods. ALL COUNTIES.

1854 **Sisyrinchium atlanticum** Bickn.　　　　　　　Atlantic Blue-eyed Grass
Rare; fields, marshes, thickets. ST: grassy places, Kuehnle s.n., 13 Jun 1938 (KE 215).

1855 **# Sisyrinchium montanum** Greene　　　　　　Northern Blue-eyed Grass
Rare; roadsides, open fields, pond margins. CU: roadside, Solon Twp., Herrick sn., 5 Jun 1955 (OS 53539). PO: roadside, Aurora Twp., Herrick s.n., 5 May 1955 (OS 53530). SU: open field, Twinsburg Twp., Herrick s.n., 20 May 1955 (OS 53064).

1856 **Sisyrinchium mucronatum** Michx.　　　　　Narrow-leaved Blue-eyed Grass
Rare; roadsides, meadows, open woods. CU: Berea, Vickers s.n., May 1891 (Youngstown State University Herbarium 4085). GE: roadside, Bainbridge Twp., Herrick s.n., 5 Jun 1955 (OS 55670).

JUNCACEAE (Rush Family)

1857 **Juncus acuminatus** Michx.
Common; damp soil along roadsides, marshes, meadows, pond margins. All counties except MO, RI.

1858 **Juncus articulatus** L.
Infrequent; wet ditches, meadows, pond margins. AB, CU, LK, LR, MD, PO, SU, RO.

1859 **Juncus balticus** Willd.　　　　　　　　　　　　　　Baltic Rush
Infrequent; fens, sedge meadows, sandy shores. AB, CU, LK, PO, ST, TR.

1860 **Juncus biflorus** Ell.　　　　　　　　　　　　　Two-flowered Rush
Frequent; dry sandy fields, waste places. CU, FA, LR, PE, PC, PO, RO, ST, SU, TR, WA.

1861 **Juncus brachycarpus** Engelm.
Rare; damp sandy soils along roadsides, creeks and lakes. KN: lake margin in quarry, Union Twp., Pusey 746, 11 Aug 1975 (KE 38105). PC: old orchard, Pickaway Twp., Bartley s.n., 28 Sep 1968 (OS 112310).

1862 **Juncus brachycephalus** (Engelm.) Buchenau
Frequent; fens, sedge meadows, marshes. AS, CL, CU, FA, HL, LK, PC, PO, RO, ST, SU.

1863 **Juncus bufonius** L. Toad Rush
Frequent; roadsides, damp open ground, cinders, waste places. AB, CU, LK, LI, LR, MH, PE, PO, ST, SU, TR, WA.

1864 **Juncus canadensis** J. Gay
Frequent; sphagnous meadows, sedge meadows, bogs. AB, CL, CU, GE, HL, LK, LI, LR, MD, PO, RI, ST, SU, WA.

1865 **Juncus diffusissimus** Buckl.
Rare; ditches, wet soils. KN: drainage ditch, Butler Twp., Pusey 833, 10 Oct 1975 (KE 38104).

1866 **Juncus dudleyi** Wieg. Dudley's Rush
Frequent; fens, sedge meadows, stream margins. AS, AB, CU, FA, HL, KN, LK, LI, PE, PO, RO, ST, SU, TR, WA.

1867 **Juncus effusus** L. Common Rush
Common; wet fields, stream and pond margins, sphagnous thickets, fen and bog margins, marshes, roadside ditches. ALL COUNTIES.

1868 **Juncus gerardii** Loisel. Black Rush, Black Grass
Infrequent; ditches. AS, CU, LK, PO, SU, TR.

1869 **Juncus greenei** Oakes & Tuckerman Greene's Rush
Rare; wet gravelly roadside banks. SU: sandy hill, Boston Twp., Andreas & Stoutamire 4367, 1 Aug 1979 (University of Akron Herbarium s.n.); hill, Boston Twp., Tomci 1278, 23 Sep 1977 (KE 41126).

1870 **Juncus marginatus** Rostk.
Infrequent; sphagnous thickets, marshes, wet fields. AB, CL, CU, LK, PO, RO.

1871 **Juncus nodosus** L.
Frequent; wet fields, ditches, marshes, bogs. AB, CU, HL, KN, LK, LR, PO, RO, SU, TR, WA.

1872 **# Juncus platyphyllus** (Wieg.) Fern. Flat-leaved Rush
Infrequent; calcareous soils along lake margins, wet meadows, seeps. CU, GE, MH, PO, ST.

1873 **Juncus secundus** Beauv. One-sided Rush
Rare; dry meadows, thickets. PC: meadow, Pickaway Twp., Bartley s.n., 28 Jun 1967 (OS 105369).

1874 **Juncus subcaudatus** (Engelm.) Coville
Rare; moist thickets, sedge meadows. RO: swamp, Harrison Twp., Bartley s.n., 9 Sep 1961 (OS 67745).

1875 **Juncus tenuis** Willd. Path Rush, Yard Rush
Common; fields, waste places, roadsides, thickets, swamps, meadows, disturbed woods. ALL COUNTIES.

1876 **Juncus torreyi** Coville
Common; roadside ditches, marshes, wet fields. All counties except CL, MH, WA.

1877 **Luzula bulbosa** (Wood) Rydb. Southern Woodrush
Rare; dry sandy upland woods. CL: lightly wooded slope, Middletown Twp., Cooperrider 5677, 21 Jun 1960 (KE 4253). MH: sandy upland woods, Milton Twp., Cooperrider 6907, 18 Jul 1960 (KE 4251). PO: sandy, mossy woods, Franklin Twp., Cooperrider 6728, 14 Jul 1960 (KE 4252). RI: Richland County, Wilkinson s.n., 29 Apr 1892 (OS 10305).

1878 **Luzula carolinae** S. Watson Evergreen Woodrush
Common; rich woods, floodplains. All counties except LI, PC.

1879 **Luzula echinata** (Small) F. J. Hermann
Common; wet fields, open woods, clearings. AS, AB, CL, FA, HL, KN, LI, MH, MO, PE, PO, RO, SU, TR, WA.

1880 **Luzula multiflora** (Retz.) Lejeune
Common; mesic woods, wooded floodplains, ravines, roadside banks, fields. ALL COUNTIES.

JUNCAGINACEAE (Arrow-grass Family)

1881 **Triglochin maritimum** L. Seaside Arrow-grass
Rare; fens, calcareous seeps, calcareous stream margins. ST: Canton, Case s.n., 22 Aug 1900 (OS 2357). SU: base of hill, Boston Twp., Tomci 1250, 23 Sep 1977 (KE 41128); fen, Green

Twp., Stoutamire s.n., 4 Jul 1969 (University of Akron Herbarium s.n.); Long's Lake, Claassen s.n., 1895 (OS 62103).

1882 **Triglochin palustre** L. Marsh Arrow-grass
Rare; fens, sedge meadows, calcareous seeps. PO: sedge meadow, Streetsboro Twp., Andreas 5620, 11 Jun 1981 (KE 44104); marl pool, Streetsboro Twp., Andreas 5385, 15 Jul 1980 (KE 43203). RO: Green Twp., Bartley & Pontius s.n., 16 Jul 1937 (OS 2360 & OS 49428). ST: fen, Jackson Twp., Andreas 5358, 7 Jul 1980 (KE 46398); marly fen, Jackson Twp., Amann 777, 23 Jul 1960 (KE 3941). SU: marl meadow, Green Twp., Stoutamire s.n, 12 Jul 1974 (University of Akron Herbarium s.n.).

1883 # **Scheuchzeria palustris** L. var. **americana** Fern. Scheuchzeria
Infrequent; sphagnous meadows, bogs. AS, AB, GE, LI, LR, PO, WA.

LEMNACEAE (Duckweed Family)

1884 **Lemna minor** L. Lesser Duckweed
Common; ponds, quiet streams, oxbows, ditches, marshes. All counties except PE.

1885 **Lemna trisulca** L. Star Duckweed
Frequent; submerged in ponds, ditches, swamps. AS, AB, FA, GE, HL, KN, LK, LI, LR, PO, ST, SU, TR, WA.

1886 **Spirodela polyrhiza** (L.) Schleiden Greater Duckweed
Common; surface of ponds, slow-moving streams, swamps. AS, AB, CL, CU, FA, GE, HL, KN, LK, LI, LR, MD, PO, RI, ST, SU, TR, WA.

1887 **Wolffia columbiana** Karsten Common Wolffia
Common; surface of marshes, ponds, quiet streams. AB, CL, FA, GE, HL, KN, LK, LI, LR, MH, MO, PE, PO, RI, RO, ST, SU, TR, WA.

1888 **Wolffia papulifera** C. H. Thompson Pointed Wolffia
Rare; submerged in stagnant waters in ponds, ditches, marshes. PO: Aurora Lake, Aurora Twp., Herrick s.n., 8 Aug 1955 (OS 52667). LR: Camden Lake, Camden Twp., Jones 71-9-28-726, 28 Sep 1971 (OS 107247).

1889 **Wolffia punctata** Griseb. Dotted Wolffia
Infrequent; ponds, slow-moving streams. FA, KN, LK, LR, PE, PO, ST.

1890 # **Wolffiella floridana** (J. D. Smith) C. H. Thompson Star Wolffiella
Infrequent; stagnant waters in ponds, ditches, bog margins. FA, GE, LI, PE, PO. Braun (1967) indicates a record for SU.

LILIACEAE (Lily Family)

1891 **Aletris farinosa** L. Unicorn-root, Stargrass
Rare; dry sandy fields, clearings. RO: Colerain Twp., Bartley & Pontius s.n, 7 Aug 1932 (OS 9020). Braun (1967) indicates a record for FA.

1892 **Allium canadense** L. Wild Garlic
Common; moist woods, wooded and grassy floodplains, meadows. ALL COUNTIES.

1893 **Allium cernuum** Roth Nodding Wild Onion
Infrequent; meadows, roadsides, prairie remnants, fens. CL, CU, LR, MH, MD, PO, RO, ST, SU.

1894 ***Allium sativum** L. Garlic
Infrequent, adventive; roadsides, pastures, meadows. LR, PC, ST, TR, WA.

1895 ***Allium schoenoprasum** var. **sibiricum** (L.) Hartman
Rare, adventive; roadsides. MH: North Jackson Twp., Hicks s.n., 30 May 1930 (OS 8957).

1896 **Allium tricoccum** Aiton Wild Leek, Ramp
Common; rich mesic woods, wooded floodplains, shaded banks. All counties except CL, RO.

1897 ***Allium vineale** L. Field Garlic
Common, naturalized; roadsides, waste areas, railroads, fields. AS, CL, CU, FA, GE, HL, KN, LK, LI, LR, MD, PE, PC, PO, ST, SU, WA.

1898 ***Asparagus officinalis** L. Garden Asparagus
Common, naturalized; fields, roadsides, pastures. AS, AB, CU, FA, GE, HL, KN, LK, LI, LR, MH, MD, MO, PE, PO, RO, ST, SU, WA.

1899 *Camassia scilloides* (Raf.) Cory   Wild Hyacinth
Common; open and shaded floodplains, marshes, thickets. AS, AB, CL, CU, FA, GE, HL,
LK, LI, LR, MD, MO, PE, PO, RI, RO, ST, SU, WA.

1900 *Chamaelirium luteum* (L.) Gray   Fairy-wand, Devil's-bit
Common; dry woods, thickets, ravine rims. AS, CU, FA, GE, HL, KN, LK, LI, LR, MD,
PO, RI, RO, ST, SU, TR, WA.

1901 # *Clintonia borealis* (Aiton) Raf.   Yellow Clintonia, Bluebead
Rare; sphagnous thickets. AB: *Sphagnum* hummocks, Richmond Twp., Andreas 5142, 6 Jun
1980 (KE 43042); Pymatuning Swamp, Richmond Twp., Davis, Dachnowski & Detmers s.n.,
11 Aug 1900 (OS 9659).

1902 *Clintonia umbellulata* (Michx.) Morong   Speckled Wood-lily
Infrequent; rich woods, ravines, stream terraces. CL, HL, PO, ST, SU, TR, WA.

1903 *Convallaria majalis* L.   Lily-of-the-valley
Infrequent, naturalized; thickets, roadsides, moist woods, abandoned homesites. CU, GE,
KN, LR, MD, PO, ST, SU.

1904 *Disporum lanuginosum* (Michx.) Nicholson   Yellow Mandarin
Common; rich woods, stream terraces, wooded ravines. All counties except PC.

1905 *Erythronium albidum* Nutt.   White Fawn Lily
Common; alluvial soils, wooded floodplains, stream terraces. ALL COUNTIES.

1906 *Erythronium americanum* Ker   Yellow Fawn Lily, Trout Lily
Common; wooded floodplains, rich woods, meadows, thickets. All counties except PC.

1907 *Hemerocallis fulva* (L.) L.   Orange Day-lily
Common, naturalized; roadsides, fields, railroads, stream margins, waste places. AS, AB, CL,
CU, FA, GE, HL, KN, LK, LI, LR, MH, MD, MO, PE, PO, ST, SU, TR, WA.

1908 *Hemerocallis lilio-asphodelus* L.   Yellow Day-lily
Hemerocallis flava L. (in Braun, 1967)
Rare, naturalized; roadsides, waste places. PO: stream bank, Brimfield Twp., Andreas 2333,
15 Jul 1978 (KE 48962). SU: beech-maple woodlot, Twinsburg Twp., Herrick s.n., 23 May
1955 (OS 54252).

1909 *Hosta lancifolia* (Thunb.) Engelm.   Narrow-leaved Plantain-lily
Rare, adventive; thickets, roadsides, waste places. ST: woods, Jackson Twp., Amann 1801,
14 Sep 1960 (KE 8318).

1910 *Hypoxis hirsuta* (L.) Coville   Yellow Stargrass
Common; dry woods, shaded eroding calcareous slopes, sphagnous thickets, fens. AS, CL,
CU, FA, GE, HL, KN, LK, LI, LR, PE, PO, RO, ST, SU, WA.

1911 *Leucojum aestivum* L.   Summer Snowflake
Rare, adventive; woodland borders, fields. SU: field edge, Northampton Twp., Tomci 6, 24
Mar 1975 (KE 41203).

1912 *Lilium canadense* L.   Wild Lily, Canada Lily
Common; wooded floodplains, roadsides, meadows, thickets. All counties except FA, RO.

1913 *Lilium michiganense* Farw.   Michigan Lily
Rare; swamps, wet woods, thickets. FA: Hocking Twp., Goslin s.n, 24 Jun 1933 (OS 8756).
PC: Washington Twp., Bartley s.n., 18 Jul 1970 (OS 105383). WA: lightly wooded marsh,
Franklin Twp., Andreas 1170, 8 Jul 1977 (KE 48953).

1914 *Lilium philadelphicum* L.   Wood Lily
Rare; fens, sphagnous thickets, stream margins. PO: bog, Mantua Twp., Cooperrider 7384,
26 Jul 1960 (KE 4681); Mantua Swamp, Mantua Twp., Webb 523, 18 Jul 1915 (KE 10629);
river margin, Kent, Bricker s.n., 20 Jun 1955 (OS 54831). RO: Tar Hollow Park, Colerain
Twp., Bartley s.n., 1 Jul 1959 (OS 64223). ST: Canton, Case s.n., 23 Jun 1899 (OS 54831).
SU: fen, Green Twp., Cusick 11824, 3 Jul 1971 (KE 28763).

1915 *Lilium superbum* L.   Turk's Cap Lily
Infrequent; wooded floodplains, wet woods. AB, CU, GE, HL, KN, LK, PO.

1916 *Lycoris squamigera* Maxim.   Spider Lily
Rare, adventive; sandy banks, waste places. WA: sandy kame, Franklin Twp., Andreas 2649,
15 Aug 1978 (KE 48953).

139

1917 **Maianthemum canadense** Desf.   Canada Mayflower
Common; mesic and wet woods, hemlock ravines, wooded floodplains, sphagnous thickets, bogs. AS, AB, CL, CU, FA, GE, HL, LK, LI, LR, MH, MD, MO, PO, RI, ST, SU, TR, WA.

1918 **Medeola virginiana** L.   Indian Cucumber-root
Common; mesic and wet woods, ravines, floodplains. All counties except MO, PC.

1919 **Melanthium virginicum** L.   Bunchflower
Rare; fens, sphagnous thickets, sedge meadows. PO: shrubby fen, Hiram Twp., Andreas & Knoop 6871, 27 Jul 1985 (KE 48613); open fen, Hiram Twp., Andreas 2217, 5 Jul 1978 (KE 40759); wet meadow, Streetsboro Twp., Cusick 10079, 20 Jun 1969 (KE 25891); sedge meadow, Mantua Twp., Stoutamire s.n., 29 Jun 1968 (University of Akron Herbarium s.n.). RI: sedge meadow, Washington Twp., McKee 8, 23 Jun 1986 (KE 49391); margin of swamp, Mansfield, Wilkinson s.n. 20 Jun 1895 (OS 9032). WA: Highland Park, Wooster, Laughlin s.n., 4 Jul 1907 (OS 74448); Highland Park, Wooster, Hopkins s.n., 4 Jun 1904 (OS 9031).

1920 ***Muscari botryoides** (L.) Miller   Grape Hyacinth
Infrequent, adventive; roadsides, ditches, creek margins. AS, AB, KN, LR, PE, RO, SU, WA.

1921 ***Narcissus pseudo-narcissus** L.   Daffodil
Infrequent, naturalized; abandoned homesites, fields, wooded floodplains, mesic woods. AS, CU, HL, LR, PE, PO, SU, WA.

1922 ***Ornithogalum umbellatum** L.   Star-of-Bethlehem
Common, naturalized; mesic and moist woods, wooded and grassy floodplains, field, meadows. AS, CU, FA, GE, HL, KN, LK, LI, LR, MD, PE, PC, PO, RO, ST, SU, TR, WA.

1923 **Polygonatum biflorum** (Walter) Ell.   Smooth Solomon's Seal
Common; dry, mesic and moist woods, rocky slopes, woodland borders. All counties except MO, RI.

1924 **Polygonatum commutatum** (Schultes f.) A. Dietr.   Large Solomon's Seal
Polygonatum canaliculatum (Muhl.) Pursh (in Braun, 1967)
Common; shaded hillsides, roadside banks, floodplains, thickets. ALL COUNTIES.

1925 **Polygonatum pubescens** (Willd.) Pursh   Hairy Solomon's Seal
Common; mesic and moist woods, rocky slopes. ALL COUNTIES.

1926 ***Scilla non-scripta** (L.) Hoffman & Link   English Bluebell
Rare, adventive; waste places, abandoned homesites. ST: moist woods, Canton Twp., Amann 162, 21 May 1960 (KE 7960).

1927 **Smilacina racemosa** (L.) Desf.   False Solomon's Seal
Common; rich mesic and moist woods, rocky slopes, roadside banks, thickets. ALL COUNTIES.

1928 **Smilacina stellata** (L.) Desf.   Starry False Solomon's Seal
Frequent; calcareous soils in floodplains, meadows, fens, thickets. AB, CL, CU, FA, HL, LK, LR, MD, MO, PO, RO, ST, SU, TR.

1929 # **Smilacina trifolia** (L.) Desf.   Three-leaved False Solomon's Seal
Once rare, now presumably extirpated from Ohio; bogs. LR: Camden Lake, Cowles s.n., 4 Jun 1892 (OC 71557); Camden Lake, Leonard s.n, 2 Aug 1887 (OC 71563). Stuckey and Roberts (1982) indicate a record for CU.

1930 **Smilax ecirrhata** (Kunth) S. Watson
Frequent; moist woods, boggy thickets, rocky slopes. AB, FA, GE, HL, LI, LR, MD, PO, RI, ST.

1931 **Smilax glauca** Walter   Greenbrier
Frequent; mesic woods, thickets, floodplains, dry slopes. AS, CU, FA, GE, HL, KN, LI, PE, PC, PO, RO, SU, WA.

1932 **Smilax herbacea** L.   Carrion-flower
Common; roadsides, wet meadows, woodland borders, marshes. All counties except MO.

1933 **Smilax hispida** Muhl.   Bristly Greenbrier
Common; bog margins, swamps, thickets. ALL COUNTIES.

1934 **Smilax rotundifolia** L.   Common Greenbrier
Common; mesic woods, thickets, fields, marshes. ALL COUNTIES.

1935 *Stenanthium gramineum* (Ker) Morong   Featherbells
Rare; mesic woods, thickets. FA: Fairfield County, Kellerman s.n., 30 Aug 1892 (Bowling Green State University Herbarium 5561). LI: Newark, Jones s.n., 26 Jun 1890 (OC 78012). ST: Plain Twp., Brown s.n., 24 Jun 1936 (OS 9048).

1936 # *Streptopus roseus* Michx.   Rose Mandarin, Twisted-stalk
Rare; moist woods, bogs. AB: bog forest near Lake Erie, Herrick s.n., 16 May 1959 (OS 63425).

1937 *Tofieldia glutinosa* (Michx.) Persoon   False Asphodel
Rare; open marl in fens. PO: marly area, Streetsboro Twp., Tandy 1513, 29 Jun 1973 (KE 39781). RO: Paxton Twp., Bartley s.n., 24 Jul 1932 (BHO 18090). ST: marl meadow, Jackson Twp., Andreas 5359, 7 Jul 1980 (KE 43157); bog, Jackson Twp., Amann 428, 26 Jun 1960 (KE 7872). SU: calcareous bog, Green Twp., Cullen s.n., 28 Jun 1974 (University of Akron Herbarium s.n.); Green Twp., Herrick s.n., 14 Jun 1959 (OS 63348).

1938 *Trillium erectum* L.   Stinking Trillium
Frequent; mesic and wet woods, ravines, wooded floodplains. AS, AB, CL, CU, GE, LK, LR, MH, MD, PE, PO, ST, SU, TR, WA.

1939 *Trillium flexipes* Raf.
Common; moist woods, wooded seeps, floodplains, boggy thickets. All counties except PC, TR.

1940 *Trillium grandiflorum* (Michx.) Salisb.   Large White Trillium
Common; rich mesic and moist woods, thickets, floodplains, pastures. ALL COUNTIES.

1941 *Trillium sessile* L.   Toad Trillium, Sessile Trillium
Common; floodplains, stream terraces, thickets, rocky slopes. ALL COUNTIES.

1942 # *Trillium undulatum* Willd.   Painted Trillium
Rare; boggy woods, hemlock swamps. AB: hemlock woods, Richmond Twp., Cusick 19408, 21 Jun 1979 (OS s.n.); hemlock woods, Richmond Twp., Bissell s.n., 13 May 1978 (CLM s.n.); hemlock woods, Rome Twp., Bissell 1979:38, 22 May 1979 (CLM 12440); boggy woods, Monroe Twp., Bissell 1974:28, 19 May 1974 (CLM 3480); Ashtabula Twp., Bartley s.n., 1933 (OS 28867); Andover Twp., Hicks s.n., 14 Jun 1931 (OS 9209).

1943 *Tulipa gesneria* L.   Tulip
Rare, adventive; roadsides, waste places. TR: roadside, Braceville Twp., Burns 374, 13 May 1977 (KE 45782).

1944 *Uvularia grandiflora* Small   Bellwort
Common; mesic and moist woods, stream terraces, thickets. ALL COUNTIES.

1945 *Uvularia perfoliata* L.   Straw-bell
Common; moist and mesic woods, thickets, stream margins. AS, AB, CL, CU, FA, GE, HL, KN, LK, LI, LR, MH, MD, PO, RI, RO, ST, SU, TR, WA.

1946 *Uvularia sessilifolia* L.   Merry-bells
Frequent; floodplains, thickets, moist woods. AS, AB, CL, CU, GE, KN, LK,LR, MH, MD, PO, RO, ST, SU, TR.

1947 # *Veratrum viride* Aiton   White Hellebore
Rare; wooded floodplains, thickets, swamps. AB: wet thicket, Richmond Twp. Andreas 5972, 2 Jul 1982 (KE 47000); floodplain, Monroe Twp., Cusick 18492, 5 Jul 1978 (OS 156960); weedy field, Lakeville, Cooperrider 5814, 23 Jun 1960 (KE 38501); Ashtabula River, Hicks 549, 30 May 1930 (OS 10572). GE: wet depression, Burton Twp., Bartolotta 761, 1 May 1976 (OS 130389). TR: floodplain, Howland Twp., Andreas 5125, 6 Jun 1980 (KE 43092).

1948 *Yucca filamentosa* L.   Spanish Bayonet
Infrequent, adventive; railroads, dry roadside banks, old fields, abandoned homesites. AS, CL, CU, HL, KN, MD, PE, PO, ST, WA.

1949 *Zygadenus glaucus* (Nutt.) Nutt.   Wand Lily, White Camass
Rare; fens, sphagnous thickets. AB: floodplain seep, Windsor Twp., Tandy 1589, 11 Jul 1973 (KE 39793). PO: fen, Streetsboro Twp., Andreas 5792, 6 Aug 1981 (KE 44107); bog, Streetsboro Twp., Tandy 1640, 14 Jul 1973 (KE 39794); bog, Mantua Twp., Cooperrider 7381, 26 Jul 1960 (KE 4950). ST: bog, Jackson Twp., Cooperrider 8545, 30 Jul 1961 (KE 4951); Canton, Case s.n., 28 Jul 1900 (OS 9038). RO: Paxton Twp., Pontius & Bartley s.n., 24 Jul 1932 (OS 9039).

NAJADACEAE (Naiad Family)

1950 **Najas flexilis** (Willd.) Rostk. & Schmidt   Slender Naiad
Frequent; quiet streams, lakes, kettle-hole ponds. AB, CU, FA, GE, LK, LI, LR, MD, PO, ST, SU, WA.

1951 **Najas gracillima** (A. Braun) Magnus   Thread-like Naiad
Once rare, now presumably extirpated from Ohio; shallow water. PO: Sandy Lake, Hopkins s.n., 3 Aug 1918 (OS 65234). WA: Doner's Lake, Selby s.n., 21 Jul 1899 (OS 2583 & OS 86667).

1952 **Najas guadalupensis** (Sprengel) Magnus
Rare; shallow waters in lakes, ponds. AB: Williamsfield Twp., Hicks s.n., 20 Aug 1928 (OS 2582). FA: shallow water, Greenfield Twp., Haynes 3470, 30 Sep 1970 (OS 105447). PC: Washington Twp., Bartley & Pontius s.n., 28 Sep 1935 (OS 2581). ST: Myer's Lake, Plain Twp., Brown s.n., 20 Sep 1936 (OS 2573).

1953 ***Najas minor** All.
Common, naturalized; ponds, lakes, quiet streams, constructed impoundments. All counties except AS, MO, RI.

ORCHIDACEAE (Orchid Family)

1954 **Aplectrum hyemale** (Willd.) Torrey   Putty-root
Infrequent; rich mesic woods. FA, LK, LR, PO, RO, ST, SU.

1955 # **Arethusa bulbosa** L.   Dragon's Mouth, Arethusa
Rare; sphagnous thickets, peaty meadows, fens. LI: Cranberry Island, Buckeye Lake, Detmers s.n., 30 May 1910 (OS 92445); cranberry marsh, Jones s.n., 16 May 1891 (OC 78194). LR: bog, Oberlin, Dascomb s.n., 188- (OC 71704). PO: bog, Mantua Twp., Stoutamire 7625, 7 Jun 1968 (University of Akron Herbarium s.n.).

1956 **Calopogon tuberosus** (L.) BSP.   Grass Pink
Calopogon puchellus (Salisb.) R. Br. (in Braun, 1967)
Frequent; fens, sedge meadows, sphagnous meadows, calcareous sands. AS, AB, CU, GE, HL, LK, LI, LR, PE, PO, ST, SU, WA.

1957 **Coeloglossum viride** (L.) Hartm.   Long-bracted Orchid
Habenaria viridis (L.) R. Br. var. *bracteata* (Willd.) Gray (in Braun, 1967 and Voss, 1972)
Rare; rocky woods, thickets. LR: rocky woods, Elyria, Day s.n., 6 Jun 1892 (OC 99174); Oberlin, Dick s.n., 21 Jun 1891 (OC 71717); river bank, McCormick s.n., 7 Jun 1888 (OC 71728). MD: Hinkley, Watson s.n., 30 May 1896 (OS 10811). PO: rich woods, Newell's Ledges, Rood 94, 16 May 1897 (KE 10591); Nelson, Webb s.n., 29 May 1897 (OS 10810). SU: Summit County, Cranz s.n., 29 May 1892 (OS 92470).

1958 **Corallorhiza maculata** Raf.   Spotted Coral-root
Frequent; dry woods, dry wooded slopes. AS, AB, CL, CU, FA, GE, LK, LI, LR, PC, PO, SU, TR, WA.

1959 **Corallorhiza odontorhiza** (Willd.) Nutt.   Small-flowered Coral-root
Frequent; dry open woods, pine plantations, thickets. AS, AB, CU, FA, HL, KN, LK, LI, PE, RO, SU.

1960 # **Corallorhiza trifida** Chatelain   Early Coral-root, Pale Coral-root
Rare; hemlock hummocks, damp woods, bogs. AB: hemlock bog, Trumbull Twp., Bissell 1978:22, 11 May 1978 (CLM 10918). GE: *Sphagnum* hummocks, Newbury Twp., Andreas & Knoop 6353, 15 Apr 1985 (KE 47940). PO: *Sphagnum* hummocks, Franklin Twp., Andreas 7296, 27 May 1988 (KE 50263); Geauga Lake, Sullivan s.n., 189- (OC 94793). Braun (1967) indicates a record for CU.

1961 **Corallorhiza wisteriana** Conrad   Spring Coral-root
Rare; moist to mesic woods. FA: Greenfield Twp., Goslin s.n., 7 May 1935 (OS 11303). RO: Green Twp., Bartley & Pontius s.n., 18 May 1932 (OS 11296 & BHO 18137).

1962 **Cypripedium acaule** Aiton   Pink Moccasin-flower
Frequent; hemlock ravines, bog margins, sphagnous thickets, dry and mesic woods. AB, CL, CU, FA, GE, LK, LI, LR, MH, PO, ST, SU, TR, WA.

1963a *Cypripedium calceolus* L. var. *parviflorum* (Salisb.) Fern.
Small Yellow Lady's Slipper
Rare; fens, thickets. PO: border of meadow, Streetsboro Twp., Andreas & Stoutamire 5098, 3 Jun 1980 (KE 43128). SU: Summit County, Claypole s.n., no date (University of Akron Herbarium s.n.).

1963b *Cypripedium calceolus* L. var. *pubescens* (Willd.) Correll
Large Yellow Lady's Slipper
Freqeuent; dry, mesic and moist woods, calcareous seeps, hemlock slopes. CU, FA, GE, LK, LI, LR, MD, PO, RO, ST, SU. Braun (1967) indicates a record for AB.

1964 *Cypripedium candidum* Willd.   White Lady's Slipper
Rare; calcareous meadows, fens. PO: Mantua Bog, Mantua Twp., Herrick s.n., 5 Jun 1970 (KE 2105); Rood s.n., 21 May 1933 (KE 36460). TR: rich woods, Phalanx, Braceville Twp., Rood s.n., 10 Jun 1936 (OS 10679).

1965 *Cypripedium reginae* Walter   Showy Lady's Slipper, Queen Lady's Slipper
Infrequent; fens, sphagnous thickets. CU, GE, LK, PO, SU. Braun (1967) indicates a record for LR.

1966 *Epipactis helleborine* (L.) Crantz   Helleborine
Infrequent, naturalized; shaded roadside banks, dry slopes. AB, CU, GE, LK, SU, WA.

1967 *Goodyera pubescens* (Willd.) R. Br.   Downy Rattlesnake Plantain
Common; rich moist woods, floodplains, stream terraces, steep slopes. All counties except MO, PC.

1968 *Goodyera repens* (L.) R. Br.
Rare; wet woods, conifer swamps (Voss, 1972). CU: Cleveland. No collector. No date. (University of Cincinnati Herbarium s.n.).

1969 # *Goodyera tesselata* Lodd   Checkered Rattlesnake Plantain
Once rare, now presumably extirpated from Ohio; dry or moist woods. AB: Phelps Run, Hicks s.n., 20 Aug 1929 (OS 11077 & OS 78730).

1970 *Isotria verticillata* (Willd.) Raf.   Whorled Pogonia
Infrequent; dry to moist woods. CL, CU, FA, GE, LK, MD, PO, RO, SU. Braun (1967) indicates a record for AB.

1971 *Liparis lilifolia* (L.) Richard   Large Twayblade
Common; pine woods, thickets, wet meadows. All counties except LR, MO, ST.

1972 *Liparis loeselii* (L.) Richard   Fen Orchid
Frequent; sphagnous thickets, fens. AB, CL, CU, GE, LK, LI, PO, RO, ST, SU, TR, WA.

1973 # *Listera cordata* (L.) R. Br.   Heartleaf Twayblade
Once rare, now presumably extirpated from Ohio; moist woods. AB: east Monroe Twp., Hicks & Chapman s.n., 16 Jun 1933 (OS 11043 & OS 11044).

1974 *Malaxis unifolia* Michx.   Green Adder's Mouth
Infrequent; dry or moist woods, swamp borders. AB, FA, KN, RO, WA.

1975 *Orchis spectabilis* L.   Showy Orchis
Common; rich moist woods, stream terraces. AS, AB, CL, CU, FA, GE, HL, KN, LK, LI, LR, MD, MO, PO, RI, RO, ST, SU, WA.

1976 # *Platanthera blephariglottis* (Willd.) Lindley   White Fringed Orchid
Habenaria blephariglottis (Willd.) Hooker (in Braun, 1967 and Voss, 1972)
Rare; *Sphagnum* bogs. AB: bog, Andover Twp., Hicks s.n., 14 Aug 1930 (OS 10860). GE: Everett Pond, Burton, Dambach s.n., 24 Jul 1937 (OS 10858); Burton, Aldrich s.n.,23 Jul 1936 (CLM 2917); Burton, Metcalf s.n., 2 Aug 1888 (OC 78213). PO: *Sphagnum* bog, Suffield Twp., Andreas 6180, 17 Aug 1983 (KE 46458); Dargen Swamps, Ravenna, Krebs s.n., 1871 (OC 94802). SU: Summit Lake swamp, Claypole s.n., 24 Jul 1889 (University of Akron Herbarium s.n.).

1977 *Platanthera ciliaris* (L.) Lindley   Yellow Fringed Orchid
Habenaria ciliaris (L.) R. Br. (in Braun, 1967 and Voss, 1972)
Rare; sandy thickets, wet depressions. RO: Colerain Twp., Bartley s.n., 7 Aug 1932 (OS 10851 & BHO 18149).

1978 *Platanthera clavellata* (Michx.) Luer   Club-spur Orchid, Green Woodland Orchid
Habenaria clavellata (Michx.) Sprengel (in Braun, 1967 and Voss, 1972)
Infrequent; boggy thickets, wet sandy woods, sphagnous hummocks. AB, GE, LK, LI, LR, PE, PO, RI, SU, TR.

143

1979 **Platanthera flava** (L.) Lindley var. **herbiola** (R. Br.) Luer   Tuberculed Rein Orchid
Habenaria flava (L.) R. Br. var. *herbiola* (R. Br.) Ames & Correll (in Braun, 1967 and Voss, 1972)
Frequent; sedge meadows, thickets, open wet woods. AB, CL, CU, LK, LI, LR, PO, ST, SU, TR, WA.

1980 **Platanthera grandiflora** (Bigel.) Lindley   Large Purple-fringed Orchid
Habenaria grandiflora (Bigel.) Lindley (in Braun, 1967 and Voss, 1972)
Once rare, now presumably extirpated from Ohio; sphagnous thickets, meadows. AB: along Pymatuning River, south Wayne Twp., Hicks s.n., 6 Aug 1928 (OS 78789); Orwell, Bogue s.n., 29 Jul 1889 (OS 19911). PO: Garrettsville, Webb s.n., 27 Jul 1913 (OS 10912).

1981 **Platanthera hookeri** (Torrey) Lindley   Hooker's Orchid
Habenaria hookeri Torrey (in Braun, 1967 and Voss, 1972)
Once rare, now presumably extirpated from Ohio; rich moist woods. GE: Little Mountain, Chardon, Kurtz s.n., 16 May 1896 (CLM 6963). MD: Whips Ledge, Watson s.n., 30 May 1896 (OS 10846). SU: Summit County, Cranz s.n., 7 Jun 1892 (OS 92534).

1982 **Platanthera hyperborea** (L.) Lindley   Tall Northern Green Orchid
Habenaria hyperborea (L.) R. Br. (in Braun, 1967 and Voss, 1972)
Once rare, now presumably extirpated from Ohio; sphagnous thickets. ST: Canton, Case s.n., 21 Jun 1901 (OS 10819). WA: Browns Lake Bog, Thomas & Walker s.n., 12 Aug 1927 (OS 10818).

1983 **Platanthera lacera** (Michx.) D. Don   Ragged Fringed Orchid
Habenaria lacera (Michx.) Lodd (in Braun, 1967 and Voss, 1972)
Common; boggy woods, thickets, dry or wet meadows. All counties except PC, RO.

1984 **Platanthera leucophaea** (Nutt.) Lindley   Prairie Fringed Orchid
Habenaria leucophaea (Nutt.) Gray (in Braun, 1967 and Voss, 1972)
Rare; calcareous meadows. WA: abandoned field and wet sedge meadow, Clinton Twp., Andreas 5990, 13 Jul 1982 (KE 46449).

1985 **Platanthera orbiculata** (Pursh) Lindley   Large Round-leaf Orchid
Habenaria orbiculata (Pursh) Torrey (in Braun, 1967 and Voss, 1972)
Frequent; rich mesic and wet woods. AS, AB, CL, CU, GE, LK, LI, LR, MD, PO, SU, TR, WA.

1986 **Platanthera peramoena** Gray   Purple Fringeless Orchid
Habenaria peramoena Gray (in Braun, 1967 and Voss, 1972)
Rare; meadows, thickets, roadside ditches. WA: Wayne County, Hopkins s.n., 15 Jul 1906 (OS 46077).

1987 **Platanthera psycodes** (L.) Lindley   Small Purple Fringed Orchid
Habenaria psycodes (L.) Sprengel (in Braun, 1967 and Voss, 1972)
Frequent; calcareous fens, wet meadows, sphagnous thickets, open wet woods. AB, CL, CU, FA, GE, LK, LI, LR, PO, RI, ST, SU, WA.

1988 # **Pogonia ophioglossoides** (L.) Ker   Rose Pogonia
Frequent; sphagnous hummocks, bogs. AS, AB, GE, HL, LI, LR, PE, PO, RI, SU, WA.

1989 **Spiranthes cernua** (L.) Richard   Nodding Ladies' Tresses
Common; fields, wet woodland borders, thickets. AS, AB, CL, CU, FA, GE, HL, KN, LI, LI, LR, MH, PE, PO, RO, ST, SU, TR, WA.

1990 **Spiranthes lacera** (Raf.) Raf. var. **gracilis** (Bigel.) Luer   Slender Ladies' Tresses
Spiranthes lacera (Raf.) Raf. (in Braun, 1967 and Voss, 1972)
Frequent; grassy slopes, pond margins, open wet woods, damp thickets. AS, AB, CU, FA, KN, LK, LI, LR, PE, PO, RI, RO, SU, TR.

1991 **Spiranthes lucida** (H. H. Eaton) Ames   Shining Ladies' Tresses
Infrequent; moist soil in meadows, thickets, open woods. CU, GE, LK, LR, MD, PO, RO, SU, TR.

1992 **Spiranthes magnicamporum** Sheviak   Great Plains Ladies' Tresses
Rare; calcareous slopes. RO: Paxton Twp., Bartley & Pontius s.n., 28 Sep 1930 (OS 10953). SU: seep, Boston Twp., Andreas 6725, 4 Oct 1985 (KE 49766); moist areas, Boston Twp., Cusick 12332, 3 Oct 1971 (KE 35708).

1993 **Spiranthes ochroleuca** (Rydb.) Rydb.
Spiranthes cernua (L.) Richard (in Braun, 1967 and Voss, 1972)
Frequent; moist roadsides, open woods, marshes. AS, AB, CL, GE, HL, LK, MH, PE, PO, RO, ST, SU, TR.

1994 **#*Spiranthes romanzoffiana*** Cham. Hooded Ladies' Tresses
Rare; fens, sedge meadows, calcareous thickets. AB: fen, Wayne Twp., Bissell 1984: 112, 14
Aug 1984 (CLM 20914); Ashtabula, South s.n., 15 Aug 1899 (OS 76775). GE: marl seep,
Burton Twp., Andreas & Knoop 6286, 27 Aug 1984 (KE 46350). ST: bog, Jackson Twp.,
Amann 1391, 21 Aug 1960 (KE 6276); Canton, Case s.n., 28 Jul 1900 (OS 10948).

1995 ***Spiranthes tuberosa*** Raf. Little Ladies' Tresses
Infrequent; dry fields, eroded slopes, disturbed open woods. AS, AB, CU, FA, HL, RO, ST,
SU, TR, WA.

1996 ***Tipularia discolor*** (Pursh) Nutt. Cranefly Orchid
Infrequent; wooded banks, boggy woods, rich mesic woods. AB, CU, MD, LR, RO.

1997 ***Triphora trianthophora*** (Sw.) Rydb. Three Birds Orchid
Infrequent; rich woods. CU, FA, LK, LI, LR, MD, RO, ST, SU, WA. Braun (1967) indicates
a record for AB.

PONTEDERIACEAE (Pickerelweed Family)

1998 ***Heteranthera dubia*** (Jacq.) MacM. Water Stargrass
Frequent; ponds, quiet streams, mudflats. AB, CU, GE, KN, LK, LI, LR, PC, PO, ST, SU,
TR, WA.

1999 ***Pontederia cordata*** L. Pickerelweed
Frequent; rivers, pond margins, marshes. AS, AB, CU, GE, HL, LK, LI, LR, MH, MD, PE,
PO, ST, SU, WA.

POTAMOGETONACEAE (Pondweed Family)

2000 **#** ***Potamogeton amplifolius*** Tuckerman Large-leaf Pondweed
Frequent; ponds, lakes. AS, AB, CL, GE, HL, KN, LR, MD, MO, PO, ST, SU, TR, WA.

2001 ***Potamogeton berchtoldii*** Fieber Berchtold's Pondweed
Infrequent; kettle-hole lakes, ponds, streams. AB, CL, GE, LR, PO, ST, TR.

2002 ***Potamogeton crispus*** L. Curly-leaf Pondweed
Common, naturalized; streams, ponds, lakes. AS, AB, CU, GE, HL, KN, LK, LI, LR, MH,
MD, PO, ST, SU, TR, WA. Braun (1967) indicates a record for FA.

2003 ***Potamogeton diversifolius*** Raf.
Infrequent; pools, shallow lakes. AB, RO, ST, SU, TR.

2004 ***Potamogeton epihydrus*** Raf. Nuttall's Pondweed
Common; ditches, oxbows, lakes, streams. AB, CL, CU, GE, HL, LK, LI, LR, MH, MD,
MO, PO, RI, RO, SU, TR, WA.

2005 ***Potamogeton foliosus*** Raf.
Common; ponds, lakes, kettle-hole lakes. AS, AB, CU, FA, GE, HL, KN, LK, LR, MO, PC,
PO, RO, ST, SU, TR, WA.

2006 ***Potamogeton friesii*** Rupr. Fries' Pondweed
Once rare, now presumably extirpated from Ohio; calcareous ponds, lakes. SU: Turkeyfoot
Lake, Case s.n., 30 Jun 1901 (OS 62516 & OS 108577).

2007 ***Potamogeton gramineus*** L. Grass-like Pondweed
Rare; slightly calcareous lakes, ponds. ST: Congress Lake, Kellerman s.n., 28 Jun 1898 (OS
2428). WA: Browns Lake, Selby 2462, 13 Oct 1900 (OS 2450 & OS 86657). Braun (1967)
indicates a record for LK.

2008 **#** ***Potamogeton hillii*** Morong Hill's Pondweed
Rare; calcareous streams, pools. AB: marshy pool, Wayne Twp., Andreas & Bissell 6203, 6
Sep 1983 (KE 46406); pond, Wayne Twp., Bissell 1982:300, 4 Sep 1982 (KE 45083); pools,
Ashtabula County, Hill s.n., no date (Field Museum of Natural History (Haynes, 1974)).
PO: Silver Creek, Garrettsville, Webb 1242, 27 Jul 1913, Gray Herbarium (Haynes, 1974).

2009 ***Potamogeton illinoensis*** Morong
Infrequent; calcareous lakes, streams, ponds. AB, CU, GE, LK, PO, ST, TR, WA.

2010 ***Potamogeton natans*** L. Floating Pondweed
Infrequent; ponds, lakes, streams. AB, FA, GE, LK, LI, LR, PE, RI, ST, SU.

2011 **Potamogeton nodosus** Poiret
Common; ponds, streams. AS, AB, CL, CU, FA, GE, KN, LK, LI, LR, MD, MO, PE, PC, PO, RI, RO, ST, SU, WA.

2012 **Potamogeton pectinatus** L.   Fennel-leaved Pondweed
Common; ponds, lakes. AS, AB, CL, CU, FA, HL, KN, LK, LI, LR, MD, PE, PC, PO, RO, ST, SU, TR, WA.

2013 **Potamogeton praelongus** Wulfen   White-stemmed Pondweed
Rare; calcareous ponds and lakes. PO: deep clear lake, Rootstown Twp., Andreas 5772, 14 Jul 1981 (KE 44085). GE: Punderson Lake, Newbury Twp., Andreas 5769, 12 Jul 1981 (KE 44086). ST: Plain Twp., Brown s.n., 12 Jul 1936 (OS 2451).

2014 **Potamogeton pulcher** Tuckerman   Spotted Pondweed
Rare; muddy lakes, peaty kettle-hole ponds. AB: small pond, Rome, Kuhn 318, 3 Aug 1946 (CLM s.n.); lake, Hicks s.n., 16 Jun 1932 (OS 2397). PO: small pond, Aurora Twp. Andreas & Spooner 5646, 16 Jun 1981 (KE 44081). SU: Singer Lake, Green Twp., Andreas 3556, 31 May 1979 (KE 42231). WA: Big Fox Lake, Selby & Duvel 199, 27 Jul 1899 (OS 86679 & OS 86680).

2015 **Potamogeton pusillus** L.   Small Pondweed
Frequent; lakes, ponds. AB, CL, GE, HL, LK, LI, MO, PE, RO, SU, WA.

2016 **Potamogeton richardsonii** (Benn.) Rydb.   Richardson's Pondweed
Rare; calcareous waters in clear lakes, ponds. PO: cold clear lake, Rootstown Twp., Andreas & Emmitt 5715, 6 Jul 1981 (KE 44082); Brady Lake, Franklin Twp., Hopkins s.n., 2 Jun 1913 (OS 73997 & OS 42632). SU: lakes, Krebs & Claassen s.n., 1890 (OS 2453). Braun (1967) indicates a record for ST.

2017 # **Potamogeton robbinsii** Oakes   Robbin's Pondweed
Rare; clear water in ponds, lakes. GE: Lake Aquilla, Claridon Twp., Bissell s.n., 20 Sep 1980 (CLM 13872). PO: Lake Pippen, Franklin Twp., Andreas & Emmitt 5683, 24 Jun 1981 (KE 44068); Steward Pond, Franklin Twp., Cooperrider & Zuppke 9856, 4 April 1966 (KE 20239); West Twin Lake, Rood 2690, 19 Jun 1953 (KE 10746). SU: Cuyahoga Falls, Kellerman s.n., 3 Jul 1898 (OS 2524).

2018 **Potamogeton spirillus** Tuckerman   Spiral Pondweed
Infrequent; ponds, lakes. AB, GE, MH, PO, RO, TR.

2019 **Potamogeton strictifolius** Benn.   Straight-leaved Pondweed
Once rare, now presumably extirpated from Ohio; lakes. Braun (1967) indicates a record for SU.

2020 # **Potamogeton vaseyi** Robbins   Vasey's Pondweed
Once rare, now presumably extirpated from Ohio; ponds, small lakes. AB: pond near Farmham, Hicks s.n., 12 Aug 1928 (OS 2511 & OS 84470). PO: Sandy Lake, Rootstown Twp., Hopkins s.n., 3 Aug 1918 (OS 98865 & OS 98863); Brady Lake, Webb & Hopkins 1352, 29 Jul 1913 (KE 10753 & OS 42628). TR: Mosquito Creek, Howland Twp., Rood & Shanks s.n., 4 Jul 1935 (OS 62106).

2021 **Potamogeton zosteriformis** Fern.   Flat-stemmed Pondweed
Infrequent; peaty waters, kettle-hole ponds, streams, slow-moving rivers. GE, LI, PE, PO, ST, SU.

RUPPIACEAE (Ditch-grass Family)

2022 ***Ruppia maritima** L.   Ditch-grass
Rare, naturalized; shallow saline waters. WA: saline pond, Rittman, Unger & Loveland 1, 26 Sep 1980 (BHO 34511).

SPARGANIACEAE (Bur-reed Family)

2023 **Sparganium americanum** Nutt.   Bur-reed
Common; muddy waters, ditches, marshes. All counties except CU, LK, PC.

2024 **Sparganium androcladum** (Engelm.) Morong   Keeled Bur-reed
Infrequent; marshes, wet thickets. AS, GE, LI, LR, SU.

2025 **Sparganium eurycarpum** Engelm.
Common; marshes, ditches, stream margins. All counties except PC.

TYPHACEAE (Cat-tail Family)

2026 *Typha angustifolia* L.   Narrow-leaved Cat-tail
Common; roadside ditches, stream margins, ponds, marshes. ALL COUNTIES. Stuckey (1987) indicates that this species may be alien.

2027 *Typha latifolia* L.   Common Cat-tail
Common; marshes, ditches, stream and lake margins, sphagnous thickets. ALL COUNTIES.

XYRIDACEAE (Yellow-eyed Grass Family)

2028 *#Xyris difformis* Chapman   Yellow-eyed Grass
Xyris caroliniana Walter (in Braun, 1967)
Rare; sphagnous mats, bog margins. GE: Burton, Metcalf s.n., 2 Aug 1888 (OC 77963); Geauga Lake, Claassen s.n., 189- (OS 10351). PO: *Sphagnum* mat, Suffield Twp., Andreas 5480, 25 Aug 1980 (KE 43099); *Sphagnum* mat, Franklin Twp., Andreas 5393, 16 Jul 1980 (KE 43098); peat swamp, Earlville, Claassen s.n., 1891 (OS 74057); Portage County, Krebs s.n., 1890 (OS 10345). SU: quaking mat, Green Twp., Karlo s.n., 5 Aug 1962 (KE 42178).

ZANNICHELLIACEAE (Horned Pondweed Family)

2029 *Zannichellia palustris* L.   Horned Pondweed
Infrequent; calcareous streams, shallow ponds. AB, CU, GE, HL, LI, PC, ST.

147

APPENDIX A—OTHER COUNTY RECORDS

The following list of 376 taxa includes additional records from the 23 counties surveyed for the flora of the Glaciated Allegheny Plateau region of Ohio, but the records are from herbarium specimens collected from localities outside the study area. Many of the vascular plant records included here from Lake County are from herbarium records labelled "near Painesville" or "Painesville", with no additional locality information. These records often are the only occurrence for the taxon within the Glaciated Allegheny Plateau region of Ohio, and possibly were collected from cultivation or were introduced into Lake County with nursery stock. These records are indicated by plus symbols "++" to the right of the county.

Taxa included in this list are from regions within the county that are included in other phytogeographical regions and these taxa may be indicative of that region. Those taxa definitive of unglaciated Ohio, as defined by Cusick and Silberhorn (1977), are indicated by the letters "ug" to the left of the county, and those taxa definitive to the Eastern Lake Plain Section of the Central Lowland Province (Anderson, 1983) are indicated by the letters "lp".

Taxa are listed by family in the order of appearance in the catalogue. An asterisk (*) preceding the scientific name indicates that the taxon is alien (non-native) to Ohio.

EQUISETACEAE
Equisetum nelsonii (A. A. Eaton) Schaffner	CU
Equisetum variegatum Schleich.	PC

ADIANTACEAE
Pellaea glabella Mett.	LI, RO

ASPLENIACEAE
Cystopteris tennesseensis Shiner	CL

OPHIOGLOSSACEAE
Botrychium simplex E. Hitchc.	AB

SCHIZAEACEAE
Lygodium palmatum (Bernh.) Swartz	CL (ug)

CUPRESSACEAE
Thuja occidentalis L.	RO

TAXODIACEAE
*Taxodium distichum (L.) Richard	CL

ACANTHACEAE
Ruellia humilis Nutt.	CU, PC

ACERACEAE
Acer rubrum L. var. tridens Wood	PE

AMARANTHACEAE
*Amaranthus cruentus L.	CU
*Amaranthus lividus L.	CU
*Amaranthus spinosus L.	LK, PC

ANACARDIACEAE

Rhus aromatica Aiton var. *arenaria* (Greene) Fern. AB (lp)
Rhus x *boreale* (Britton) Greene AB

ARISTOLOCHIACEAE

Aristolochia serpentaria L. PE

ASCLEPIADACEAE

Asclepias hirtella (Pennell) Woodson LR, RO
Asclepias sullivantii Engelm. PC
Asclepias viridiflora Raf. CL, LK
Cynanchum nigrum (L.) Persoon CU, LK

BETULACEAE

Betula alba L. LK
Betula nigra L. RO (ug)

BORAGINACEAE

Amsinckia lycopsoides (Lehm.) Lehm. LK (++)
Anchusa officinale L. LK (++)
Asperugo procumbens L. LK (++)
Heliotropium indicum L. LK (++)
Lithospermum caroliniense (Walter) MacM. AB (lp)
Lithospermum latifolium Michx. PC
Lycopsis arvensis L. LK
Myosotis stricta Link LK
Onosmodium molle Michx.
    var. *hispidissimum* (Mackenz.) Cronq. PC

BUXACEAE

Pachysandra terminalis Sieb. & Zucc. CU

CAMPANULACEAE

Campanula rotundifolia L. RO
Campanula trachelium L. LR, PC
Lobelia puberula Michx. RO (ug)
Lobelia spicata Lam. var. *leptostachys* RO (ug)
    (DC.) Mackenz. & Bush

CAPRIFOLIACEAE

Lonicera maackii (Rupr.) Maxim. CU, LK
Lonicera prolifera (Kirchn.) Rehd. PC
Triosteum angustifolium L. RO
Viburnum lantana L. LK

CARYOPHYLLACEAE

Arenaria stricta Michx. var. *stricta* CU
Holosteum umbellatum L. LK, PC
Silene caroliniana Walter var. *pensylvanica* (Michx.) Fern. CL, LK
Silene cserei Baumg. CU
Silene rotundifolia Nutt. RO (ug)

CELASTRACEAE

Euonymus europeaeus L. LK
Euonymus fortunei (Turcz.) Hand-Maz. CU

CHENOPODIACEAE

Chenopodium leptophyllum Nutt. CL, LK, LR
Chenopodium polyspermum L. LK (++)
Chenopodium vulvaria L. LK
Corispermum nitidum Kit. CU
Suaeda calceoliformis (Hooker) Moq. LK

149

CISTACEAE

Helianthemum bicknellii Fern.	AB
Helianthemum canadense (L.) Michx.	AB

COMPOSITAE

Ambrosia psilostachya DC.	LK
Antennaria solitaria Rydb.	FA
Antennaria virginica Stebbins	CL
Anthemis nobilis L.	LK
Anthemis tinctoria L.	LK
Arnoseris minima (L.) Schweigger & Koerte	LK (++)
Artemisia abinthium L.	LK, LR
Artemisia annua L.	CU, LK
Artemisia biennis Willd.	CU, LK
Artemisia caudata Michx.	AB, CU (lp)
Artemisia pontica L.	LK
Aster drummondii Lindley	PC
Aster dumosus L.	LR, RO
Aster pilosus Willd. var. *demotus* Blake	LK
Aster subulatus Michx.	LK
Boltonia asteroides Gray var. *latisquama* (Gray) Cronq.	LK (++)
Calendula officinale L.	HL
Carduus nutans L.	CU
Centaurea cyanus L.	LK, PC
Centaurea scabiosa L.	LK (++)
Cirsium pitcheri (Torrey) T. & G.	CU
Conyza ramosissima Cronq.	PC
Crepis tectorum L.	PC
Elephantopus carolinianus Raeuschel	FA (ug)
Eupatorium serotinum Michx.	CU
Eupatorium rotundifolium L.	FA (ug)
Helianthus x *laetiflorus* Persoon	AB
Helianthus maximiliani Schrader	LK
Helianthus mollis Lam.	AB, LK
Helianthus occidentalis Riddell	AS
Hieracium canadense Michx.	CU, LK (lp)
Hieracium trailii Greene	AB
Lapsana communis L.	CU
Leontodon taraxacoides (Vill.) Merat	LK
Liatris aspera Michx.	FA
Matricaria chamomilla L.	CU, LK
Matricaria maritima L. var. *agrestris* (Knaf.) Wilmott	LK
Onopordum acanthium L.	LK, WA
Parthenium integrifolium L.	CU, LK
Picris hieracioides L.	CU, LR
Prenanthes racemosa Michx.	AB
Rudbeckia fulgida Aiton	PE, PC
Rudbeckia grandiflora (Sweet) DC.	CU
Senecio sylvaticus L.	LK (++)
Solidago remota (Greene) Friesner	CU, LK (lp)
Solidago sempervirens L.	CU, LK
Tagetes erecta L.	AB, LR
Taraxacum laevigatum (Willd.) DC.	LK

CONVOLVULACEAE

Cuscuta cephalanthii Engelm.	AB
Cuscuta glomerata Choisy	PC
Ipomoea lacunosa L.	FA (ug)

CRASSULACEAE

Sedum acre L.	LK
Sedum sarmentosum Bunge	HL

CRUCIFERAE

Alyssum saxatile L.	LK (++)
Brassica alba (L.) Rabenh.	AB
Cakile edentula (Bigelow) Hooker	AB, CU, LK, LR (lp)
Cardamine rotundifolia Michx.	HL, LK
Coronopus didymus (L.) Sm.	CU, LK
Dentaria heterophylla Nutt.	HL, RO
Descurainia sophia (L.) Prantl	PC
Diplotaxis muralis (L.) DC.	AB, CU, LK
Diplotaxis tenuifolia (L.) DC.	AB, CU
Erysimum repandum L.	PC
Lepidium perfoliatum L.	LK
Neslia paniculata (L.) Desv.	LK (++)
Rorippa sessiliflora (Nutt.) Hitchc.	PC
Sisymbrium loeselli L.	PC
Thlaspi perfoliatum L.	PE

CUCURBITACEAE

Citrullus lanatus (Thunb.) Matsumura & Nakai	LR
Cucumis melo L.	AB, HL
Cucurbita maxima Duchesne	HL
Cucurbita pepo L.	HL

ELATINACEAE

Elatine americana (L.) Nutt.	LK

ERICACEAE

Arctostaphylos uva-ursi (L.) Sprengel	AB (lp)
Calluna vulgaris (L.) Hull	LK
Ledum groenlandicum	LK
Oxydendrum arboreum (L.) DC.	FA, PE, RO (ug)
Rhododendron calendulaceum (Michx.) Torrey	FA (ug)
Rhododendron maximum L.	FA (ug)

EUPHORBIACEAE

Acalypha gracilens Gray	LK
Croton capitatus Michx.	PC
Euphorbia helioscopia L.	LK (++)
Euphorbia obtusata Pursh	FA
Euphorbia platyphylla L.	AB, CU, LK
Euphorbia polygonifolia L.	AB, CU, LK, LR (lp)
Mercurialis annua L.	LK
Ricinus communis L.	AB

FAGACEAE

Querucs x *hillii* Trel.	AB
Quercus x *bushii* Sarg.	RO
Quercus stellata Wang.	PE, PC (ug)

FUMARIACEAE

Fumaria parviflora Lam.	LK (++)

GENTIANACEAE

Centaurium erythraea Raf.	AB, LK
Gentiana crinita Froel.	LK

GERANIACEAE

Erodium cicutarium (L.) L'Her.	CU, LK
Geranium columbianum L.	LK
Geranium dissectum L.	LK

151

GUTTIFERAE

Hypericum ellipticum Hooker | LK
Hypericum hypericoides (L.) Crantz | PE (ug)

HIPPOCASTANACEAE

Aesculus octandra Marsh. | PE (ug)

LABIATAE

*Acinos thymoides (L.) Moench | LK (++)
Cunila origanoides (L.) Britton | FA, HL, PE, RO
*Galeopsis tetrahit L. | LK
*Lycopus europeaus L. | LK, CU
*Mentha longifolia (L.) Hudson | CU, AB, LK
*Mentha x piperita L. var. citrata (Ehrh.) Briq. | LK
*Mentha rotundifolia (L.) Hudson | CU, LK
*Monarda punctata L. var. villicaulis (Pennell) Shinners | CU
*Perilla frutescens (L.) Britton | HL
Pycnanthemum pycnanthemoides (Leavenw.) Fern. | FA, PE (ug)
Salvia lyrata L. | RO (ug)
Satureja arkansana (Nutt.) Briq. | RO
*Satureja hortensis L. | LK (++)
Scutellaria ellipitca Muhl. var. hirsuta (Short) Fern. | LI, RO (ug)
Scutellaria ovata Hill | PC
Scutellaria parvula Michx. var. leonardii (Epling) Fern. | AB
Scutellaria saxatilis Riddell | CL (ug)
*Stachys aspera Michx. | LI, LR
Stachys nuttallii Shuttlew. | AS, RO (ug)
Synandra hispidula (Michx.) Britton | PE, RO

LEGUMINOSAE

*Anthyllis vulneraria L. | LK (++)
Astragalus canadensis L. | AB
Astragalus neglectus (T. & G.) Sheldon | AB, CU, LK, LR
*Baptisia australis (L.) R. Br. | LK (++)
Baptisia lactea (Raf.) Thieret | PC
Cladrastris lutea (Michx. f.) K. Koch | PC, RO
Desmodium illinoense (Michx.) MacM. | PC
Desmodium obtusum (Willd.) DC. | FA
*Genista tinctoria L. | LK
Lathyrus japonicus Willd. | AB, CU, LK (lp)
*Lathyrus pratensis L. | LK
*Lathyrus tuberosus L. | LK
Lathyrus venosus Willd. | LK (++)
*Lespedeza cuneata (Dumont) G. Don | LK, LR
Lupinus perennis L. | AB, LK
Phaseolus polystachios (L.) BSP. | LK
Psoralea psoralioides (Walter) Cory | KN
Trifolium reflexum L. | LK (++)
*Vicia hirsuta (L.) S. F. Gray | LK
*Vicia sativa L. | LK, LR
*Vicia tetrasperma (L.) Schreber | CU, FA, LK

LENTIBULARIACEAE

Utricularia vulgaris L. | LK

LINACEAE

*Linum perenne L. | AB, LK

LYTHRACEAE

Ammannia coccinea Rottb. | PC
Rotala ramosior (K.) Koehne | RO

152

MALVACEAE

Malva alcea L.   LK, LR
Malva sylvestris L.   CU, LR, PC
Malva verticillata L. var. *crispa* L.   LR

MENYANTHACEAE

Nymphoides peltata (Gmel.) Kuntze   LK

MYRICACEAE

Myrica pensylvanica Loisel.   AB

NYCTAGINACEAE

Mirabilis albida (Walter) Heirmel   CU
Mirabilis hirsuta (Pursh) MacM.   AB

NYMPHAEACEAE

Brasenia schreberi Gmel.   PE

OLEACEAE

Fraxinus quadrangulata Michx.   PC

ONAGRACEAE

Ludwigia polycarpa Short & Peter   CU, LK
Oenothera fruticosa L. var. *fruticosa*   LK (++)

PAPAVERACEAE

Macleaya cordata (Willd.) R. Br.   LK (++)
Papaver dubium L.   RO
Papaver somniferum L.   HL

PASSIFLORACEAE

Passiflora lutea L.   FA, LI, PC

PLANTAGINACEAE

Plantago psyllium L.   CU, LK

POLEMONIACEAE

Collomia linearis Nutt.   PC
Ipomopsis rubra (L.) Wherry   LK (++)
Phlox glaberrima L. var. *triflora* (Michx.) Reveal & Broome   FA
Phlox stolonifera Sims   FA (ug)

POLYGALACEAE

Polygala incarnata L.   FA (ug)
Polygala paucifolia Willd.   CL
Polygala polygama Walter   AB, CU

POLYGONACEAE

Polygonum ramosissimum Michx.   LK
Rumex conglomeratus Murr.   LK
Rumex maritimus L.   RO

PORTULACACEAE

Montia perfoliata (Donn) Howell   LK (++)
Portulaca oleracea L.   PE

PRIMULACEAE

Dodecatheon meadia L.   CU, RO
Lysimachia vulgaris L.   LK

153

RANUNCULACEAE

*Clematis dioscoreifolia Levl. & Vaniot	FA
Clematis viorna L.	LI, PE, RO
Delphinium exaltatum Aiton	AB, PC
Ranunculus fascicularis Muhl.	CL, PC, RI
*Ranunculus ficaria L.	CU, LK
*Ranunculus pusillus Poiret	RO
Trollius laxus Salisb.	FA

RESEDACEAE

*Reseda alba L.	CU, PC

RHAMNACEAE

Ceanothus herbaceus Raf.	LK (++)

ROSACEAE

*Cotoneaster pyracantha (L.) Spach	LK
Porteranthus trifoliatus (L.) Britton	LK
Potentilla anserina L.	AB, CU, LK, LR (lp)
Potentilla paradoxa Nutt.	AB, LK, LR
*Prunus mahaleb L.	LK
*Rosa gallica L.	LK (++)
*Rosa rugosa Thunb.	LK
*Rubus laciniatus Willd.	AB, CU
*Rubus phoenicolasius Maxim.	AB, LK, LR
*Sanguisorba minor Scop.	LK (++)
Spiraea alba Duroi var. latifolia (Aiton) Dippel	LK (++)

RUBIACEAE

Houstonia purpurea L.	RO
*Sherardia arvensis L.	LK

SALICACEAE

*Salix caprea L.	LK (++)

SAXIFRAGACEAE

*Ribes grossularia L.	CU, LK
*Ribes odoratum Wendl. f.	CU
Sullivantia sullivantii (T. & G.) Britton	RO (ug)

SCROPHULARIACEAE

*Antirrhinum majus L.	AB
Aureolaria laevigata (Raf.) Raf.	FA (ug)
*Digitaria purpurea L.	CU, LK
*Digitaria lutea L.	CU
*Kickxia elatine (L.) Dumort.	LK
*Kickxia spuria (L.) Dumort.	LK (++)
*Mazus japonicus (Thunb.) Kuntze	PC
Penstemon laevigatus Aiton	RO
Veronica catenata Pennell	PC
*Veronica chamaedrys L.	CU, LK
*Veronica longifolia L.	CU, LK

SOLANACEAE

*Datura inoxia Miller	CU, LK
*Nicandra physalodes (L.) Gaertner	LI, RO
*Nicotiana alata Link & Otto	FA

TILIACEAE

Tilia heterophylla Vent.	CL, PE, PC, RO

UMBELLIFERAE

*Aethusa cynapium L.	LK
Anthriscus sylvestris (L.) Hoffm.	PE
Apium graveolens L.	CU
Chaerophyllum procumbens (L.) Crantz var. shortii T. & G.	PE
Hydrocotyle umbellata L.	AB
*Scandix pecten-veneris L.	LK (++)

URTICACEAE

*Urtica urens L.	LK (++)

VALERIANACEAE

Valeriana pauciflora Michx.	LR, PC, RO

VERBENACEAE

Verbena x engelmannii Moldenke	LR

VIOLACEAE

Viola rafinesquii Greene	LI
Viola x bissellii House	HL

VITACEAE

*Ampelopsis brevipedunculata (Maxim.) Trautv.	LK (++)
Ampelopsis cordata Michx.	RO (ug)
Vitis labrusca L.	CU

ZYGOPHYLLACEAE

*Tribulus terrestris L.	CU

BUTOMACEAE

*Butomus umbellatus L.	LK

COMMELINACEAE

Tradescantia subaspera Ker	RO

CYPERACEAE

Carex albolutescens Schwein.	LK
Carex argyrantha Tuckerman	HL, LK
Carex aquatilis Wahlenb.	AB, CL, LK, LR (lp)
Carex bebbii (Bailey) Fern.	LR
Carex brevior (Dewey) Mackenz.	CL, LK
Carex careyana Dewey	PE
Carex conoidea Willd.	LK
Carex eburnea Boott	LK
Carex garberi Fern.	AB
Carex haydenii Dewey	LR
Carex laevenworthii Dewey	LK
Carex muskingumensis Schwein.	HL
Carex projecta Mackenz.	RO
Carex retroflexa Muhl.	CL
Carex rugosperma Mackenz.	AB
Carex striatula Michx.	LI (ug)
Carex suberecta (Olney) Britton	PE, RO
Carex texensis (Torrey) Bailey	PC
Cyperus ovularis (Michx.) Torrey	RO
Cyperus schweinitzii Torrey	AB, CU (lp)
Eleocharis compressa Sulliv.	RO
Eleocharis quadrangulata (Michx.) R. & S.	LI
Eleocharis wolfii Gray	RO (ug)
Fimbristylis autumnalis (L.) R. & S.	FA
Hemicarpha micrantha (Vahl) Pax	AB
Rhynchospora globularis (Chapman) Small	RO

155

Scirpus purshianus Fern.	LK
Scirpus saximontanus Fern.	PC
Scirpus torreyi Olney	LK
Scleria triglomerata Michx.	AB
Scleria verticillata Willd.	PC

DIOSCOREACEAE

Dioscorea quaternata (Walter) J. F. Gmel.	CL, RO (ug)

GRAMINEAE

Agrostis elliotiana Schultes	CU
Aira caryophyllea L.	LK (++)
Alopecurus carolinianus Walter	CU, RO
Alopecurus pratensis L.	LK
Aegilops cylindrica Host	PC
Ammophila breviligulata Fern.	AB, CU, LK, LR (lp)
Andropogon elliottii Champ.	PE (ug)
Aspera spica-venti (L.) Beauv.	CU, LK
Beckmannia syzigachne (Steudel) Fern.	CU
Bouteloua curtipendula (Michx.) Torrey	LK
Bromus arvensis L.	CU, PE
Bromus briziformis Fischer & Meyer	CU
Bromus mollis L.	CU, LK, PE
Calamovilfa longifolia (Hooker) Scribner	CU
Cynosurus cristatus L.	CU
Diarrhena americana Beauv.	LR
Deschampsia caespitosa (L.) Beauv.	AB, LK
Digitaria filiformis (L.) Koeler	RO (ug)
Eragrostis pilosa (L.) Beauv.	LK
Festuca dertonensis (All.) Aschers. & Graebn.	LK, RO (++)
Heleochloa schoenoides (L.) Roemer	CU
Hordeum distichon L.	LK (++)
Hordeum pusillum Nutt.	PC
Muhlenbergia curtisetosa (Scribner) Bush	HL
Panicum calliphyllum Ashe	LK (++)
Panicum depauperatum Muhl.	CU, LK
Panicum flexile (Gattinger) Scribner	AB, CU
Panicum laxiflorum Lam.	PE, RO (ug)
Panicum longifolium Torrey	RO (ug)
Panicum villosissimum Nash	AB
Paspalum dilatatum Poiret	LK (++)
Paspalum laeve Michx.	FA, PE
Poa bulbosa L.	CU
Sporobolus cryptandrus (Torrey) Gray	AB, CU, LK, LR (lp)
Triplasis purpurea (Walter) Champ.	AB, CU, LK, LR (lp)
Triticum aestivum L.	PE
Uniola latifolia Michx.	RO (ug)

IRIDACEAE

Sisyrinchium atlanticum Bickn.	RO

JUNCACEAE

Juncus alpinus Vill.	AB, CU, LK (lp)
Juncus brachycarpus Engelm.	HL

JUNCAGINACEAE

Triglochin martimum L.	LK

LILIACEAE

Aletris farinosa L.	AB, LK
Leucojum aestivum L.	LK
Smilacina trifolia (L.) Desf.	LK
Tofieldia glutinosa (Michx.) Persoon	RO
Trillium cernuum L.	LK (++)
Veratrum viride Aiton	LK

ORCHIDACEAE

Aplectrum hyemale (Willd.) Torrey	PE
Orchis spectabilis L.	PE
Platanthera ciliaris (L.) Lindley	LK
Platantera hookeri (Torrey) Lindley	LK
Platanthera peramoena Gray	FA, PE, RO

POTAMOGETONACEAE

Potamogeton diversifolius Raf.	PE
Potamogeton praelongus Wulfen	AB
Potamogeton richardsonii (Benn.) Rydb.	AB, CU, LK
Potamogeton tennesseensis Fern.	CL

RUPPIACEAE

*Ruppia maritima L.	LK

XYRIDACEAE

Xyris difformis Chapman	LK

APPENDIX B—DEFINITIVE TAXA OF
THE GLACIATED ALLEGHENY
PLATEAU REGION OF OHIO

The following list of 151 native vascular plants includes those taxa that have their geographical distribution in Ohio primarily limited to the Glaciated Allegheny Plateau region. Approximately 50% of these vascular plants occur in wetlands, especially bogs and fens as described by Andreas (1985). Eighteen percent are typical components of the flora of the hemlock-eastern white pine-hardwood community described in Braun (1950), and 7% occur on eroding slopes communities as described in Anderson (1982). The remaining 25% are found primarily in a variety of forested communities including beech-maple forests, hemlock slopes, oak-hickory forests and mixed mesic forests. Forty-four of these 151 definitive taxa are known within the region from only one or two collections. Apparently, in Ohio, these taxa always have been rare.

Taxa are listed by family in the order of appearance in the catalogue. The number of the taxon in the catalogue is provided in parentheses after the scientific name. A plus sign (+) to the left of the scientific name indicates that the taxon was listed by Schaffner (1932) as indicative of the "northeastern phytogeographic region."

ISOETACEAE
Isoetes echinospora Dur. (1) Spiny-spored Quillwort
+ *Isoetes engelmannii* A. Braun (2) Appalachian Quillwort

LYCOPODIACEAE
Lycopodium inundatum L. (7) Bog Clubmoss
Lycopodium tristachyum Pursh (12) Ground Cedar

EQUISETACEAE
+ *Equisetum sylvaticum* L. (21) Woodland Horsetail

ASPLENIACEAE
Dryopteris x boottii (Tuckerman) Underw. (37) Boott's Wood Fern
Dryopteris clintoniana (D. C. Eaton) Dowell (38) Clinton's Wood Fern
Gymnocarpium dryopteris (L.) Newman (46) Oak Fern

BLECHNACEAE
Woodwardia virginica (L.) Small (53) Virginia Chain Fern

OPHIOGLOSSACEAE
+ *Botrychium lanceolatum* (Gmel.) Angstr. (57) Lance-leaved Grape-fern
Botrychium multifidum (Gmel.) Rupr.
  var. *intermedium* (D. C. Eaton) Farw. (59) Leathery Grape-fern

THELYPTERIDACEAE
+ *Thelypteris phegopteris* (L.) Slosson (70) Long Beech Fern

PINACEAE
Larix laricina (DuRoi) K. Koch (74) Tamarack

AQUIFOLIACEAE
Nemopanthus mucronatus (L.) Trel. (121) Mountain-holly

ARALIACEAE
Aralia hispida Vent. (122) Bristly Sarsaparilla

158

BETULACEAE
Alnus rugosa (DuRoi) Sprengel (155) Speckled Alder

BORAGINACEAE
Cynoglossum virginianum L. var. *boreale* (Fern.) Cooperrider (173a) Northern Wild Comfrey

CALLITRICHACEAE
Callitriche palustris L. (196) Water-starwort

CAPRIFOLIACEAE
Diervilla lonicera Mill. (214) Bush Honeysuckle
+ *Linnaea borealis* var. *americana* (Forbes) Rehd. (215) Twinflower
Lonicera canadensis Bart. (217) Fly Honeysuckle
Lonicera oblongifolia (Goldie) Hooker (222) Swamp Fly Honeysuckle
Sambucus pubens Michx. (229) Red-berried Elder
Viburnum alnifolium Marsh. (236) Hobble-bush
Viburnum opulus L. var. *americana* Aiton (241b) Highbush Cranberry
Viburnum recognitum Fern. (244) Northern Arrow-wood

CISTACEAE
Lechea intermedia Leggett (314) Round-fruited Pinweed
Lechea leggettii Britton & Hollick (315) Leggett's Pinweed

COMPOSITAE
Aster acuminatus Michx. (342) Mountain Aster
Aster junciformis Rydb. (351) Swamp Aster
Bidens discoidea (T. & G.) Britton (377) Stick-tight
Gnaphalium viscosum HBK. (443) Macoun's Everlasting
Megalodonta beckii (Torrey) Greene (495) Water-marigold

CORNACEAE
+ *Cornus canadensis* L. (573) Bunchberry
Cornus rugosa Lam. (577) Roundleaf Dogwood

DROSERACEAE
Drosera intermedia Hayne (661) Spathulate-leaved Sundew

ELAEAGNACEAE
Shepherdia canadensis (L.) Nutt. (666) Buffalo-berry

ERICACEAE
+ *Andromeda glaucophylla* Link (667) Bog-rosemary
Chamaedaphne calyculata (L.) Moench (668) Leatherleaf
+ *Gaultheria hispidula* (L.) Bigelow (673) Creeping Snowberry
+ *Ledum groenlandicum* Oeder (677) Labrador-tea
+ *Pyrola secunda* L. (683) One-sided Wintergreen
Vaccinium angustifolium Aiton (685) Low Sugarberry
Vaccinium myrtilloides Michx. (688) Velvet-leaf Blueberry
Vaccinium oxycoccos L. (689) Small Cranberry

GENTIANACEAE
Gentiana clausa Raf. (738) Closed Gentian

GUTTIFERAE
Hypericum boreale (Britton) Bickn. (754) Northern St. John's-wort
Hypericum ellipticum Hooker (757) Few-flowered St. John's-wort
Hypericum virginianum L. var. *virginianum* (768a) Marsh St. John's-wort

LABIATAE
Pycnanthemum muticum (Michx.) Persoon (828) Blunt Mountain Mint

LENTIBULARIACEAE

Utricularia intermedia Hayne (936) Flat-leaved Bladderwort
Utricularia minor L. (937) Lesser Bladderwort

MYRICACEAE

+ *Myrica pensylvanica* Loisel. (973) Bayberry

NYMPHAEACEAE

Brasenia schreberi Gmel. (976) Watershield

ONAGRACEAE

Epilobium strictum Muhl. (998) Simple Willow-herb

OXALIDACEAE

Oxalis montana Raf. (1017) White Wood-sorrel

POLYGONACEAE

Polygonum cilinode Michx. (1055) Mountain Bindweed

PORTULACACEAE

Claytonia caroliniana Michx. (1079) Carolina Spring Beauty

PRIMULACEAE

Trientalis borealis Raf. (1095) Star-flower

RANUNCULACEAE

+ *Aconitum noveboracense* Gray (1096) Northern Monkshood
Coptis trifolia (L.) Salisb. (1110) Goldthread
+ *Trollius laxus* Salisb. (1135) Spreading Globe-flower

RHAMNACEAE

Rhamnus alnifolia L'Her. (1138) Alder-leaved Buckthorn

ROSACEAE

Agrimonia striata Michx. (1146) Hairy Agrimony
Amelanchier laevis Wieg. (1148) Allegheny Serviceberry
+ *Dalibarda repens* L. (1166) Dew-drop
Geum rivale L. (1176) Water Avens
+ *Potentilla palustris* (L.) Scop. (1191) Marsh Five-fingers
Prunus pensylvanica L. f. (1200) Pin Cherry
Sorbus decora (Sarg.) Schneider (1207) Mountain-ash

RUBIACEAE

Galium labradoricum (Wieg.) Wieg. (1242) Bog Bedstraw
Galium palustre L. (1247) Marsh Bedstraw
Galium trifidum L. (1251) Three-cleft Bedstraw

SALICACEAE

Salix candida Fluegge (1274) Hoary Willow
Salix pedicillaris Pursh (1283) Bog Willow
Salix serissima (Bailey) Fern. (1288) Autumn Willow

SARRACENIACEAE

Sarracenia purpurea L. (1291) Pitcher-plant

SAXIFRAGACEAE

Ribes glandulosum Grauer (1302) Skunk Currant
Ribes hirtellum Michx. (1304) Smooth Gooseberry
Ribes triste Pallas (1307) Swamp Red Currant

UMBELLIFERAE
Conioselinum chinense (L.) BSP. (1389) Hemlock Parsley
+ *Hydrocotyle americana* L. (1396) American Pennywort
+ *Hydrocotyle umbellata* L. (1398) Navelwort

VALERIANACEAE
Valeriana uliginosa (T. & G.) Rydb. (1422) Swamp Valerian

VIOLACEAE
Viola hastata Michx. (1445) Halberd-leaved Violet
Viola rotundifolia Michx. (1456) Round-leaved Violet

ARACEAE
Arisaema stewardsonii Britton (1481) Swamp Jack-in-the-pulpit
+ *Calla palustris* L. (1482) Wild Calla

CYPERACEAE
Carex alata T. & G. (1493) Broad-winged Sedge
Carex arctata Boott (1498) Drooping Wood Sedge
Carex argyrantha Tuckerman (1499) Silvery Sedge
Carex bebbii (Bailey) Fern. (1504) Bebb's Sedge
Carex brunnescens (Persoon) Poiret (1507) Brownish Sedge
Carex canescens L. (1509) Glaucous Sedge
Carex cephalantha (Bailey) Bickn. (1512) Little Prickly Sedge
Carex cephaloidea Dewey (1513) Thin-leaved Sedge
Carex debilis Michx. var. *rudgei* Bailey (1526) Weak Sedge
Carex deweyana Schwein. (1528) Dewey's Sedge
Carex emmonsii Dewey (1532) Emmon's Sedge
Carex folliculata L. (1536) Long Sedge
Carex haydenii Dewey (1544) Hayden's Sedge
Carex howei Mackenz. (1548) Howe's Sedge
Carex leptonervia Fern. (1562) Nerveless Sedge
Carex limosa L. (1563) Mud Sedge
Carex louisianica Bailey (1564) Louisiana Sedge
Carex oligosperma Michx. (1574) Few-seeded Sedge
Carex ormostachya Wieg. (1575) Stiff Broad-leaved Sedge
Carex pallescens L. (1576) Pale Sedge
Carex pedunculata Muhl. (1577) Long-stalked Sedge
Carex projecta Mackenz. (1584) Necklace Sedge
Carex radiata (Wahlenb.) Dewey (1585) Radiate Sedge
Carex rostrata Stokes (1588) Beaked Sedge
Carex scabrata Schwein. (1590) Rough Sedge
Carex seorsa Howe (1592) Weak Bog Sedge
Carex trisperma Dewey (1609) Three-seeded Sedge
Eriophorum virginicum L. (1645) Tawny Cotton-sedge
Eriophorum viridi-carinatum (Engelm.) Fern. (1646) Green Cotton-sedge
Rhynchospora alba (L.) Vahl (1649) Weak Beak-rush
Scirpus expansus Fern. (1656) Woodland Wool-grass

ERIOCAULACEAE
+ *Eriocaulon septangulare* With. (1666) White Buttons

GRAMINEAE
Cinna latifolia (Trev.) Griseb. (1711) Broad-leaved Northern Wood-reed
Danthonia compressa Aust. (1716) Flattened Wild Oat-grass
Deschampsia flexuosa (L.) Beauv. (1719) Wavy Hair-grass
Glyceria acutiflora Torrey (1746) Sharp-glumed Manna-grass
Glyceria canadensis (Michx.) Trin. (1747) Rattlesnake Grass
Glyceria grandis S. Watson (1748) Tall Manna-grass
Glyceria melicaria (Michx.) F. T. Hubbard (1749) Long Manna-grass
Oryzopsis asperifolia Michx. (1774) Large-leaved Mountain Rice Grass
Trisetum pensylvanicum (L.) Beauv. (1838) Swamp Oat-grass

HYDROCHARITACEAE
Elodea nuttallii (Planchon) St. John (1844) Nuttall's Waterweed

IRIDACEAE
Sisyrinchium montanum Greene (1855) Northern Blue-eyed Grass

JUNCACEAE
Juncus platyphyllus (Wieg.) Fern. (1872) Flat-leaved Rush

JUNCAGINACEAE
Scheuchzeria palustris L. var. *americana* Fern. (1883) Scheuchzeria

LEMNACEAE
Wolffiella floridana (J. D. Smith) C. H. Thompson (1890) Star Wolffiella

LILIACEAE
+ *Clintonia borealis* (Aiton) Raf. (1901) Bluebead
Smilacina trifoliata (L.) Desf. (1929) Three-leaved False Solomon's Seal
Streptopus roseus Michx. (1936) Twisted-stalk
+ *Trillium undulatum* Willd. (1942) Painted Trillium
Veratrum viride Aiton (1947) White Hellebore

ORCHIDACEAE
Arethusa bulbosa L. (1955) Dragon's Mouth Orchid
Corallorhiza trifida Chatelain (1960) Early Coral-root
Goodyera tesselata Lodd (1969) Checkered Rattlesnake Plantain
Listera cordata (L.) R. Br. (1973) Heartleaf Twayblade
+ *Platanthera blephariglottis* (Willd.) Lindley (1976) White-fringed Orchid
Pogonia ophioglossoides (L.) Ker (1988) Rose Pogonia
+ *Spiranthes romanzoffiana* Cham. (1994) Hooded Ladies'-tresses

POTAMOGETONACEAE
Potamogeton amplifolius Tuckerman (2000) Large-leaved Pondweed
Potamogeton hillii Morong (2008) Hill's Pondweed
Potamogeton robbinsii Oakes (2017) Robbin's Pondweed
Potamogeton vaseyi Robbins (2020) Vasey's Pondweed

XYRIDACEAE
Xyris difformis Chapman (2028) Yellow-eyed Grass

APPENDIX C—FAMILIES ARRANGED ACCORDING TO THE ENGLERIAN SYSTEM

The vascular plant families included in the catalogue for the Glaciated Allegheny Plateau region are listed below, arranged according to the Englerian system (Fernald, 1950). Departures from Fernald, such as the merging of two families into one or the separation of one family into two or more families, are noted in parentheses. Explanations for departures from Fernald are discussed in the "Explanation of the Catalogue."

Equisetaceae
Lycopodiaceae
Selaginellaceae
Isoetaceae
Ophioglossaceae
Osmundaceae
Polypodiaceae (including here the Adiantaceae, Aspleniaceae, Blechnaceae, Dennstaedtiaceae, Thelypteridaceae)
Salviniaceae
Taxaceae
Pinaceae (including here the Cupressaceae, Taxodiaceae)
Typhaceae
Sparganiaceae
Zosteraceae (including here the Potamogetonaceae, Ruppiaceae, Zannichelliaceae)
Najadaceae
Juncaginaceae
Alismataceae
Hydrocharitaceae
Gramineae
Cyperaceae
Araceae
Lemnaceae
Xyridaceae
Eriocaulaceae
Commelinaceae
Pontederiaceae
Juncaceae
Liliaceae
Dioscoreaceae
Amaryllidaceae (included here within the Liliaceae)
Iridaceae
Orchidaceae
Saururaceae
Salicaceae
Myricaceae
Juglandaceae
Corylaceae (=Betulaceae)
Fagaceae
Ulmaceae
Moraceae
Cannabinaceae (=Cannabaceae)
Urticaceae
Santalaceae
Loranthaceae (=Viscaceae)
Aristolochiaceae
Polygonaceae
Chenopodiaceae

Saxifragaceae
Hamamelidaceae
Platanaceae
Rosaceae
Leguminosae
Linaceae
Oxalidaceae
Geraniaceae
Zygophyllaceae
Rutaceae
Simaroubaceae
Polygalaceae
Euphorbiaceae
Callitrichaceae
Limnanthaceae
Anacardiaceae
Aquifoliaceae
Celastraceae
Staphyleaceae
Aceraceae
Hippocastanaceae
Balsaminaceae
Rhamnaceae
Vitaceae
Tiliaceae
Malvaceae
Guttiferae
Cistaceae
Violaceae
Passifloraceae
Cactaceae
Thymelaeaceae
Elaeagnaceae
Lythraceae
Nyssaceae
Melastomataceae
Onagraceae
Haloragaceae
Araliaceae
Umbelliferae
Cornaceae
Pyrolaceae (included here within the Ericaceae)
Ericaceae
Primulaceae
Ebenaceae
Oleaceae
Gentianaceae (including here the Menyanthaceae)
Apocynaceae
Asclepiadaceae
Convolvulaceae
Polemoniaceae

163

Amaranthaceae
Nyctaginaceae
Phytolaccaceae
Aizoaceae
Portulacaceae
Caryophyllaceae
Ceratophyllaceae
Nymphaeaceae
Ranunculaceae
Berberidaceae
Menispermaceae
Magnoliaceae
Annonaceae
Lauraceae
Papaveraceae (including here the Fumariaceae)
Capparidaceae (=Capparaceae)
Cruciferae
Resedaceae
Sarraceniaceae
Droseraceae
Crassulaceae

Hydrophyllaceae
Boraginaceae
Verbenaceae
Labiatae
Solanaceae
Scrophulariaceae
Orobanchaceae
Lentibulariaceae
Acanthaceae
Phyrmaceae
Plantaginaceae
Rubiaceae
Caprifoliaceae
Valerianaceae
Dipsacaceae
Cucurbitaceae
Campanulaceae
Compositae

REFERENCES

Adams, William. 1982. Pteridophytes. Pages 22-25, *in* Endangered and Threatened Plants of Ohio. Tom S. Cooperrider, ed. Ohio Biol. Surv. Biol. Notes No. 16. Columbus, Ohio. 92 p.

Aiken, Susan G. 1981. A conspectus of *Myriophyllum* (Haloragaceae) in North America. Brittonia 33: 57-69.

Aldrich, John W. 1943. Biological survey of the bogs and swamps in northeastern Ohio. Amer. Midl. Naturalist 30: 346-402.

Amann, Joyce E. 1961. A survey of the vascular plants of Stark County, Ohio. M.A. Thesis, Kent State University, Kent, Ohio. 230 p.

Anderson, Dennis M. 1982. Plant communities of Ohio: a preliminary classification and description. Ohio Department of Natural Areas and Preserves, Ohio Department of Natural Resources, Columbus, Ohio. 183 p.

_____ **1983.** The natural divisions of Ohio. Nat. Areas J. 3: 23-32.

Anderson, Sandra M. 1969. The vascular plant flora of Medina County, Ohio. M.A. Thesis, Kent State University, Kent, Ohio. 92 p.

Andreas, Barbara K. 1970. The herbaceous Gentianales of Ohio. M.A. Thesis, Kent State University, Kent, Ohio. 143 p.

_____ **1980.** The flora of Portage, Stark, Summit, and Wayne Counties, Ohio. Ph.D. Dissertation, Kent State University, Kent, Ohio. 680 p.

_____ **1985.** The relationship between Ohio peatland distribution and buried river valleys. Ohio J. Sci. 85: 116-125.

Andreas, Barbara K. and Gary Bryan. 1990. The vegetation of three *Sphagnum*-dominated kettle-hole bogs in Ohio. Ohio J. Sci. 90: In press.

Andreas, Barbara K. and Tom S. Cooperrider. 1979. The Apocynaceae of Ohio. Castanea 44: 238-241.

_____ **1980.** The Asclepiadaceae of Ohio. Castanea 45: 51-55.

_____ **1981.** The Gentianaceae and Menyanthaceae of Ohio. Castanea 46: 102-108.

Arbak, Zelia E. and Will H. Blackwell, Jr. 1982. The Chenopodiaceae of Ohio. Castanea 47: 284-297.

Banks, Phillip O. 1970. General geology of northeastern Ohio. Pages 1-8, *in* Guide to the Geology of Northeastern Ohio. Northern Geological Society No. 64. Phillip O. Banks and Rodney M. Feldmann, eds. Department of Geology, Case Western Reserve University. Cleveland, Ohio. 164 p.

Bayer, Randall J. and G. Ledyard Stebbins. 1982. A revised classification of *Antennaria* (Asteraceae: Inuleae) of the eastern United States. Syst. Bot. 7: 300-313.

Benninghoff, William S. 1964. The prairie peninsula as a filter barrier to postglacial plant migration. Proc. Indiana Acad. Sci. 72: 116-124.

Bentz, Gregory D. 1969. The Rhinanthoideae of Ohio. M.A. Thesis, Kent State University, Kent, Ohio. 95 p.

Bentz, Gregory D. and Tom S. Cooperrider. 1978. The Scrophulariaceae subfamily Rhinanthoideae of Ohio. Castanea 43: 145-154.

Bigelow, John M. 1841. Florula Lancastriensis or a catalog of nearly all the flowering and felicoid plants growing naturally within the limits of Fairfield county with notes on such as are medicinal. Proc. Medical Convention of Ohio at Columbus, May, 1841: 49-79.

Bissell, James K. 1984. A currant new to Ohio: *Ribes triste* Pall. Ohio J. Sci. 84: 268-269.

Bissell, James K., Brian C. Parsons, and Thomas A. Yates. 1985. A new locality for *Trollius laxus* Salisb. ssp. *laxus* in Ohio. Rhodora 87:459-461.

Black, Robert A. and Richard N. Mack. 1977. *Tsuga canadensis* in Ohio: Synecological and phytogeographical relationships. Vegetatio 31: 11-19.

Blackwell, Will H., Jr. 1970. The Lythraceae of Ohio. Ohio J. Sci. 70: 346-352.

Braun, E. Lucy. 1950. Deciduous forest of Eastern North America. Blakiston Co., Philadelphia, Pennsylvania. 596 p.

—————— 1955. The phytogeography of unglaciated eastern United States and its interpretation. Bot. Rev. 21: 297-367.

—————— 1961. The woody plants of Ohio. The Ohio State University Press, Columbus, Ohio. 362 p.

—————— 1967. The Monocotyledoneae [of Ohio]: Cat-tails to Orchids. The Ohio State University Press, Columbus, Ohio. 464 p.

Brockett, Bruce L. 1969. The Primulaceae of Ohio, and a study of *Lysimachia* x *producta* (Gray) Fern. M.A. Thesis, Kent State University, Kent, Ohio. 98 p.

Brockett, Bruce L. and Tom S. Cooperrider. 1983. The Primulaceae of Ohio. Castanea 48: 37-40.

Burns, James F. 1980. The flora of Trumbull County, Ohio. M.S. Thesis, Kent State University, Kent, Ohio. 197 p.

—————— 1986. The Polygalaceae of Ohio. Castanea 51: 137-143.

Chuey, Carl F. 1972. The herbaceous angiosperm flora of Mahoning County, Ohio. Ohio J. Sci. 72: 30-49.

Cline, Robert J. 1977. The vascular plants of Perry County, Ohio. M.S. Thesis, Kent State University, Kent, Ohio. 201 p.

Cooperrider, Tom S. 1961. Ohio floristics at the county level. Ohio J. Sci. 61: 318-320.

—————— (ed.). 1982. Endangered and threatened plants of Ohio. Ohio Biol. Surv. Biol. Notes No. 16, Columbus, Ohio. 92 p.

—————— 1984a. Ohio's herbaria and the Ohio Flora Project. Ohio J. Sci. 84 (4): 189-196.

—————— 1984b. Some species mergers and new combinations in the Ohio flora. Michigan Bot. 23: 164-168.

—————— 1985a. Checklist and distribution maps for Ohio Flora, Volume 3. Dicotyledons. Linaceae through Campanulaceae. Reprographic manuscript, Kent State University, Kent, Ohio. 112 p.

—————— 1985b. *Thaspium* and *Zizia* (Umbelliferae) in Ohio. Castanea 50: 116-119.

—————— 1986a. The genus *Phlox* (Polemoniaceae) in Ohio. Castanea 51: 145-148.

—————— 1986b. *Viola* x *brauniae* (*Viola rostrata* x *V. striata*). Michigan Bot. 25: 107-109.

Cooperrider, Tom S. and Robert F. Sabo. 1969. *Stachys hispida* and *S. tenuifolia* in Ohio. Castanea 34: 432-435.

Crabbe, James A., A. C. Jermy, and John T. Mickel. 1975. A new generic sequence for the Pteridophyte herbarium. Fern Gaz. 11: 141-162.

Cronquist, Arthur. 1973. Basic Botany. Harper and Row, New York, New York. 536 p.

—————— 1980. Vascular flora of southeastern United States. Volume 1. Asteracaeae. The University of North Carolina Press, Chapel Hill, North Carolina. 261 p.

Cronquist, Arthur, Armen Takhtajan, and Walter Zimmerman. 1966. On the higher taxa of Embryobionta. Taxon 15: 129-134.

Crowl, Gordon S. 1937. A vegetation survey of Ross County. M.S. Thesis, The Ohio State University, Columbus, Ohio. 176 p.

Cruden, Robert W. 1962. The Campanulaceae of Ohio. Ohio J. Sci. 62: 142-149.

166

Cummin, James W. 1959. Buried river valleys in Ohio. Ohio Water Plan Inventory Report No. 10, Division of Water, Ohio Department of Natural Resources. Columbus, Ohio. 2 maps.

Curtis, John T. 1959. The vegetation of Wisconsin. Univ. Wisconsin Press, Madison, Wisconsin. 657 p.

Cusick, Allison W. 1983. *Spergularia* (Caryophyllaceae) in Ohio. Michigan Bot. 22: 69-71.

_____ **1984.** *Carex praegracilis*: a halophytic sedge naturalized in Ohio. Michigan Bot. 23: 103-106.

Cusick, Allison W. and Gene M. Silberhorn. 1977. The vascular plants of unglaciated Ohio. Ohio Biol. Surv. Bull. N.S. 5(4). 157 p.

Cusick, Allison W. and K. Roger Troutman. 1978. The prairie survey project, a summary of data to date. Ohio Biol. Surv. Info. Circ. No. 10, Columbus, Ohio. 60 p.

Dachnowski, Alfred P. 1912. Peat deposits in Ohio: their origin, formation and uses. Geol. Surv. Ser. 4, Bull. No. 16, Columbus, Ohio. 424 p.

Delcourt, Paul A. and Hazel R. Delcourt. 1981. Vegetation maps for eastern North America: 40,000 yr B. P. to the present. Pages 123-165 *in* Geobotany II. Robert C. Romans, ed. Plenum Press. New York, New York. 263 p.

Division of Geological Survey. n.d. Physiographic sections of Ohio. Ohio Department of Natural Resources, Columbus, Ohio. 1 map.

Division of Natural Areas and Preserves. 1988. Rare species of native Ohio wild plants. 1988-89 status list. Ohio Department of Natural Resources, Columbus, Ohio. 19 p.

Easterly, Nathan W. 1964. Distribution patterns of Ohio Cruciferae. Castanea 29: 164-172.

Emmitt, David P. 1981. The vascular plants of Ashland County, Ohio. M.S. Thesis, Kent State University, Kent, Ohio. 276 p.

Fenneman, Nevin M. 1938. Physiography of eastern United States. McGraw Hill Book Co., New York, New York. 714 p. + 7 plates.

Fernald, Merritt L. 1950. Gray's manual of botany. 8th Ed. American Book Co., New York, New York. 1632 p.

Fisher, T. Richard. 1966. The genus *Silphium* in Ohio. Ohio J. Sci. 66: 259-263.

Flint, Richard F. 1971. Glacial and quaternary geology. John Wiley and Sons, Inc. New York, New York. 892 p.

Forsyth, Jane L. 1970. A geologist looks at the natural vegetation map of Ohio. Ohio J. Sci. 70: 180-191.

Fuller, J. Osborn. 1965. Bedrock geology of the Garrettsville quadrangle. Ohio Div. Geol. Surv. Report of Investigations No. 54, Ohio Department of Natural Resources, Columbus, Ohio. 26 p.

Gardner, George D. 1972. A regional study of landsliding in the lower Cuyahoga River Valley, Ohio. M.S. Thesis, Kent State University, Kent, Ohio. 85 p.

Gleason, Henry A. and Arthur Cronquist. 1963. Manual of vascular plants of northeastern United States and adjacent Canada. D. Van Nostrand Company, Princeton, New Jersey. 810 p.

Goldthwait, Richard P. 1959. Scenes in Ohio during the last Ice Age. Ohio J. Sci. 9: 193-216.

Goldthwait, Richard P., George W. White, and Jane L. Forsyth. 1967. Glacial map of Ohio. U.S. Geol. Surv. and Ohio Div. Geol. Surv., Map I-316. Columbus, Ohio. 1 map.

Gordon, Robert B. 1966. Natural vegetation map of Ohio at the time of the earliest land surveys. Ohio Biol. Surv., Columbus, Ohio. 1 map.

_____ **1969.** The natural vegetation of Ohio in pioneer days. Ohio Biol. Surv. Bull. N.S. 3(2). 109 p.

Hardin, E. Dennis. 1985. *Geranium bicknellii* extant to Ohio. Castanea 50: 123.

Hardin, E. Dennis and Warren A. Wistendahl. 1983. The effects of floodplain trees on herbaceous vegetation patterns, microtopography and litter. Torrey Bot. Club Bull. 110: 23-30.

Hauser, Edward J. P. 1963. The Dipsacaceae and Valerianaceae of Ohio. Ohio J. Sci. 63: 26-30.

——————— 1964. The Rubiaceae of Ohio. Ohio J. Sci. 64: 27-35.

——————— 1965. The Caprifoliaceae of Ohio. Ohio J. Sci. 65: 118-129.

Hawver, William D. 1961. The vascular flora of Geauga County, Ohio. M.A. Thesis, Kent State University, Kent, Ohio. 94 p.

Haynes, Robert R. 1974. A revision of North American *Potamogeton* subsection *Pusilli* (Potamogetonaceae). Rhodora 76: 564-649.

Heffner, George A. 1939. Vegetation survey of an area in central Ohio at the edge of the edge of the Allegheny Plateau. M.S. Thesis, The Ohio State University, Columbus, Ohio. 49 p.

Hicks, Lawrence E. 1933. The original forest vegetation and the vascular flora of Ashtabula County, Ohio. Ph.D. Dissertation, The Ohio State University, Columbus, Ohio. 211 p.

International Code of Botanical Nomenclature. 1972. Utrecht, The Netherlands. 426 p.

Kerrigan, Warren J., Jr. and Will H. Blackwell, Jr. 1973. The distribution of Ohio Araliaceae. Castanea 38: 168-170.

Knoop, Jeffrey D. and Barbara K. Andreas. 1987. Vegetational survey of Sinking Creek Fen, Clark County, Ohio. Ohio J. Sci. 87(2): 4.

Lafferty, Michael B., (ed.). 1979. Ohio's natural heritage. Ohio Academy of Science, Columbus, Ohio. 324 p.

Lawrence, George H. M. 1951. Taxonomy of vascular plants. Macmillan Company, New York, New York. 823 p.

Lellinger, David B. 1985. A field manual of the ferns and fern allies of the United States and Canada. Smithsonian Institution Press, Washington, D.C. 389 p.

Loconte, Henry and Will H. Blackwell, Jr. 1984. Berberidaceae of Ohio. Castanea 49: 39-43.

Luer, Carlyle A. 1975. The native orchids of the United States and Canada excluding Florida. New York Botanical Gardens, New York, New York. 361 p.

McCance, Robert M., Jr. and James F. Burns, (eds.). 1984. Ohio endangered and threatened vascular plants: Abstracts of state-listed taxa. Div. Natural Areas and Preserves, Ohio Department of Natural Resources, Columbus, Ohio. 635 p.

McCready, George A. 1968. The Scrophularioideae of Ohio and a study of *Chelone* chromosome numbers. M.A. Thesis, Kent State University, Kent, Ohio. 125 p.

McCready, George A. and Tom S. Cooperrider. 1978. The Scrophulariaceae subfamily Scrophularioideae of Ohio. Castanea 43: 76-86.

——————— 1984. The Geraniaceae of Ohio. Castanea 49: 138-141.

Mickel, John T. 1979. How to know the ferns and fern allies. William C. Brown Co., Dubquque, Iowa. 229 p.

Miller, Lynne D. 1976. The Violaceae of Ohio. M.S. Thesis, Kent State University, Kent, Ohio. 203 p.

Miller, Marvin E. 1969. Climatic guide for selected locations in Ohio. Div. of Water, Ohio Department of Natural Resources, Columbus, Ohio.

Noelle, Helen J. and Will H. Blackwell, Jr. 1972. The Cactaceae of Ohio. Castanea 37: 119-124.

168

O'Connor, Mildred A. and Will H. Blackwell, Jr. 1974. Taxonomy and distribution of Ohio Cistaceae. Castanea 39: 228-239.

Pell, William F. and Richard N. Mack. 1977. The *Fagus grandifolia - Acer saccharum - Podophyllum peltatum* association in northeastern Ohio. Bot. Gaz. 138: 64-70.

Pringle, James S. 1967. Taxonomy of *Gentiana*, section *Pneumonanthae*, in eastern North America. Brittonia 19: 1-32.

Pusey, Paul L. 1976. The vascular plants of Knox County, Ohio. M.A. Thesis, Kent State University, Kent, Ohio. 221 p.

Radford, Albert E., William C. Dickison, Jimmy R. Massey, and C. Ritchie Bell. 1974. Vascular plant systematics. Harper & Row, New York, New York. 891 p.

Rau, Jon L. 1970. Pennsylvanian system in northeast Ohio. Pages 1-8, *in* Guide to the Geology of Northeastern Ohio. Northern Geological Society No. 64. Phillip O. Banks and Rodney M. Feldmann, eds. Department of Geology, Case Western Reserve University, Cleveland, Ohio. 164 p.

Riehl, Terrence E. and Irwin A. Ungar. 1980. *Spergularia marina*, a new species record for the flora of Ohio. Ohio J. Sci. 80: 36.

Roberts, Marvin L. and Ronald L. Stuckey. 1974. Bibliography of theses and dissertations on Ohio florisitics and vegetation in Ohio colleges and universities. Ohio Biol. Surv. Info. Circ. No. 7, Columbus, Ohio. 92 p.

Ruffner, James A. 1980. Climate of the states. 2nd Ed. Gate Research Co., Detroit, Michigan. 1175 p.

Sabo, Robert F. 1965. The Labiatae of Ohio. M.A. Thesis, Kent State University, Kent, Ohio. 168 p.

Sampson, Homer C. 1930a. Succession in the swamp forest formation in northern Ohio. Ohio J. Sci. 30: 340-357.

——————— 1930b. The mixed mesophytic forest community of northeastern Ohio. Ohio J. Sci. 30: 358-367.

Schaffner, John H. 1932. Revised catalog of Ohio vascular plants. Ohio Biol. Surv. Bull. 25, The Ohio State University, Columbus, Ohio. 87-215.

Sears, Paul B. 1926. The natural vegetation of Ohio. II. The Prairies. Ohio J. Sci. 26: 128-146.

Shupe, Lloyd M. 1930. The original vegetation of Pickaway County, Ohio. M.S. Thesis, The Ohio State University, Columbus, Ohio. 69 p.

Silberhorn, Gene M. 1970a. The flora of the unglaciated Allegheny Plateau of southeastern Ohio. Ph.D. Dissertation, Kent State University, Kent, Ohio. 371 p.

——————— 1970b. A distinct phytogeographic area of Ohio: the southeastern Allegheny Plateau. Castanea 35: 277-292.

Spencer, J.W. 1894. A review of the history of the Great Lakes. Amer. Geol. 14: 289-301.

Spooner, David M. 1983. The northern range of eastern mistletoe, *Phoradendron serotinum* (Viscaceae), and its status in Ohio. Torry Bot. Club Bull. 110: 489-193.

Spooner, David M. and John S. Shelly. 1983. The national historical distribution of *Platanthera peramoena* (A. Gray) A. Gray (Orchidaceae) and its status in Ohio. Rhodora 85: 55-64.

Stout, Wilbur E., Karl Ver Steeg, and G. F. Lamb. 1943. Water in Ohio. Ohio Div. Geol. Surv. Bull. 44, Ohio Department of Natural Resources, Columbus, Ohio. 694 p.

Stuckey, Ronald L. 1987. *Typha angustifolia* in North America: a foreigner masquerading as a native. Ohio J. Sci. 87(2):4.

Stuckey, Ronald L. and Guy L. Denny. 1981. Prairie fens and bog fens in Ohio: Florisitic similarities, differences and geographical affinities. Pages 1-33, *in* Geobotany II. Robert C. Romans, ed. Plenum Press. New York, New York. 263 p.

Stuckey, Ronald L. and Marvin L. Roberts. 1982. Monocotyledons. Pages 27-47, *in* Endangered and Threatened Plants of Ohio. Tom S. Cooperrider, ed. Ohio Biol. Surv. Biol. Notes No. 16. Columbus, Ohio. 92 p.

Szmuc, Eugene J. 1970a. The Devonian system. Pages 9-22, *in* Guide to the Geology of Northeastern Ohio. Northern Geological Society No. 64. Phillip O. Banks and Rodney M. Feldmann, eds. Department of Geology, Case Western Reserve University. Cleveland, Ohio. 164 p.

—————— **1970b.** The Mississippian system. Pages 23-67, *in* Guide to the Geology of Northeastern Ohio. Northern Geological Society No. 64. Phillip O. Banks and Rodney M. Feldmann, eds. Department of Geology, Case Western Reserve University. Cleveland, Ohio. 164 p.

Tandy, Lewis W. 1976. Vascular plants of four natural areas of northeastern Ohio. M.S. Thesis, Kent State University, Kent, Ohio. 140 p.

Thompson, Isabel. 1939. Geographical affinities of the flora of Ohio. Amer. Midl. Naturalist 21: 730-751.

Thorne, Robert F. 1968. Synopsis of a putatively phylogenetic classification of the flowering plants. Aliso 6: 57-66.

Transeau, Edgar N. 1935. The prairie peninsula. Ecology 16: 423-437.

Tyrrell, Lucy E. 1987. A floristic survey of buttonbush swamps in Gahanna Woods State Nature Preserve, Franklin County, Ohio. Michigan Bot. 26:29-38.

Ungar, Irwin A. and David G. Loveland. 1982. *Ruppia maritima*, new for the flora of Ohio. Ohio J. Sci. 82: 68.

Valley, Karl and Tom S. Cooperrider. 1966. The Orobanchaceae of Ohio. Ohio J. Sci. 66: 264-265.

Ver Steeg, Karl. 1946. The Teays river. Ohio J. Sci. 46: 297-307.

Voss, Edward G. 1972. Michigan flora, Pt. 1. Gymnosperms and Monocots. Cranbrook Institute Sci. Bull. No. 55, Bloomfield Hills, Michigan. 488 p.

—————— **1985.** Michigan flora, Pt. 2. Dicots (Saururaceae-Cornaceae). Cranbrook Institute of Science, Bull. No. 59 and the University of Michigan Herbarium, Ann Arbor, Michigan. 724 p.

Weishaupt, Clara G. 1971. Vascular plants of Ohio, 3rd Ed. Kendall/Hunt Publishing Company, Dubuque, Iowa. 292 p.

White, George W. 1934. Drainage history of north-central Ohio. Ohio J. Sci. 34: 365-382.

—————— **1982.** Glacial geology of northeastern Ohio. Div. Geol. Surv. Bulletin No. 68, Ohio Div. Geol. Surv., Ohio Department of Natural Resources, Columbus, Ohio. 75 p.

Wilson, Hugh D. 1972. The vascular plants of Holmes County, Ohio. M.A. Thesis, Kent State University, Kent, Ohio. 273 p.

—————— **1974.** Vascular plants of Holmes County, Ohio. Ohio J. Sci. 74: 277-281.

Winslow, John D. 1957. The stratigraphy and bedrock topography of Portage County, Ohio, and their relation to ground water resources. Ph.D. Dissertation, University of Illinois, Urbana, Illinois. 88 p. + 6 plates.

INDEX TO FAMILIES, GENERA, AND SPECIES

Family names are in block capitals. Synonyms are italicized. Catalogue numbers of the taxa are presented in the right column.

172

Name	No.	Name	No.	Name	No.	Name	No.
diandra	1529	stricta	1600	fontanum	250	arvense	402
digitalis	1530	suberecta	1601	nutans	251	discolor	403
eburnea	1531	swanii	1602	viscosum	252	muticum	404
emmonsii	1532	tenera	1603	*vulgatum*	250	plattense	405
emoryi	1533	tetanica	1604	CERATOPHYLLACEAE	293-294	pumilum	406
festucacea	1534	texensis	1605	Ceratophyllum		vulgare	407
flava	1535	torta	1606	demersum	293	CISTACEAE	312-319
folliculata	1536	tribuloides	1607	echinatum	294	Citrullus	
frankii	1537	trichocarpa	1608	Cercis		lanatus	653
glaucodea	1538	trisperma	1609	canadensis	863	Cladium	
gracilescens	1539	tuckermanii	1610	Chaenomeles		mariscoides	1619
gracillima	1540	typhina	1611	lagenaria	1153	Claytonia	
granularis	1541	umbellata	1612	Chaenorrhinum		caroliniana	1079
gravida	1542	vesicaria	1613	minus	1317	virginica	1080
grayi	1543	virescens	1614	Chaerophyllum		Clematis	
haydenii	1544	viridula	1615	procumbens	1386	dioscoreifolia	1107
hirsutella	1545	vulpinoidea	1616	Chamaedaphne		virginiana	1108
hirtifolia	1546	willdenowii	1617	calyculata	668	Cleome	
hitchockiana	1547	woodii	1618	Chamaelirium		hassleriana	211
howei	1548	Carpinus		luteum	1900	*spinosa*	211
hyalinolepis	1549	caroliniana	163	Chelidonium		Clinopodium	
hystricina	1550	Carum		majus	1020	vulgare	795
incomperta	1502	carvi	1385	Chelone		Clintonia	
interior	1551	Carya		glabra	1318	borealis	1901
intumescens	1552	cordiformis	782	CHENOPODIACEAE	295-311	umbellulata	1902
jamesii	1553	glabra	783	Chenopodium		Clitoria	
lacustris	1554	laciniosa	784	album	296	mariana	864
laevivaginata	1555	ovalis	785	ambrosioides	297	CLUSIACEAE	754-768
lanuginosa	1556	ovata	786	botrys	298	Coeloglossum	
lasiocarpa	1557	tomentosa	787	capitatum	299	viride	1957
laxiculmis	1558	CARYOPHYLLACEAE	245-286	glaucum	300	Collinsia	
laxiflora	1559	Cassia		hybridum	301	verna	1319
leavenworthii	1560	chamaecrista	859	leptophyllum	302	Collinsonia	
leptalea	1561	*fasciculata*	859	murale	303	canadensis	796
leptonervia	1562	hebecarpa	860	stanleyanum	304	Comandra	
limosa	1563	marilandica	861	urbicum	305	umbellata	1290
louisianica	1564	nictitans	862	vulvaria	306	Commelina	
lupuliformis	1565	Castanea		Chimaphila		communis	1485
lupulina	1566	dentata	713	maculata	669	diffusa	1486
lurida	1567	Castilleja		umbellata	670	COMMELINACEAE	1485-1490
mesochorea	1568	coccinea	1316	Chorispora		COMPOSITAE	320-554
molesta	1569	Catalpa		tenella	619	Conioselinum	
muhlenbergii	1570	bignonioides	168	Chrysanthemum		chinense	1389
muskingumensis	1571	ovata	169	balsamita	394	Conium	
normalis	1572	speciosa	170	leucanthemum	395	maculatum	1390
oligocarpa	1573	Caulophyllum		maximum	396	*Conobea*	
oligosperma	1574	thalictroides	151	parthenium	397	*multifida*	1326
ormostachya	1575	Ceanothus		Chrysogonum		Conopholis	
pallescens	1576	americanus	1137	virginianum	398	americana	1011
pedunculata	1577	CELASTRACEAE	287-292	Chrysopsis		Conringia	
pensylvanica	1578	Celastrus		camporum	399	orientalis	620
plantaginea	1579	scandens	287	Chrysosplenium		Consolida	
platyphylla	1580	Celtis		americanum	1293	ambigua	1109
praegracilis	1581	occidentalis	1378	Cichorium		Convallaria	
prairea	1582	Cenchrus		intybus	400	majalis	1903
prasina	1583	longispinus	1709	Cicuta		CONVOLVULACEAE	555-570
projecta	1584	*pauciflorus*	1709	bulbifera	1387	Convolvulus	
radiata	1585	Centaurea		maculata	1388	arvensis	559
retroflexa	1586	cyanus	388	Cimicifuga		*pellitus*	556
rosea	1587	dubia	389	racemosa	1106	*sepium*	557
rostrata	1588	jacea	390	Cinna		*spithamaeus*	558
sartwellii	1589	maculosa	391	arundinacea	1710	Conyza	
scabrata	1590	nigra	392	latifolia	1711	canadensis	408
scoparia	1591	solstitialis	393	Circaea		ramosissima	409
seorsa	1592	Centaurium		alpina	990	Coptis	
shortiana	1593	pulchellum	736	alpina x C. lutetiana	991	*groenlandica*	1110
sparganioides	1594	Centunculus		intermedia	991	trifolia	1110
sprengelii	1595	minimus	1083	lutetiana	992	Corallorhiza	
squarrosa	1596	Cephalanthus		*quadrisulcata*	992	maculata	1958
sterilis	1597	occidentalis	1235	Cirsium		odontorhiza	1959
stipata	1598	Cerastium		altissimum	401	trifida	1960
straminea	1599	arvense	249			wisteriana	1961

174

175

178

181

INDEX TO COMMON NAMES

Catalogue numbers are shown in the right-hand column.

Burnet
 Canada 1227
Bush-clover 892
Creeping 898
Japanese 899
Nuttall's 896
Pubescent 894
Trailing 897
Butter-and-eggs 1327
Buttercup Family 1096-1135
Buttercup
 Allegheny 1119
 Bulbous 1121
 Creeping 1128
 Kidney-leaf 1117
 Small-flowered 1125
 Swamp 1123, 1130
 Tall 1118
 Water-
  White 1124
  Yellow 1122
Butterfly
 -pea 864
 -weed 138
Butternut 788
Buttonbush 1235
Buttonweed 1236

Cabbage 604
Cactus Family 198
Calla
 Wild 1482
Caltrop Family 1470
Caltrop 1470
Camass
 White 1949
Campion
 Red 258
 Rose 257
 Starry 275
 White 274
 Bladder- 277
Cancer-root 1011
 One-flowered 1013
Candytuft 634
Canker-root 1325
Caper Family 211-213
Caper-spurge 704
Caraway 1385
Cardinal Flower 202
Carpet-weed Family 94
Carpet-weed 94
Carrion-flower 1932
Carrot
 Wild 1392
Cashew Family 105-111
Cat's Ear 472
Cat-tail Family 2026-2027
Cat-tail
 Common 2027
 Narrow-leaved 2026
Catalpa
 Chinese 169
 Southern 168
Catchfly
 Carolina 270
 Forking 272
 German 260
 Night-flowering 273
 Sleepy 268
 Sweet William 269
 Von Cserei's 271

Catfoot 440
Catnip 822
Cedar
 Ground 12
 Red 72
 White 73
Celandine 1020
Centaury 736
Chaffweed 1083
Chain Fern Family 52-53
Chain Fern
 Netted 52
 Virginian 53
Chamomile
 Corn 330
 Garden 332
 Stinking 331
 Yellow 333
Charlock 601
Cheeses 961
Cherry
 Choke 1204
 Fire 1200
 Ground 1364
 Hairy 1366
 Perfumed 1199
 Pin 1201
 Sour 1198
 Sweet 1197
 Wild 1203
Chervil 1386
 Wild 1391
Chestnut 713
Chickweed
 Common 284
 Field 249
 Forked 262, 263
 Giant 261
 Great 285
 Jagged 256
 Mouse-eared 250
 Nodding 251
 Water 195
Chicory 400
Chokeberry
 Black 1150
 Purple 1151
Cigar-tree 170
Cinquefoil
 Field 1195
 Shrubby 1189
 Silvery 1186
 Tall 1187
 Upright 1193
Clammyweed 213
Clearweed 1418
Cleavers 1237
Clintonia
 Yellow 1901
Clover
 Alsike 922
 Buffalo 925
 Crimson 923
 Hop 919
  Little 921
  Low 920
 Rabbit-foot 918
 Red 924
 White 926
 Yellow 919
Clubmoss Family 3-12

Clubmoss
 Bog 7
 Flat-branch 4
 Rock 11
 Shining 8
 Trailing 3
 Tree 10
Coffee
 Wild 233
Coffee-tree
 Kentucky 885
Cohosh
 Black 1106
 Blue 151, 152
Coltsfoot 548
 Sweet 495
Columbine
 Garden 1104
 Wild 1103
Columbo
 American 744
Comfrey
 Common 194
 Prickly 193
 Wild 173
Compass-plant 515
Coneflower 506
 Cut-leaf 508
 Prairie 505
 Purple 419
 Three-lobed 509
Coolwort 1417
Coontail 293
Copperleaf
 Hornbeam-leaved 692
Coral-root
 Early 1960
 Small-flowered 1959
 Spotted 1958
 Spring 1961
Coralberry 232
Coreopsis
 Tall 413
Corn
 Indian 1841
 -cockle 245
 -salad 1424, 1426
Cosmos 414
Costmary 394
Cotton-grass
 Green 1646
 Tawny 1645
Cottonweed
 Slender 104
Cottonwood 1264
 Swamp 1267
Cow-cockle 286
Cow-cress 636
Cow-parsnip 1395
Cow-wheat 1329
Cowbane 1401
Cowherb 286
Cowslip 1105
Crab
 Narrow-leaved 1180
 Wild 1181
Cranberry
 European 241
 Highbush 241
 Large 687
 Small 689

Cranesbill
 Blood-red 753
 Dove's-foot 750
Cress
 Garden- 638
 Hoary 618
 Mouse-ear 586
Croton
 Northern 694
Crowfoot
 Bristly 1126
 Cursed 1129
Crown-vetch 865
Cuckoo-flower 259, 616
Cucumber
 Prickly 656
Cucumber-root
 Indian 1918
Cucumber-tree 952
Cudweed
 Low 442
 Purple 441
Culver's-root 1356
Cup-plant 516
Cuphea
 Clammy 945
Currant
 Black 1300
 Buffalo 1305
 Garden 1306
 Red 1306
 Skunk 1302
 Swamp
  Red 1307
Custard-apple Family 112
Cypress Family 71-73

Daffodil 1921
Daisy
 English 372
 Ox-eye 395
 Shasta 396
Dame's-rocket 633
Dandelion
 Common 544
 Dwarf 475
 Red-seeded 543
Day-lily
 Orange 1907
 Yellow 1908
Dayflower 1485
Deerberry 691
Deptford-pink 253
Devil's-bit 1900
Dewberry 1218
 Small 1219
Dewdrop 1166
Ditch-grass Family 2022
Ditch-grass 2022
Dock
 Blunt-leaved 1076
 Pale 1072
 Swamp 1078
  Whorled 1078
 Water 1077
 Yellow 1074
Dodder
 Common 563
 Five-angled 564
 Flax 562
 Hazel 561
 Prairie 560

184

186

187